NOW YOU'RE TALKING!

ALL YOU NEED TO GET YOUR FIRST HAM RADIO LICENSE

Second Edition

Edited by:

Larry D. Wolfgang, WR1B
Jim Kearman, KR1S
Joel P. Kleinman, N1BKE

Production staff:

Deborah Strzeszkowski, Typesetting and Layout
Joe Shea, Production Assistant
David Pingree, N1NAS, Technical Illustrations
Steffie Nelson, KA1IFB, Proofreading
Jodi Morin, KA1JPA, Assistant Production Supervisor
Sue Fagan, Cover Design
Michelle Bloom, WB1ENT, Production Supervisor

Published by:
The American Radio Relay League
Newington, CT 06111-1494

This book may be used for Novice and Technician exams given beginning July 1, 1993. The book may be used for exams given until July 1, 1997. (This ending date assumes no FCC rules changes to the licensing structure or privileges for these two license classes, which would force the VEC Question Pool Committee to modify these question pools.)

For your convenience, there is a blank FCC Form 610 in the back of *Now You're Talking!*. Novice license exams are now being given by VEC-certified Volunteer Examiners, and all VE Teams should have a supply of the Forms when you go for your exam. Only Forms with a revision date of November 1993 or later may be used. For additional copies of the latest Form, write to: Form 610, ARRL, 225 Main Street, Newington, CT 06111-1494. Please enclose a self-addressed, stamped envelope.

Foreword

Amateur Radio is a very exciting avocation to many people all over the world. It provides a medium through which people in all walks of life can share a common interest. Amateur Radio also has many faces and can be seen differently by each person. To some just the ability to talk to people of diverse cultures is what drew them to Amateur Radio. For some it provides an important communications link home from remote parts of the earth (and beyond). The Amateur Radio Service enjoys access to the radio waves with very few restrictions. This has given many electronics experimenters an opportunity to dabble at the forefront of communications technology that they would not otherwise have.

I call it an avocation because to many of us it is much more than just a hobby. My own entry into Amateur Radio was in 1963 at the age of 12. Little did I know that it would play such an important role in my future life. During my college career I was employed by the university planetarium as a technician. The hands-on experience with soldering iron and voltmeter that I acquired in my Amateur Radio activities got me that job. Later when applying for a job teaching physics at a technical college, I was turned down until they found out I was a "ham." The ability to teach practical electronics learned through Amateur Radio, as well as physics, gave me the edge over the other candidates. The practical technical experience I have gained through Amateur Radio has continued to serve me well throughout my professional career.

The climax of my Amateur Radio activities (so far) has been the operation of SAREX-2 from aboard the space shuttle *Columbia*. Between December 2 and December 10, 1990 I was able to communicate with over 100 people on voice, and nearly 1000 on packet radio! Imagine the excitement of talking to Amateur Radio operators all over the world from earth orbit. It was indeed the ultimate DXpedition!

The Federal Communications Commission and the ARRL have now provided even more ways than ever before to enter this continually changing "hobby." With the new laws in place you can choose an entry-level license that best suits your current level of expertise. Depending upon your interests, you can also choose a license that either stresses the technical or traditional aspects of Amateur Radio. Whatever direction you decide to take you will enter a worldwide fraternity that will never cease to provide new challenges. Best of luck, and see you on the air!

73,

Ron

Ronald A. Parise, WA4SIR

Preface

Welcome to the exciting world of Amateur Radio! You are about to join the more than two million people in every corner of the globe who call themselves radio amateurs, or "hams." They have earned the very special privilege of being able to communicate directly with one another, by radio, without regard to the geographic and political barriers that so often interfere with our understanding of the world. Whether across town or across the sea, hams are always looking for new friends. So wherever you may happen to be, you are probably near someone—perhaps a whole club—who would be glad to help you get started.

Whether you want to learn Morse code and enter Amateur Radio as a Novice licensee or choose to skip the code and start with the Technician license, you'll find plenty of Amateur Radio activity to keep you busy. You'll also find plenty of friendly folks who are anxious to help you get started. Amateur Radio has many interesting areas to explore. You may be interested in one particular aspect of the hobby now, but be willing to investigate new avenues occasionally. You'll discover a world of unlimited potential!

Most of the active radio amateurs in the United States are members of the American Radio Relay League. ARRL has been the hams' own organization for over 75 years. We provide training materials and other services, and represent our members nationally and internationally. *Now You're Talking!* is just one of the many ARRL publications for all levels and interests in Amateur Radio. You don't need a ham license to join. If you're interested in ham radio, we're interested in you. It's as simple as that!

David Sumner, K1ZZ
Executive Vice President

Newington, Connecticut
March 1993

Contents

Two Tracks To Your First Ham Radio License

Until recently, the Novice license was the entry route for most newcomers to Amateur Radio. The Novice license written exam covers basic theory and regulations. It also includes a 5-word-per-minute (WPM) Morse code test. Novices are allowed to operate code on four high-frequency (HF) bands, where worldwide communication is possible without satellites or repeater stations. Novices are also allowed to use voice on one HF band and on two higher-frequency bands used mostly for local communications. Other privileges include radioteletype and computer-to-computer communications on certain bands. The Technician class license was usually the next step up for many radio amateurs.

The Technician license has now become the favorite entry-level license for most newcomers. In fact, today there are more Technician licensees than any other license class. Effective February 14, 1991, the FCC removed the Morse-code requirement for obtaining a Technician license.

The Technician exam includes the Novice written exam and a slightly more technical exam covering electronics and radio theory along with questions about the Federal Communications Commission (FCC) rules and regulations.

Technician licensees have full VHF and UHF privileges granted to higher-class licensees, but *not* the HF privileges granted to Novices. Those Novice-band privileges can be earned, however, by any code-free Technician licensee who passes a 5-WPM code test. (Previously licensed Technician-class operators still have Novice HF privileges because they passed a code test to earn their license.)

Any holder of a new Technician license who passes the code test is issued a *Certificate of Successful Completion of Examination* (CSCE) and is immediately granted Novice operating privileges on the HF bands. Novices who upgrade to Technician within two years of the expiration of their Novice licenses are also automatically granted Novice privileges when they pass the Technician exam.

Change to Novice Testing Procedures

Effective July 1, 1993, all Novice exams are given through the Volunteer Examiner Coordinator (VEC) system. All Amateur Radio exams are now administered by certified Volunteer Examiners through the VEC system. The Novice testing procedure is now identical to the Technician and higher-class license-testing program. The current Novice and Technician exam procedures are explained on page 2-10.

When you are ready to take a Novice exam, contact your local ham radio club, class instructor or ARRL Headquarters for information about exam opportunities near you.

WANT MORE INFORMATION?

Looking for more information about ham radio in your local area? Interested in taking a ham radio class? Ready for your license exam? Call 1-800-32 NEW HAM (1-800-326-3942). You'll receive, at no charge, a list of ham radio clubs, instructors and examiners in your local area.

How To Use This Book

You're about to begin an exciting adventure: a fun-filled journey into the world of Amateur Radio. *Now You're Talking!* is *the* study guide to help you reach your goal. This book introduces you to basic radio theory in an easily understood style. You'll also learn Federal Communications Commission (FCC) rules and regulations—just as the FCC requires. You will learn enough to pass your Novice or Technician written exam with ease. But we in the American Radio Relay League care much more about you and your future in Amateur Radio than to stop there! We won't abandon you once your new "ticket" is in hand.

Now You're Talking! is also a pathway to a successful beginning after you've passed your license exam. In this book you'll find the practical knowledge needed to become an effective communicator. We all hope you will take enough pride in the achievement of earning your license to be a considerate communicator as well.

But that's all a few weeks and a few pages down the line. To ensure your success and head you in the right direction toward your first on-the-air contact, here's how to use this book to your best advantage.

Now You're Talking! has been designed and written by a staff with a great deal of experience, backed by decades of Amateur Radio tradition. This book provides everything you'll need to learn—and understand—what you should know to operate an Amateur Radio station. We explain all the rules and regulations required of Novice and Technician candidates, not only to pass the test, but to operate properly (legally) once your license is hanging proudly in your shack.

SELF STUDY OR CLASSROOM USE?

We designed *Now You're Talking!* both for self study and for classroom use. An interested student will find this book complete, readable and easy to understand. Read carefully, and test yourself often as you study. Before you know it, you'll be ready to pass that exam!

Why deprive yourself of the company of fellow beginners and the expertise of those "old-timers" in your hometown, though? *Now You're Talking!* goes hand-in-hand with a very effective ARRL-sponsored training program run by over 6000 volunteer instructors throughout the United States. If you would like to find out about a local class, just contact the Educational Activities Department at ARRL Headquarters—we'll be happy to assist.

Hams are very social animals who derive a great deal of pleasure from helping a newcomer along the way. The most effective learning situation is often the one you share with others. There are knowledgeable people to turn to when you have a question or problem. You can practice the Morse code with fellow students, and quiz one another on basic electronics concepts. It doesn't matter if you're studying on your own or joining a class. Use *Now You're Talking!* to study for your Novice or Technician exam and you'll be on the air in no time at all.

USING THIS BOOK

Now You're Talking! leads you from one subject to the next in a logical sequence that builds on the knowledge learned in earlier sections. It presents the material in easily digested and well-defined "bite-sized" sections. You will be directed to turn to the question pool in Chapter 12 (and 13 if you're preparing for the Technician exam) as you complete a section of the material.

This review will help you determine if you're ready to move on. It will also highlight those areas where you need a little more study. In addition, this approach takes you through the entire question pool. By the time you complete the book, you will be familiar with all the questions used to make up your test. Please take the time to follow these instructions. Believe us, it's better to learn the material correctly the first time than to rush ahead, ignoring weak areas and unresolved questions.

Every page of this book presents information you'll need to pass the exam and become an effective operator. Pay attention to diagrams, photographs, sketches and captions; they contain a wealth of information you should know. You'll also find a few anecdotes and "mini-articles" (called Sidebars) that will help put the tradition of Amateur Radio in perspective. Our roots go back to the beginning of the 20th Century, and our community service continues even as you read this.

As you study *Now You're Talking!*, you'll find sections titled *Not For Technicians Only* in several chapters. These sections contain information needed for the Technician exam. If you are studying for a Novice license you may want to skip those sections. While you won't find any exam questions from this material on your Novice exam, you may find it interesting and helpful to read. Like it says, these sections are ***Not for Technicians Only!***

Start at the beginning. Chapter 1 summarizes the fun you will have when you earn your license and join the thousands of other active radio amateurs. Chapter 2 explains the need for international and national regulation. It describes the regulating bodies and the relevant sections of Part 97 of the FCC Rules and Regulations.

Chapter 3 breaks basic radio theory into well-defined sections. Each section builds on the previous material to explain the theory you'll need to answer the questions in Chapters 12 and 13.

Study these sections one at a time. Be sure to follow the instructions at specific places in the text that direct you to study the actual test questions. This is the best way to determine how well you understand the most important points. The text explains the theory in straightforward terms, so you shouldn't have any problems. Don't be afraid to ask for help if you don't understand something, though.

If you're participating in an official ARRL-registered class, you'll have the chance to ask the experts for help. Ask your instructor about anything you are having difficulty with. You may also find it helpful to discuss the material with your fellow students. If you're not in a class and run into snags, don't despair! The Educational Activities Department at ARRL Headquarters will be happy to put you in touch with an Amateur Radio operator in your area who can help answer your questions.

Chapters 4 and 5 cover some basic components and circuits. You will also learn how to connect various pieces of equipment to form an Amateur Radio station.

Chapter 6 offers some guidelines to help you select the equipment you'll need to set up your own radio station. Chapter 7 gives you all the information about antennas that you'll need to pass your exam and get started in Amateur Radio. Full construction details are provided for several simple antennas suitable for Novice or Technician operation.

Chapter 8 will help you assemble an effective Amateur Radio station. After your license arrives from the FCC and your station is set up, Chapter 9 leads you through those first few contacts. How do you know when to operate on what band? Where do you turn to make contact with a specific foreign country or faraway state? How do you establish contact in the first place and what do you say? What sorts of activities will you want to become involved with on the air? Chapter 9 has the operating information to make you feel at ease on the air. Chapter 10 will help you identify and solve the problems you are most likely to encounter as a new ham.

All these chapters cover information necessary to pass the exam and the practical information you'll need to make your ham radio experience enjoyable. Study the material presented in this book and follow the instructions to review the exam questions. You'll cover small sections of the text and a few questions at a time. Review the related sections if you have any difficulty, and then go over the questions again. In this way you will soon be ready for your exam. Before you know it, you'll be on the air!

Most people learn more when they are actively involved in the learning process. Turning to the questions and answers when directed in the text helps you be actively involved. Fold the edge of the question-pool pages to cover the answer key and page-reference information. Check your answers after you study each group of questions, and review the appropriate text for any questions you get wrong. Paper clips make excellent place markers to help you keep your spot in the text and the question pool.

If You Decide to Learn the Code

Chapter 11 lets you in on a little secret. The Morse code is not only easy to learn; it can be so enjoyable that you may become "addicted." You should read this chapter before beginning to learn the Morse code. People try to learn the code by many different methods. The code cassettes in *Your Introduction to Morse Code* teach you the Morse code at your own pace. You should rewind the tape to review any characters that give you difficulty. The technique used with these tapes also makes it easy to increase your speed once you've learned the code.

After reading Chapter 11, begin studying the code and theory in "parallel"; that is, split your study time between daily sessions with the code and study sessions with the text. Move ahead at a pace that allows you to master the material. You do have to keep moving, however. Don't allow yourself to stay on one section too long, even if you have to come back and review later.

THE NOVICE TEST

The FCC requires three licensed hams to give a Novice exam. If you're taking a class, you'll have no problem in locating Volunteer Examiners (VEs) to administer the code test and written exam. Who may give the Novice test? Sections 97.513 and 97.515 of the FCC rules list the qualifications for the three certified VEs who administer a Novice test. Novice examiners must:

1) hold a current General, Advanced or Amateur Extra class operator license issued by the FCC;

2) be accredited by the coordinating Volunteer Examiner Coordinator (VEC).

3) be at least 18 years of age;

4) not be related to the applicant;

5) not own a significant interest in, or be an employee of, any company or other organization engaged in the manufacture or distribution of equipment used in connection with Amateur Radio transmissions, or in preparation or distribution of any publication used in preparation for

obtaining an Amateur Radio license. An employee who can demonstrate that he or she does not normally communicate with that part of an organization engaged in such manufacture or publishing is eligible to be a volunteer examiner, however;

6) never had their Amateur Radio station or operator's license revoked or suspended.

The code test requires you to show your ability to send and receive the Morse code at 5 words per minute. Here is a simple way to estimate code speed: Five characters make one word, with punctuation and numerals counting as two characters each. You must know the 26 letters, 10 numerals and the basic punctuation marks: comma, period, question mark, double dash (= also called $\overline{BT}$) and fraction bar (/ also called $\overline{DN}$). You must also know some common procedural signals (prosigns): $\overline{AR}$ (the + sign, also used to mean "over" or "end of message") and $\overline{SK}$ ("clear; end of contact"). For further details on the code test, see Chapter 11.

For the written exam, you must answer 30 questions about Amateur Radio rules, theory and practice. Your test questions will be drawn from the pool of more than 300 questions. The question pool printed in Chapter 12 was put into effect on July 1, 1993. The question pool includes the multiple-choice answers released by the Volunteer Examiner Coordinator Question Pool Committee.

The question pool is divided into nine subelements. Each subelement is further divided into blocks, with one question to come from each block. Your test should include one question from each of the 30 blocks. Your examiners may select the questions, or the coordinating VEC or a test supplier may provide a printed exam. The questions must be used exactly as printed (and as reproduced in Chapter 12 of this book).

The passing grade for an amateur license is 74%, so you will need 22 or more correct answers out of the 30 questions on your exam to pass the test. (Another way of putting this is that you may have as many as eight incorrect answers and still pass!)

THE TECHNICIAN TEST

Since January 1, 1985, all US amateur exams except the Novice have been administered under the Volunteer-Examiner Program. Novice exams now also come under the regulations involving Volunteer-Examiner Coordinators (VECs). (This change became effective July 1, 1993.)

To qualify for a Technician-class license, you must pass Elements 2 and 3A. *You do not have to pass Element 1A* (5-WPM code test). If you already hold a valid Novice license, then you have credit for passing Elements 1A and 2 and will not have to retake those elements.

The Element 3A exam consists of 25 questions taken from a pool of more than 250. The question pools for all amateur examinations are maintained by a Question Pool Committee selected by the Volunteer Examiner Coordinators. The FCC allows Volunteer Examiners to select the questions for an amateur exam, but they must use the questions exactly as they are released by the VEC Question Pool Committee. The questions and answers on your exam will be exactly as they are printed in Chapter 13.

BEFORE YOU TAKE YOUR TEST

Before you go to take your exam, fill out an FCC Form 610. (Chapter 2 has full details about how to fill out this form.) After you complete the test, your examiners grade it. When you pass all the exam elements required for the license, the examiners complete the Administering VE's Report on the front of the form and the Certification section on the back, indicating that you passed all required elements.

Applications for all license classes go first to the admin-

istering VEC, then to the FCC. Now comes the hard part! It may take as long as twelve weeks for your license to arrive in the mail. The FCC staff processes your application along with the thousands of other 610 forms they receive each month.

If you would like:
• help in locating someone to administer a test
• a schedule of Volunteer Examiner test sessions in your area

• a list of nearby examiners and clubs,

contact the ARRL Educational Activities Department at 800-32-NEW HAM or write EAD, ARRL, 225 Main St, Newington, CT 06111-1494.

Give *Now You're Talking!* a chance to guide you the way it was intended—by following these instructions. You'll soon be joining us on the air. Each of us at the American Radio Relay League Headquarters and the entire ARRL membership wishes you the very best of success. We are all looking forward to that day in the not-too-distant future when we hear your signal on the ham bands.

73 (best regards) and good luck!

Chapter 1

Discovering Amateur Radio

H ow did you first hear about Amateur Radio? That evening news report about the Amateur Radio operators who relayed messages after Hurricane Andrew? The funny-looking antenna in your neighbor's yard? A birthday greeting via Amateur Radio from your uncle who lives clear across the country? Or maybe you heard about hams who talked to astronauts during a space-shuttle mission.

Obviously, you know something about amateur (ham) radio; you have this book. But you want to know more. *Now You're Talking!* will introduce you to the wonderful hobby of Amateur Radio. It will answer your many questions on the subject, and lead you to your first license.

WHAT CAN I DO AS A HAM?

Ham radio offers so much variety, it would be hard to describe all its activities in a book twice this size! Most of all,

ham radio gives you a chance to meet other people who like to communicate. That's the one thing all hams have in common. You can communicate with other hams on a simple hand-held radio that fits in your pocket. Thanks to "repeater stations" (*repeaters*) operated by local ham clubs, your range with a tiny "hand-held" may be 50 miles or more.

Repeaters also make it possible for mobile ham stations to keep in touch. Mobile operating is very popular during commuting hours. There's almost always someone to talk to on a repeater. Most people would like to do something to help their communities. Someday, you may spot another motorist who needs help, and use the repeater telephone link to call the police. Or, your repeater club may provide communications for a parade, foot race, or even during an emergency like a flood. Best of all, you'll be meeting other people nearby who are as excited about Amateur Radio personal communications as you are.

Imagine talking to a missionary operating a battery-powered station deep in the Amazon jungle, or a sailor attempting an around-the-world solo journey! You can talk to hams all over the world in many ways. The most popular way is by bouncing your signal off the ionosphere, a layer in the upper atmosphere. Other hams like to use the OSCAR satellites. OSCAR means Orbiting Satellite Carrying Amateur Radio. Hams have designed and built over 20 OSCARs since 1961. If you like a challenge, you can even bounce your signals off the moon. It's possible to contact other hams in over 100 countries by this method, as strange as it sounds!

How do you talk to these other hams? Well, you can use voice of course, but there are other ways as well. Maybe you've tried computer-to-computer conversations over the telephone lines. As a ham, you can have similar conversations with other hams around the world. The best part is, you don't have to pay for the call!

The earliest radio communication was done with Morse code. You don't have to know the code to become a ham anymore, but many hams still enjoy using this funny language. To them, the beeping of Morse code is like listening to a favorite song.

Maybe you're interested in photography or video. Many hams have television equipment, including color. Hams led the way in *slow-scan television*, which lets you send color photographs, slides and artwork to other hams thousands of miles away.

How about radio-controlled models? Yes, many hams enjoy using exclusive ham frequencies to fly gliders and powered planes, or pilot graceful sailboats. The no-code amateur license is all you need to join them.

If you like to work on electronic circuits, ham radio gives you the chance to build your own transmitters and receivers, and actually use them to talk to other people. Even if you'd just like to plug in a radio and go on the air, you might enjoy building some part of your station, from scratch or from a kit.

Unlike shortwave or scanner listening, Amateur Radio doesn't make you sit on the sidelines. When East and West Germany became one country, hams didn't have to learn about it from the TV news: they talked directly to hams in both halves of the country! Instead of listening to police calls on Halloween, you can join your club's "Goblin Watch," and help make the streets safer for Trick-or-Treaters.

Ham radio operators are proud of their hobby. That's why you see so many license plates with ham call signs. Communicating on the air isn't all we do, though. We get together at club meetings, hamfests and conventions. To welcome newcomers we sponsor thousands of classes each year to help anyone who's interested join our great hobby.

TWO PATHS TO CHOOSE FROM

There are five classes of Amateur Radio license in the United States. All US amateur licenses have 10-year terms and are renewable. Licenses are issued by the Federal Communications Commission (FCC). There is no license fee.

Although you can take the exams and start right out at the highest class, most beginners enter Amateur Radio with either the Novice or Technician license.

There are several differences between the two entry-level licenses, in exams and operating privileges. They are described in the following sections. To pass the Novice license exam, you're required to know Morse code, but there's no code exam for the Technician license. No matter which path you choose, this book has all the information you need to pass the written exams.

The Novice License

To receive a Novice license you have to pass a 30-question written exam in basic electronics theory, FCC regulations and operating practices. You also have to copy the international Morse code at 5 words per minute (WPM).

Novice licensees are granted a wide range of privileges on several amateur bands. Some of these bands are mainly useful for local communications, but the "high-frequency" (HF) Novice bands offer the ability to talk to other radio amateurs around the world, without the use of satellites (we'll tell you more about satellite communications later). The Morse code and many abbreviations such as "Q signals" are understood internationally. This makes the code ideal for communicating with hams in foreign countries. Plenty of stations in other countries operate in the Novice bands. Quite a few Novices have even contacted hams in more than 100 different countries. Chapter 2 has more information about Novice-license privileges.

Novice exams are given by three Volunteer Examiners with General, Advanced or Extra-class licenses. The examiners must be accredited by a Volunteer Examiner Coordinator (VEC). For help learning the code we recommend *Your Introduction to Morse Code*, a package of two audio cassettes that teach you the code, letter by letter. When you know the code, the tapes give you practice at 5 WPM, the same speed used on the Novice exam.

The Technician License

If you aren't interested in learning the Morse code right now, the Technician license is for you. To receive a Technician license you have to pass the same 30-question written exam as for the Novice license, plus an additional 25-question exam. The additional exam covers the same subjects, but in more detail.

Technician licensees are granted full amateur privileges on the extensive VHF, UHF and microwave amateur bands, including Amateur Radio satellite bands. Yes, with a Technician license you can talk to other amateurs around the world via satellite! Technician licensees can use the popular 2-meter band, where thousands of amateur-owned repeater stations provide solid communications over large areas.

If you start out as a Technician and then decide you'd also like to use the high-frequency Novice bands, all you have to do is pass a 5-WPM Morse code exam. Your new privileges will begin the day you pass the exam. There is enough Morse code activity on the Technician bands to help you if you decide to learn the code. Chapter 2 has more information about Technician-license privileges.

The Technician license exam (and all other amateur license exams) are given by teams of Volunteer Examiners. Thousands of exam sessions are held each year, so you shouldn't have to travel far or wait long to take one. Write to ARRL for information on exam sessions near you.

WE COME FROM ALL WALKS OF LIFE

Having fun communicating and experimenting—that's what Amateur Radio is all about. That's why people from all walks of life become hams. Young or old, we all enjoy the thrill of meeting and exchanging ideas with people from across town or from the other side of the earth. The excitement of building a new project or getting a circuit to work properly is almost beyond description.

Carm Prestia is a diamond-in-the-rough police sergeant, patrolling a bustling university town tucked away in the mountains of Pennsylvania. By night, he packs a .38-caliber Police Special to protect thousands of his fellow townspeople. By day, he wields a soldering iron in pursuit of the world's greatest hobby: Amateur Radio.

"I love to talk to people on the radio," Carm explains as he unstraps his portable radio from his uniform belt. "I talk all night at work, but I still go home and fire up the ham gear."

Carm's radio room (hams all over the world affectionately call this room their "shack") is in a corner of his basement. His equipment table holds a transmitter for sending and a receiver for listening. His radio gear works with an antenna outside, above his backyard. Carm can talk with a friend in the next town one minute and with a ham halfway around the world in Australia the next.

Each Amateur Radio station has its own distinctive call sign. The Federal Communications Commission (FCC) issued Carm his call sign, WB3ADI.

Ham radio operators are so proud of their call signs that the two often become inseparable in the minds of friends. Barry, K7UGA, of Arizona has worked (talked with) thousands of hams on the air. Many of them didn't know that his last name is Goldwater or that he was a United States senator.

AGE IS NO BARRIER

Age is no barrier to getting a ham license and joining in the fun. There are hams of all ages, from five years to more than 80 years. Samantha Fisher received her Technician license at age 9. Known to her ham friends as KA1VBQ, she enjoys talking on 10 meters.

Then there's Luke Ward, KC4UJS, of Alexandria, Virginia, who was licensed at age 7. Luke upgraded to General at age 8, and hopes to experiment with "moonbounce"—reflecting VHF/UHF signals off the moon to communicate over long distances.

Like many hams, Dan Broniatowski, N8QWT, enjoys chasing DX from his Cleveland Heights, Ohio, station. Dan was licensed at the age of 10, and received his General license less than a year later.

Of course, young people aren't the only ones who are active radio amateurs. Involved in many hobbies, Evelyn Fox of Merrimac, Wisconsin, plays contract bridge with her AARP (American Association of Retired Persons) group on the 40-meter band.

Evelyn was over 75 years old when she became interested in Amateur Radio. That didn't stop her from taking on the job of learning radio theory and the international Morse code. She found the theory a pleasant challenge. She joined a club, attended its classes, and became WB9QZA. Not bad for someone who knew nothing about electronics when she first got started.

Bill Bliss, KD6AHF received his Novice license about 3 weeks before his birthday. So what, you say? He was 82 years old on that birthday! Bill says, "Some think that a bit old to start ham radio, but I have the rest of my life to enjoy it."

Herb Kline is further proof that it's never too late to become a ham. Herb, KB2QGD, of Adams, New York passed his Novice exam just before his 83rd birthday and upgraded to Technician Plus a few weeks later. With plans to upgrade to General, Herb likes to operate CW. He is also learning about computers.

"BREWING IT" AT HOME

Ham radio operators pop up in some of the least expected places. Dr Peter Pehem, 5Z4JJ, is one of Africa's flying doctors. He works out of a small village on the north slope of Mount Kilimanjaro in Kenya. Pete has been bitten by an OSCAR bug, but can't do anything about it while on medical duty. But he can and does attack it with great pleasure when he's off duty.

Somebody gave Pete an old radiotelephone, a vacuum tube and some coaxial cable. The doctor added empty aspirin tins and a quartz crystal from his airplane radio. Right out there in the African bush, he fired up a home-made transmitter, built on the aspirin tins. Then he talked to the world through OSCAR, the Orbiting Satellite Carrying Amateur Radio.

Pete proved something with his home-made gear: You don't need a shack full of the latest commercial equipment to have fun on the air. New hams find this out every day. Tom Giugliano, WA2GOQ, of Brooklyn, New York, contacted 26 states using a pre-World War II transmitter and receiver. He used a simple home-made wire antenna. Other hams have bridged the oceans to contact hams in Europe and Japan using simple equipment running less than 1 watt of power. As a Novice, you'll be permitted to run 200-watts output on the HF bands. That's more than enough to contact other hams around the world. Technicians who have passed a 5-WPM code test have the same privileges. On the VHF, UHF and microwave bands, Technicians are allowed to use 1500-watts output, although that much power is rarely needed or used.

There was a time, many years ago, when no commercial equipment was available. The earliest hams, beginning more than 75 years ago, tried to find more efficient ways of communicating with each other. All early radio sets were home brew (homebuilt) and were capable only of communication over several miles. Some transmitters were nothing more than a length of copper wire wrapped around an oatmeal box, attached to a few other basic parts and a wire antenna. Often the transmissions were one-way, with one transmitting station broadcasting to several receive-only stations. Over the years, hams have continually looked for ways to transmit farther and better. They are constantly developing and advancing the state of the art in their quest for more effective ways of communicating.

PEERING BACK THROUGH TIME

It all started on a raw December day in 1901. Italian inventor and experimenter Guglielmo Marconi launched the Age of Wireless from an abandoned barracks at St John's, Newfoundland. He listened intently for a crackling series of buzzes, the letter S in international Morse code, traversing the 2000 miles from Cornwall, England. That signal was the culmination of years of experimentation. News of Marconi's feat stimulated hundreds of electrical hobbyists to build their own "wireless" equipment. They became the first hams.

Later, Marconi set up a huge station at Cape Cod that was unlike anything today's ham has experienced. Marconi's 3-foot-diameter spark-gap rotor fed 30,000 watts of power to a huge antenna array suspended from four 200-foot towers on the dunes at South Wellfleet, Massachusetts.

By 1914, Marconi had set up a station and antennas for daily transmission across the Atlantic. At the same time, Amateur Radio operators all over America were firing up their own homebuilt transmitters. Soon, several hundred amateurs across the country joined Hiram Percy Maxim in forming the American Radio Relay League (ARRL), based in Hartford, Connecticut. These amateurs set up a series of "airborne trunk lines" through which they could relay messages from coast to coast. If you're interested in learning more about the history of Amateur Radio, you'll enjoy *200 Meters and Down* by Clinton B. DeSoto. *Fifty Years of ARRL* is also an interesting account of the ARRL's first half century. Both books are available directly from ARRL.

There were more hams experimenting all the time. Commercial broadcasting stations began to spring up after World War I. This brought a great deal of confusion to the airwaves. Congress created the Federal Radio Commission in 1927 to unravel the confusion and assign specific frequencies for specific uses. Soon amateurs found themselves with their own frequency bands.

Continued experimentation over the years has brought us tubes and transistors. Equipment has grown smaller and more sophisticated. In the early days of radio communications, equipment was large and heavy. Sometimes it would take a roomful of equipment to accomplish what can now be done with the circuitry in a tiny box.

The state-of-the-art in amateur gear has come a long way since the days of "Old Betsy," Hiram Percy Maxim's own spark-gap transmitter. Spark-gap transmitters were the very first type of radio transmitters used in the early 1900s.

WE PITCH IN WHEN NEEDED

Traditionally, amateurs have served their countries in times of need. During wartime, amateurs have patriotically taken their communications skills and technical ability into the field. During natural disasters, when normal channels of communication are interrupted, hams provide an emergency communications system. Practically all radio transmitters operating in the United States provide some public service at one time or another. Hams don't wait for someone to ask for their help; they pitch in when needed. They provide communications on March of Dimes walk-a-thons, help plug the dikes when floods threaten and warn of approaching hurricanes. Hams bring assistance to sinking ships, direct medical supplies into disaster areas, and search for downed aircraft and lost children.

Amateur Radio operators recognize their responsibility to provide these public-service communications. They train in various ways to be effective communicators in times of trouble. Every day, amateurs relay thousands of routine messages across the country. They send many of these messages through "traffic nets" devoted to developing the skill of sending and receiving messages efficiently. (A net is a gathering of hams on a single frequency for some specific purpose. In this case the net's purpose is to "pass traffic"—relay messages.) This daily operation helps prepare hams for real emergencies. Also, each June thousands of hams across the country participate in the ARRL Field Day. They set up portable stations, including antennas, and use emergency power. Relaying messages and operating on Field Day help hams to test their emergency communications capabilities and to help them identify problems that could arise in the event of a disaster.

Public service has been a ham tradition since the very beginning. Whether it's a walk-a-thon, the Olympic Torch Run or the aftermath of a hurricane, hams are always there to help with communications, which they provide at no cost to the group involved.

MAYDAY—WE'RE GOING DOWN

In September 1991, a Pacific storm caught two boats, the *Molly Sue* and the *Dauntless*, eastbound from Hawaii to California. The *Molly Sue* and the *Dauntless* had intended to travel together, but the *Dauntless*, a 65-foot schooner, outran the other boat by 500 miles in the storm, out of maritime radio range.

The storm's winds ripped the *Molly Sue*'s sails and broke her halyard, injuring a crewman in the process—injuries that appeared serious. Alerted by hams operating on the California-Hawaii net, with whom the *Molly Sue*'s skipper was in contact, the Coast Guard diverted a freighter to help. Unfortunately, the *Molly Sue* needed new sails. A local ham picked

up a set of sails from the skipper's home in San Diego, and delivered them to the Coast Guard. The Coast Guard parachuted the sails to the *Molly Sue*.

Meanwhile, the *Dauntless*, back in radio range, reported she was taking on water and running out of fuel. The net relayed this word to the Coast Guard, and the *Dauntless* was towed to San Francisco. The *Molly Sue*, fitted with new sails, arrived safely in San Diego.

An FCC official commended the hams on the net for their "shining example of the communications skills, dedication and public spirit of the amateur service."

IN QUAKES, HURRICANES AND FLOODS: HAMS ARE THERE

In October 1989, a major earthquake struck the San Francisco area. Thousands of inquiries came in to ARRL Headquarters and to individual hams across the country. Friends, relatives and business associates worried about people in the affected area.

Why did they choose to ask Amateur Radio operators to help them? Over the decades, hams have volunteered their services in times of emergency to relay vital information to and from stricken areas. With local power and telephone communications knocked out, San Francisco hams went to work. Using amateur stations that fit in briefcases, they relayed messages between residents and their worried relatives and friends outside the stricken area. These tiny stations included a hand-held Amateur Radio transceiver, a lap-top computer and a radio modem called a TNC. Local hams went back to their regular routine of working and spending time with their families only after workers restored regular communications channels.

When towns and cities in the Fargo, North Dakota and Moorhead, Minnesota area were threatened by flooding, members of the Red River Radio Amateur Club were ready. Portable transceivers in hand, radio amateurs patrolled sand-bag dikes and riverbanks, reporting on the flood's progress. When one dike started leaking at the bottom and was undermining the sand bags, Amateur Radio was used to call in the Army Corps of Engineers.

In 1992, Hurricane Andrew chewed across southern Florida, then took aim at Louisiana. The city of Homestead, Florida, and the adjacent air base were almost totally destroyed. Despite the damage to their own homes, local hams provided essential communications in the first hours after Andrew passed. In Andrew's aftermath, amateurs from other states volunteered their time and equipment to aid relief efforts. A few weeks later, Hurricane Iniki struck Hawaii. Again, radio amateurs played key communications roles when commercial communications systems were destroyed.

Radio amateurs help out even when the emergency is outside the US. In September 1989, Hurricane Hugo did terrible damage to island nations in the Caribbean, and Puerto Rico. Most of the islands were completely cut off from the outside world; all normal means of communication were destroyed. Through volunteer Amateur Radio networks, shipments of relief and medical supplies were coordinated. Island residents were able to contact loved ones on the mainland and other islands. Amateur Radio "jump teams" even sped to the affected areas to help restore communications.

NO BARRIERS TO ENJOYING HAM RADIO

Amateur Radio holds no roadblocks for people with disabilities. Many people who are unable to walk, see or talk are able to enjoy their Amateur Radio hobby, conversing with friends in their home town or across the world. Some local ham clubs even take classes to a home to help a person with a disability discover ham radio.

Although unable to leave his bed, Otho Jarman was able to earn his ham license. Bill Haney, WA6CMZ, helped Otho learn the code and pass his Novice exam through such a club program. Bill spent one hour each week teaching Otho the code, and in seven weeks he passed his Novice exam. Other members of the Barstow (California) Amateur Radio Club helped Otho put his station together. They obtained equipment for him and constructed antennas.

Soon, Otho, WB6KYM, was on the air talking with hams in Mozambique, Nicaragua and Puerto Rico. He could communicate with the world from his bed. Sixteen years earlier, at age 22, Otho had broken his spine when he dove into a reservoir to rescue a drowning child. Although interested in Amateur Radio for years, he had not had an opportunity to learn about it until the club came along. Now he monitors local frequencies from 7 AM to 10 PM, often just chatting or giving directions to motorists.

"Amateur Radio can take a disabled person out of his living room, out of his bed or out of his wheelchair, and put him in the real world," Otho says.

The Courage Center, of Golden Valley, Minnesota, sponsors the HANDI-HAM system to help people with physical disabilities obtain amateur licenses. The system provides materials and instruction to persons with disabilities interested in obtaining ham licenses. The Center also provides information to other hams, "verticals," who wish to help people with disabilities earn a license.

NOW YOU'RE TALKING!

Once you've earned your license, you can join the thousands of other hams on the air. Then you'll begin to experience the thrill of Amateur Radio firsthand. As a ham, you can contact other hams in your town and around the world. You can talk into a microphone to use either single sideband (SSB) or frequency modulation (FM) to communicate. If you want, you can experience the thrill of using the international Morse code (called "CW" by hams). You can even communicate by packet radio or radioteletype using a computer. You can do all this on parts of special frequency bands set aside by the FCC for Amateur Radio operators.

For your first contact, you'll probably tune around looking for a station calling "CQ" (calling for any station to make contact). Perhaps you'll even try calling a CQ on your own. Suddenly you'll hear your own call coming back! It's hard to describe the excitement. Someone else is sending your call

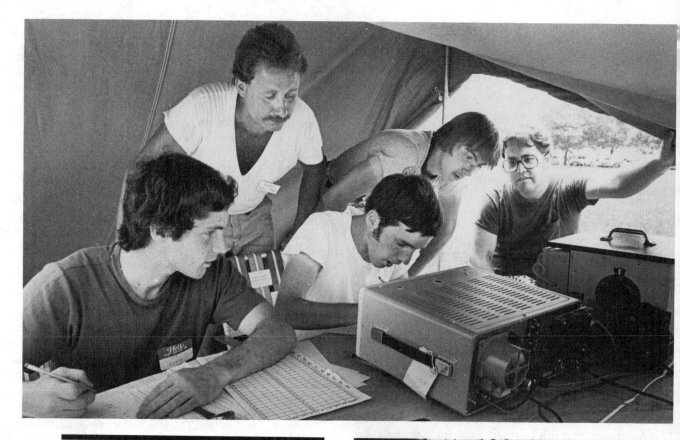

sign back to let you know that they hear you and want to make contact.

Each time you send a CQ, you'll wonder who will answer. It could be a ham in the next town, the next state or clear across the country. The whole world is full of hams to talk to.

There are many wonderful things you can do as a ham. After you have been on the air for a while, you'll be known as one of the regulars on the band. It's surprising how many people from all over you'll recognize and who will recognize you. Many a fast friendship has developed through repeated on-the-air contacts.

Soon you'll be collecting contacts with different states and exchanging QSL cards (postcards) with other hams you've talked to. These special cards commemorate each contact. They also serve as proof of the contact as you begin working toward some of the awards issued by the American Radio Relay League. [The Worked All States (WAS) award is one popular example.] You'll learn more about the special language and abbreviations that hams use as you study. Chapter 9 has a list of common "Q signals" used by most hams.

There is a certain intrigue about DX (long-distance communication) that catches many hams. Talking to hams from other lands can be quite an experience. After all, foreign hams are people just like you who enjoy finding out about other people and places! Also, hams in other countries often speak enough English to carry on a limited conversation, so you'll have little problem there. Whether you start as a Novice or a Technician, Amateur Radio offers many opportunities to contact DX stations.

If you enjoy a little competition, perhaps you'll like on-

the-air contesting. The object of a contest is to work as many people in as many different areas as possible in a certain time. Each year, the American Radio Relay League sponsors a "Novice/Technician Roundup" contest. In the Roundup, you have the chance to contact old friends and make new ones. You might work some new states or countries, and if you use Morse code you will increase your code speed. You are certain to improve your general operating skills and ability. Most of all, though, you will have fun.

OTHER MODES

Novice- and Technician-class licensees may use just about every type of communication available to Amateur

There are many hobbies within the hobby of Amateur Radio. Some hams collect QSL cards and awards. Others love the thrill of talking to rare DX stations in foreign lands, while still others enjoy exchanging pictures by slow-scan television. You'll find one or more of these pastimes enjoyable, and in time you'll probably try most of them.

Radio operators. You should become familiar with these modes. In addition to voice and Morse code, you may want to investigate some of the less traditional modes.

With slow-scan television (SSTV), hams send still photos to each other, one frame at a time. It takes about 8 seconds for the bright band of light to creep down the screen to make a complete picture. (Your home TV makes 30 complete pictures per second.) SSTV pictures are more like those shots of the moon or Saturn that you may have seen transmitted from space. SSTV pictures may be transmitted around the world via shortwave ham transmitters. Amateur Radio operators were the first to flash TV snapshots of Mars to foreign countries.

Facsimile (fax) is a means of sending drawings, charts, maps and graphs. You can even play games over the air by transmitting fax pictures of each move.

Using the "digital" modes, a ham can type out a message and send it over the air to a friend's station. Even if the friend is away, her radioteletype system can receive and hold the message until she returns. Early digital systems used mechanical machines cast off by news services. Today, many hams use personal computer systems. These display the message silently on a TV screen rather than using roll after roll of paper with noisy, clacking typewriter keys. There is a growing number of modern computer-controlled systems capable of relaying messages and storing them for later reception by the intended ham.

HAM SATELLITES

Hams use satellites to communicate in voice, code, radioteletype and packet radio around the world.

Some schools use the OSCARs to instruct students in science and math. No license is needed to listen, so many students across the country have eavesdropped on Amateur Radio transmissions through an OSCAR. A receiver and an antenna are all you need to introduce students to the exciting world of space technology.

Hams from many countries worked in a joint effort to build several OSCARs. Weighing less than your living-room TV and powered by sun-charged batteries, OSCARs retransmit hams' "uplink" signals down to other earth-bound stations.

WHAT ARE YOU WAITING FOR?

You can operate in any of these exciting modes when you become an Amateur Radio operator. Take the time to explore the adventure of ham radio!

SUBELEMENT N1—COMMISSION'S RULES
[10 exam questions—10 groups]

N1A Basis and purpose of amateur service and definitions.

N1B Station/Operator license.

N1C Novice control operator frequency privileges.

N1D Novice eligibility, exam elements, mailing addresses, US call sign assignment and life of license.

N1E Novice control operator emission privileges.

N1F Transmitter power on Novice sub-bands and digital communications (limited to concepts only).

N1G Responsibility of licensee, control operator requirements.

N1H Station identification, points of communication and operation, and business communications.

N1I International and space communications, authorized and prohibited transmissions.

N1J False signals or unidentified communications and malicious interference.

SUBELEMENT N2—OPERATING PROCEDURES
[2 exam questions—2 groups]

N2A Choosing a frequency for tune-up, operating or emergencies; understanding the Morse code; RST signal reports; Q signals; voice communications and phonetics.

N2B Radio teleprinting; packet; repeater operating procedures; special operations.

SUBELEMENT N3—RADIO WAVE PROPAGATION
[1 exam question—1 group]

N3A Radio wave propagation, line of sight, ground wave, sky wave, sunspots and the sunspot cycle, reflection of VHF/UHF signals.

SUBELEMENT N4—AMATEUR RADIO PRACTICES
[4 exam questions—4 groups]

N4A Unauthorized use prevention, lightning protection, and station grounding.

N4B Radio frequency safety precautions, safety interlocks, antenna installation safety procedures.

N4C SWR meaning and measurements.

N4D RFI and its complications.

SUBELEMENT N5—ELECTRICAL PRINCIPLES
[4 exam questions—4 groups]

N5A Metric prefixes, i.e. pico, micro, milli, centi, kilo, mega, giga.

N5B Concepts of current, voltage, conductor, insulator, resistance, and the measurements thereof.

N5C Ohm's Law (any calculations will be kept to a very low level—no fractions or decimals) and the concepts of energy and power, and open and short circuits.

N5D Concepts of frequency, including AC vs DC, frequency units, AF vs RF and wavelength.

SUBELEMENT N6—CIRCUIT COMPONENTS
[2 exam questions—2 groups]

N6A Electrical function and/or schematic representation of resistor, switch, fuse, or battery.

N6B Electrical function and/or schematic representation of a ground, antenna, transistor, or a triode vacuum tube.

SUBELEMENT N7—PRACTICAL CIRCUITS
[2 exam questions—2 groups]

N7A Functional layout of transmitter, transceiver, receiver, power supply, antenna, antenna switch, antenna feed line, impedance matching device, SWR meter.

N7B Station layout and accessories for telegraphy, radiotelephone, radioteleprinter or packet.

SUBELEMENT N8—SIGNALS AND EMISSIONS
[2 exam questions—2 groups]

N8A Emission types, key clicks, chirps or superimposed hum.

N8B Harmonics and unwanted signals, equipment and adjustments to help reduce interference to others.

SUBELEMENT N9—ANTENNAS AND FEED LINES
[3 exam questions—3 groups]

N9A Wavelength vs antenna length.

N9B Yagi parts, concept of directional antennas, and safety near antennas.

N9C Feed lines, baluns and polarization via element orientation.

SUBELEMENT T1—COMMISSION'S RULES
[5 exam questions—5 groups]

T1A Station control, frequency privileges authorized to the Technician class control operator, term of licenses, grace periods and modifications of licenses.

T1B Emission privileges for Technician class control operator, frequency selection and sharing, transmitter power.

T1C Digital communications, station identification, ID with CSCE.

T1D Correct language, Phonetics, Beacons and Radio Control of model craft and vehicles.

T1E Emergency communications; broadcasting; permissible one-way, satellite and third party communication; indecent and profane language.

SUBELEMENT T2—OPERATING PROCEDURES
[3 exam questions—3 groups]

T2A Repeater operation, courteous operation.

T2B Simplex operations, Q signals, RST signal reporting, repeater frequency coordination.

T2C Distress calling and emergency drills and communications— operations and equipment, Radio Amateur Civil Emergency Service (RACES).

SUBELEMENT T3—RADIO-WAVE PROPAGATION
[3 exam questions—3 groups]

T3A Ionosphere, Ionospheric regions, solar radiation.

T3B Ionospheric absorption, causes and variation, maximum usable frequency.

T3C Propagation, including ionospheric, tropospheric, line-of-sight, scatter propagation, and Maximum Usable Frequency.

SUBELEMENT T4—AMATEUR RADIO PRACTICES
[4 Exam Questions— 4 groups]

T4A Electrical wiring, including switch location, dangerous voltages and currents.

T4B Meters, including volt, amp, multi, peak-reading, RF watt and placement, and ratings of fuses and switches.

T4C Marker generator, crystal calibrator, signal generators and impedance-match indicator.

T4D Dummy antennas, S-meter, exposure of the human body to RF.

SUBELEMENT T5—ELECTRICAL PRINCIPLES
[2 exam questions—2 groups]

T5A Definition of resistance, inductance, and capacitance and unit of measurement, calculation of values in series and parallel.

T5B Ohm's Law.

SUBELEMENT T6—CIRCUIT COMPONENTS
[2 Exam Questions—2 groups]

T6A Resistors, construction types, variable and fixed, color code, power ratings, schematic symbols.

T6B Schematic symbols—inductors and capacitors, construction of variable and fixed, factors affecting inductance and capacitance, capacitor construction.

SUBELEMENT T7—PRACTICAL CIRCUITS
[1 exam question—1 group]

T7A Practical circuits.

SUBELEMENT T8—SIGNALS AND EMISSIONS
[2 exam questions—2 groups]

T8B RF carrier, modulation, bandwidth and deviation.

SUBELEMENT T9—ANTENNAS AND FEED LINES
[3 exam questions—3 groups]

T9A Parasitic beam and non-directional antennas.

T9B Polarization, impedance matching and SWR, feed lines, balanced vs unbalanced (including baluns).

T9C Line losses by line type, length and frequency, RF safety.

KEY WORDS

Amateur operator—A person holding a written authorization to be the control operator of an amateur station.

Amateur service—A radiocommunication service for the purpose of self-training, intercommunication and technical investigations carried out by amateurs, that is, duly authorized persons interested in radio technique solely with a personal aim and without pecuniary interest. (*Pecuniary* means payment of any type, whether money or other goods.)

Amateur station—A station licensed in the amateur service, including necessary equipment, used for amateur communication.

Broadcasting—Transmissions intended to be received by the general public, either direct or relayed.

Control operator—An amateur operator designated by the licensee of a station to be responsible for the transmissions of an amateur station.

Emission—The transmitted signal from an amateur station.

Emission privilege—Permission to use a particular emission type (such as Morse code or voice).

False or deceptive signals—Transmissions that are intended to mislead or confuse those who may receive the transmissions. For example, distress calls transmitted when there is no actual emergency are false or deceptive signals.

Frequency bands—A group of frequencies where amateur communications are authorized.

Frequency privilege—Permission to use a particular group of frequencies.

Malicious (harmful) interference—Intentional, deliberate obstruction of radio transmissions.

MAYDAY—From the French *m'aidez* (help me), MAYDAY is used when calling for emergency assistance in voice modes.

Operator license—The portion of an Amateur Radio license that gives permission to operate an amateur station.

Peak envelope power (PEP)—The average power of a signal at its largest amplitude peak.

SOS—A Morse code call for emergency assistance.

Space station—An amateur station located more than 50 km above the Earth's surface.

Station license—The portion of an Amateur Radio license that authorizes an amateur station at a specific location. The station license also lists the call sign of that station.

Third-party communications—Messages passed from one amateur to another on behalf of a third person.

Third-party communications agreement—An official understanding between the United States and another country that allows amateurs in both countries to participate in third-party communications.

Third-party participation—The way an unlicensed person can participate in amateur communications. A control operator must ensure compliance with FCC rules.

Unidentified communications or signals—Signals or radio communications in which the transmitting station's call sign is not transmitted.

Chapter 2

The Radio Spectrum: A Valuable Resource

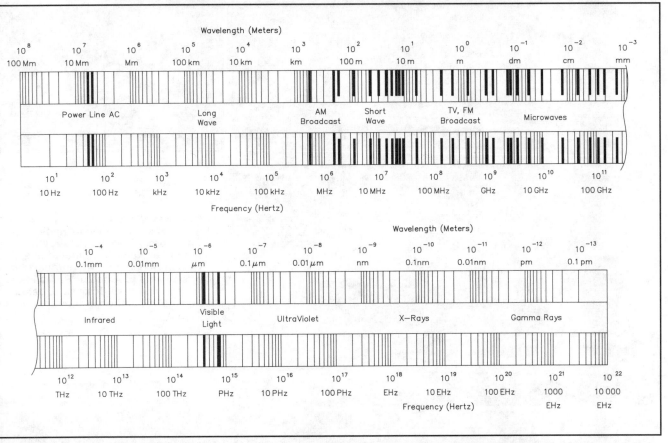

Figure 2-1—The electromagnetic spectrum, showing the range from the 60-hertz ac power line in your home, through the radio frequencies, X-rays and gamma rays. Amateur Radio operations use a small, but significant, portion of these frequency bands. The amateur bands are represented as short bars on the graph.

hen you tune an AM or FM broadcast radio to your favorite station, you select a specific spot on the tuning dial. There are many stations spread across that dial. Each radio station occupies a small part of the entire range of "electromagnetic waves." Other parts of this range, or spectrum, include microwaves, X-rays and even infrared, ultraviolet and visible light waves.

Figure 2-1 shows the electromagnetic spectrum from below the radio range all the way through the X-rays. Amateur Radio occupies only a small part of the total available space; countless users must share the electromagnetic spectrum.

You may be thinking: "Who decides where Amateur Radio frequencies will be, and where my favorite FM broad-

cast station will be?" That's a good question, and the answer has several parts.

Radio signals travel to distant corners of the globe, so there must be a way to prevent total chaos on the bands. The International Telecommunication Union (ITU) has the important role of dividing the entire range of communications frequencies among those who use them. Many radio services have a need for communications frequencies. These services include commercial broadcast, land mobile and private radio (including Amateur Radio). ITU member nations decide which radio services will be given certain bands of frequencies, based on the needs of the different services. This process takes place at ITU-sponsored World Administrative Radio Conferences (WARCs).

Nearly every country on earth has its share of active Amateur Radio operators. Above left, a direction-finding exercise in China; above, a ham in Rome; left, two enthusiastic newcomers to Amateur Radio in the African nation of Botswana; below, an amateur station in the Kingdom of Tonga.

WHY IS THERE AMATEUR RADIO?

In the case of Amateur Radio, the ITU has long recognized that hams are invaluable in times of emergency or disaster. At an ITU WARC held in 1979, Amateur Radio gained several new bands of frequencies. Mighty praise indeed!

The ITU makes these allocations on an international basis. The Federal Communications Commission (FCC) decides the best way to allocate frequency bands to those services using them in the US. The FCC is the United States governing body when it comes to Amateur Radio. In fact, an entire part of the FCC Rules is devoted to the amateur service—Part 97. As your amateur career develops, you'll become more familiar with Part 97.

To gain the privilege of sending a radio signal over the airwaves in the US, you must pass a license exam. To earn a Novice license, you will need to pass a 30-question exam covering basic radio theory and FCC regulations, as well as a Morse code exam. To earn a Technician license, you will need to pass two written exams—the 30-question Novice written test plus a more comprehensive 25-question written exam. There is no Morse code requirement for the Technician exam. Whether you choose to learn the code and try for the Novice license or work toward the code-free Technician license, this book will prepare you to earn your first Amateur Radio license.

THE FIVE PRINCIPLES

In Section 97.1 of its Rules, the FCC describes the basis and purpose of the amateur service. It consists of five principles:

(a) Recognition and enhancement of the value of the amateur service to the public as a voluntary noncommercial communication service, particularly with respect to providing emergency communications.

Probably the best-known aspect of Amateur Radio is our ability to provide life-saving emergency communications. Normal communications channels often break down during hurricanes, earthquakes, tornadoes, airplane crashes and other disasters. Amateur Radio is frequently the first available means of contact with the outside world from the

---

In the days before the giant eruption that demolished vast areas around the Mount St Helens volcano in Washington, volunteer amateurs monitored its behavior. Hams are ready to do whatever they can when someone needs their communications services. Because of the suddenness and extent of that eruption, two hams lost their lives while helping to monitor the volcano.

affected area. Red Cross and other civil-defense agencies rely heavily on the services of volunteer radio amateurs.

One of the more noteworthy aspects of Amateur Radio is its noncommercial nature. In fact, amateurs may not accept any form of payment for operating their radio stations. (There is one limited exception to this rule, which we will explain later.) This means that hams make their services available free of charge. This is true whether they are assisting a search-and-rescue operation in the Sierra Nevada, relaying health-and-welfare messages from a disaster-stricken Caribbean island or providing communications assistance at the New York City Marathon. Talk about value!

Why do hams work so hard if they can't be paid? It gives them an immense feeling of personal satisfaction! It's like the good feeling you get when you lend a hand to an elderly neighbor, only on a much grander scale. Hams operate their stations only for personal satisfaction and enjoyment. They don't talk about business matters on the air. (FCC rules don't allow business communications over Amateur Radio.)

(b) Continuation and extension of the amateur's proven ability to contribute to the advancement of the radio art.

In the early days of radio there were no rules, but in 1912 Congress passed a law to regulate the airwaves. Amateurs had to keep to a small range of frequencies (known as "short waves"). There they would remain "out of the way"—everyone knew that radio waves couldn't travel very far at those frequencies. Ha! Amateurs soon overcame the restrictions. They were among the first to experiment with radio propagation, the study of how radio waves travel through the atmosphere. When vacuum tubes became available, amateurs began to develop much-improved radio communication circuits.

Today, the traditions and spirit in Amateur Radio remain. Amateurs continue to experiment with state-of-the-art technologies. Advancement of the radio art takes a major portion of an amateur's energies. The FCC promotes this amateur experimentation and technical development by establishing rules that are consistent with amateur techniques.

Section 97.1 of the amateur rules continues:

(c) Encouragement and improvement of the amateur service through rules which provide for advancing skills in both the communications and technical phases of the art.

Along with some of the technical aspects of the service, amateurs also hold special training exercises in preparation for communications emergencies. Simulated Emergency Tests and Field Days, where amateurs practice communicating under emergency conditions, are just two ways amateurs sharpen their operating skills.

The Commission's rules even specify a "service within a service" called the Radio Amateur Civil Emergency Service, or RACES. RACES provides amateur communications assistance to federal, state and local civil defense in times of need.

(d) Expansion of the existing reservoir within the amateur radio service of trained operators, technicians and electronics experts.

Self-training, station operating standards, intercommunication and technical investigation are all important parts of the amateur service. We need more amateurs who are experienced in communications methods, because they are a national resource to the public.

(e) Continuation and extension of the amateur's unique ability to enhance international goodwill.

Hams are unique, even in this time of worldwide jet travel. They journey to the far reaches of the earth and talk with amateurs in other countries every day. They do this simply by walking into their ham shacks. International peace and coexistence are very important today. Amateurs represent their countries as ambassadors of goodwill. Amateur-to-amateur communications often cross the cultural boundaries between societies. Amateur Radio is a teacher in Lincoln, Nebraska, trading stories with the headmaster of a boarding school in a London suburb. It is a tropical-fish hobbyist learning about fish in the Amazon from a missionary stationed in Brazil. Amateur Radio is a way to make friends with other people everywhere.

These five principles provide the basis and purpose for the amateur service, as set down by the FCC. These principles place a large responsibility on the amateur community—a responsibility you will share. It is the Commission's duty to ensure that amateurs are able to operate their stations properly, without interfering with other radio services. All amateurs must pass an examination before the FCC will issue a license authorizing amateur station operation. It's a very serious matter.

While the FCC expects you to operate your station properly, the construction of your station is up to you. There are no FCC requirements for amateur station construction.

[At this point you should turn to Chapter 12 and study the questions that cover this material in the Novice question pool. Before you turn to Chapter 12, though, it may be helpful to understand a bit about the numbering system used for these questions. The numbers match the study guide or syllabus printed at the end of Chapter 1. The syllabus forms a type of outline, and the numbering system follows this outline format. There are nine subelements, labeled N1 through N9. This chapter covers Subelement N1 and part of Subelement N2, so all the question references will be to questions with numbers that begin "N1" or "N2."

Each subelement is further divided into question groups. These groups represent the number of questions to be selected from each subelement. There are ten questions on the exam from Subelement N1, Commission's Rules. So there are 10 question groups labeled N1A through N1J. Individual questions end with a two-digit number, such as N1A01, N1B11, N1E13 and so on.]

[You should study questions N1A01 through N1A06 in Chapter 12 now. If you have difficulty with any of these questions, review the material in this section.]

THE AMATEUR SERVICE

The FCC defines some important terms in Section 97.3 of the amateur rules. The **amateur service** is "A radio communication service for the purpose of self-training, intercommunication and technical investigations carried out by amateurs, that is, duly authorized persons interested in radio technique solely with a personal aim and without pecuniary interest." *Pecuniary* means related to money or other payment. In other words, you can't be paid for operating your station or providing a communications service for some individual or group. Amateurs learn various communications skills on their own, and carry out technical experiments in electronics and radio principles.

An **amateur operator** is a person holding a valid license to operate an amateur station. In the United States, the Federal Communications Commission issues these licenses. A US amateur license allows you to operate wherever the amateur service is regulated by the FCC. An amateur operator performs communications in the amateur service.

How does the FCC define an **amateur station**? "A station licensed in the amateur service, including the apparatus necessary for carrying on radiocommunications." The person operating an amateur station has an interest in self-training, intercommunication and technical investigations or experiments.

An Amateur Radio license is really two licenses in one—an **operator license** and a **station license**. The operator license is one that lets you operate a station within your authorized privileges on amateur-service frequencies. The station license authorizes you to have an amateur station. It also lists the call sign that identifies that station. Figure 2-2 shows an actual Amateur Radio license.

One piece of paper includes both licenses. You won't even be able to tell which part is which license. It is important to realize that your license includes both of these parts, however.

The operator license lists your license class and gives you the authority to operate an amateur station. The station license includes the address of your primary, or main, Amateur Radio station. The station license also lists the call sign of your station. It is your written authorization for an amateur station.

You must have your original operator license, or a photocopy, in your possession any time you are operating an amateur station. You are using the *operator license* portion when you serve as the control operator of a station. It is a good idea to carry a copy of your license in your wallet or purse at all times. That way you are sure to have it if a chance to operate arises.

You must also have your original station license, or a photocopy of it, in your station any time someone operates

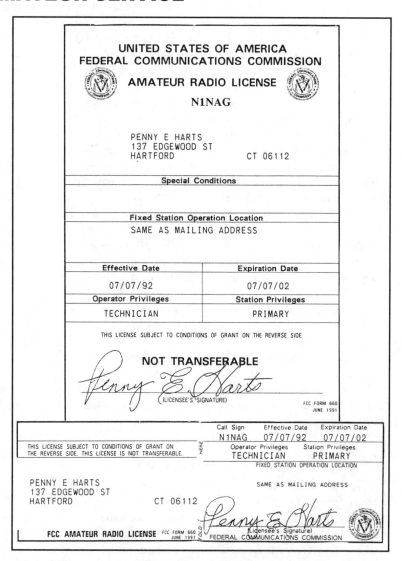

Figure 2-2—An FCC Amateur Radio license consists of two parts— the station license and the operator's license.

the station. The station license can be in the control operator's possession or you can have it on display in the station.

Amateur licenses are printed on a laser printer, and issued in two parts (Figure 2-2). One part is small enough to carry with you; the other can be framed and displayed in your shack. This means you can carry your license *and* display it! Laser ink can smear, so it's a good idea to have your wallet copy laminated. If the small part is too large for your wallet, you can make a reduced-size copy on a photocopier. If you carry a copy, you can leave your original safely at home. Although you can legally carry a copy, your original license must be available for inspection by any US government official or FCC representative, however. Don't lose the original license!

The FCC issues all licenses for a 10-year term. You should always renew your license for another 10 years, about 60 to 90 days before the present one expires. If your license is issued in November 1993, it expires in November 2003. It's a good idea to form the habit of looking at the expiration date on your license every now and then. You'll be less likely to forget to renew the license in time.

If you do forget to renew your license, you have up to two years to apply for a new license. After the two-year grace period, you will have to take the exam again. Your license is not valid during this two-year grace period, however. You may not operate an amateur station with an expired license. All the grace period means is that the FCC will renew the license if you apply during that time.

If your license is lost, mutilated or destroyed, request a new one from the FCC. A Form 610 isn't required; a letter will do. Be sure to explain why you are requesting a new license. (You don't have to explain that your dog ate your license—just that it was destroyed.) If you move or change your name, you will need to modify your license. Notify the FCC of the new information on an FCC Form 610 (available from ARRL). Then FCC will be able to write to you if necessary.

[Turn to Chapter 12 and study questions N1A07, N1A08, N1A09, N1B01, N1B03, N1B04, N1B06 and N1B10. Also study questions N1D05, N1D06, N1D10 and N1D11. Review this section if you have difficulty with any of these questions.]

THE CONTROL OPERATOR

A **control operator** is an "amateur operator designated by the licensee of a station to be responsible for the transmissions from that station to assure compliance with the FCC Rules." In effect, the control operator operates the Amateur Radio station. Only a licensed ham may be the control operator of an amateur station. If another licensed radio amateur operates your station with your permission, he or she assumes the role of control operator.

If you let another amateur with a higher class license than yours control your station, he or she may use any operating privileges allowed by the higher class license. On the other hand, if you are the control operator at the station of another amateur who has a higher class of license than yours, you can use only the privileges allowed by your license.

Any amateur operator may designate another licensed operator as the control operator, to share the responsibility of station operation. The FCC holds both the control operator and the station licensee responsible for proper operation of the station. If you are operating your own Amateur Radio station, then you are the control operator at that time.

A control operator must be present at the station control point whenever a transmitter is operating. (A control operator can control any number of transmitters.) This means that you may not allow an unlicensed person to operate your radio transmitter while you are not present. There is one time when a transmitter may be operated without a control operator being present, however. Some types of stations, such as repeater stations, may be operated by automatic control. In this case there is no control operator at the transmitter control point.

Your Novice or Technician license authorizes you to be the control operator of an amateur station in the Novice or Technician frequency bands. This means you can be the control operator of your own station or someone else's station. In either case, you are responsible to the FCC for the proper operation of the station.

A Canadian citizen who holds a Canadian amateur license can be a control operator of a US amateur station. So can a ham from another country who has a reciprocal operating permit issued by the FCC. Otherwise, the control operator must hold a valid US amateur license.

If you allow another licensed ham to operate your station, you are still responsible for its proper operation. You are always responsible for the proper operation of your station. Your primary responsibility as the station licensee is to ensure the proper operation of your station. It isn't necessary to use a station logbook. If you don't, though, it will not be possible to show when someone else has been the control operator of your station. Whenever you give someone else permission to be the control operator at your station, enter the other person's name and call sign, and the time, date and frequencies used, in your logbook.

[Before you go on to the next section, you should turn to Chapter 12 and study questions N1A10, N1B02, N1B05, N1B07 and N1G01 through N1G11. If you have difficulty with any of these questions, review the material in this section.]

AMATEUR LICENSE CLASSES

There are five kinds, or levels, of amateur license. They vary in degree of knowledge required and **frequency privileges** granted. Higher class licenses have more comprehensive examinations. In return for passing a more difficult exam you earn more frequency privileges (frequency space and modes of operation).

The first step is either the Novice or Technician license. The FCC issues these "beginner's" licenses to those who demonstrate the ability to operate an Amateur Radio transmitter safely and properly.

An applicant for a Novice license must show a basic proficiency in Morse code by passing a test at 5 words per minute (WPM). The written part of the exam covers some very basic radio fundamentals and knowledge of a few rules and regulations. With a little study and some common sense, you'll soon be ready to pass the Novice exam.

An applicant for a Technician license must pass a slightly more difficult written test, but there is no Morse code requirement.

Anyone (except an agent or representative of a foreign government) is eligible to qualify for an Amateur Radio operator license. There is no age requirement. To hold an Amateur Radio station license, you must have a valid operator license. (Remember, both licenses are printed on the same piece of paper.)

A Novice or Technician license gives you the freedom to develop operating and technical skills through on-the-air experience. These skills will help you upgrade to a higher class of license, with additional privileges.

As a Novice, you can communicate with other amateur stations in the exotic reaches of the world. You can provide public service through emergency communications and message handling. You can enhance international goodwill just like operators with higher license classes. You also can use repeater stations on three popular amateur bands.

As a Technician, you can use a wide range of frequency bands—all amateur bands above 30 MHz, in fact. You'll use repeaters, packet radio and orbiting satellites to relay your signals over a wider area. By passing the 5-WPM Morse code test you also will gain Novice privileges below 30 MHz.

Later, you will probably want to earn greater operating privileges. As you gain operating experience, you will prepare for a higher-class license—the gateway to more frequencies and modes.

There is a real thrill to having more frequencies available to talk with other amateurs on the far side of the globe. That is a powerful incentive to upgrade to the General class license (the next step up from Technician). You will have to pass a 13-WPM Morse code exam and another theory and rules exam to earn the General class license. The General class license gives voice privileges on eight high-frequency (HF) bands. These bands typically carry signals over great distances. For this reason, the General class license is very popular.

As you progress and mature in Amateur Radio, you will develop specialized interests in such exciting modes as amateur television and satellite communication. You'll also want even more privileges, and that is where the Advanced and Amateur Extra licenses come in. To obtain one of these, you must be prepared to face exams on the more technical aspects of the hobby. For the Amateur Extra license, the top-of-the-line, there is also a 20-word-per-minute code test.

To qualify for a higher-class license, you must pass the theory exam for each level up to that point. For example, suppose you want to go from Novice directly to Amateur Extra. You will have to take the Technician, General and Advanced theory exams before you take the Amateur Extra test. (You can just take the 20-WPM code test, however, without passing the 13-WPM General class code test.) In fact, you could take the Novice through Extra exams all at once.

[You should turn to Chapter 12 now and study questions N1B08, N1B09, N1B11, N1D01 and N1D02. Review the material in this section if you have difficulty with any of these questions.]

THE NOVICE LICENSE

A Novice license allows you to operate on portions of six Amateur Radio bands. We normally identify these bands by specifying the frequency range they cover or by listing the wavelength in meters. (You will learn more about the relationship between frequency and wavelength later in this book.) Table 2-1 lists the frequency range for each of the Novice bands. This table also serves as a comparison be-

Table 2-1

Amateur Operator Licenses†

Class	Code Test	Written Examination	Privileges
Novice	5 WPM (Element 1A)	Novice theory and regulations. (Element 2)	Telegraphy on 3675-3725, 7100-7150 and 21,100-21,200 kHz with 200 watts PEP output maximum; telegraphy, RTTY and data on 28,100-28,300 kHz and telegraphy and SSB voice on 28,300-28,500 kHz with 200 W PEP max; all amateur modes authorized on 222-225 MHz, 25 W PEP max; all amateur modes authorized on 1270-1295 MHz, 5 W PEP max.
Technician		Novice theory and regulations; Technician-level theory and regulations. (Elements 2 and 3A)*	All amateur privileges above 50.0 MHz. Technician-class licensees who have passed a 5-WPM code test also have HF Novice privileges.
General	13 WPM (Element 1B)	Novice theory and regulations; Technician and General theory and regulations. (Elements 2, 3A and 3B)	All amateur privileges except those reserved for Advanced and Amateur Extra class; see Table 2-2.
Advanced	13 WPM (Element 1B)	All lower exam elements, plus Advanced theory. (Elements 2, 3A, 3B and 4A)	All amateur privileges except those reserved for Amateur Extra class; see Table 2-2.
Amateur Extra	20 WPM (Element 1C)	All lower exam elements, plus Extra-class theory. (Elements 2, 3A, 3B, 4A and 4B)	All amateur privileges.

† A licensed radio amateur will be required to pass only those elements that are not included in the examination for the amateur license currently held.

*If you hold a valid Technician-class license issued before March 21, 1987, you also have credit for Element 3B. You must be able to prove your Technician license was issued before March 21, 1987 to claim this credit.

tween Novice and Technician privileges and those given to higher class licensees.

Frequency Privileges

When operating, you must stay within your assigned frequency bands. Novice operators may transmit on portions of six bands in the radio spectrum. Amateurs usually refer to these frequency bands by their wavelength. Novices may operate in the 80, 40, 15 and 10-meter bands, and in the 1.25-meter (222-MHz) and 23-centimeter (1270-MHz) bands. These bands are further divided into subbands for the different classes of license. Novice operators have **frequency privileges** (or permission to operate) on these subbands:

3675-3725 kHz in the 80-meter band

(5167.5 kHz—Alaska only, emergency communications, single-sideband voice emissions only)

7100-7150 kHz in the 40-meter band*

21,100-21,200 kHz in the 15-meter band

28,100-28,500 kHz in the 10-meter band

222-225 MHz in the 1.25-meter band*

1270-1295 MHz in the 23-centimeter band

*In ITU Region 2 only. ITU Region 2 comprises all of North America, the Caribbean and the Eastern Pacific, including Hawaii. Contact ARRL Headquarters for more information.

We use the metric system of measurement in electronics. Chapter 3 includes a full explanation of the metric system and terms like *kilo, mega* and *centi* that we have used here. For now it is important that you know that *kilohertz* and *megahertz* are measures of the frequency of a radio signal. The *hertz* (Hz) is the basic unit of frequency. Kilo means thousand and mega means million. We can list any of the amateur frequency bands in either kilohertz or megahertz. For example, the 80-meter Novice band can be written either as 3.675 to 3.725 MHz, or 3675 to 3725 kHz.

[You must memorize these Novice frequency privileges. Before you go on to the next section, turn to Chapter 12 and study questions N1C01 through N1C11. Review these frequencies often as you study the remaining material in this book.]

Emission Privileges

Amateur operators transmit a wide variety of signals. These include Morse code, radioteletype, several types of voice communications and even television pictures. An **emission** is any radio-frequency (RF) signal from a transmitter.

There is a system for describing the various types of signals (or emissions) found on the amateur bands. Different modes are given identifiers, called emission types.

You should be familiar with the various emission types. The FCC lists the **emission privileges** for each license class by giving the emission types each may use. An emission

privilege is the FCC permission to use a particular emission type, such as Morse code or single-sideband phone. As a Novice licensee, you will be permitted to use all of the emission types (except pulse) on at least one frequency band. The emission types defined by the FCC are:

- CW—Morse code telegraphy.
- Data—Computer communications modes, often called *digital communications* because digital computers are used.
- Image—Television and facsimile communications.
- MCW—(Tone-modulated CW) Morse code telegraphy using a keyed audio tone.
- Phone—Speech (voice) communications.
- Pulse—Communications using a sequence of controlled signal variations.
- RTTY—Direct-printing telegraphy communications (received by automatic techniques). Since digital computers are often used on RTTY, these signals are also often called *digital communications*.
- SS—Spread-spectrum communications in which the signal energy is spread across a wide bandwidth.
- Test—Transmissions containing no information.

On 80, 40 and 15 meters, Novices may transmit only Morse code (CW). The transmitter produces this Morse code signal by *keying* (switching on and off) the signal from a continuous-wave (CW) transmitter. On 10 meters, Novices may use CW from 28.1 to 28.5 MHz. Novice operators may also transmit radioteletype (RTTY) and data from 28.1 to 28.3 MHz, and single-sideband phone from 28.3 to 28.5 MHz. On the frequencies 222.1-223.91 MHz and 1270-1295 MHz, Novices may use all emissions authorized to higher-class licensees on these bands, including FM phone and digital modes, such as packet radio. Table 2-2 summarizes the amateur band limits and operating modes for each license class.

Hams often describe radioteletype (RTTY) and packet radio or other data emissions as *digital communications*. These are signals intended to be received and printed or displayed on a computer screen automatically. Information transferred directly from one computer to another is an example of digital communications.

[You have to memorize these Novice emission privileges. You should turn to Chapter 12 now and study questions N1E01 through N1E14. Review the material in this section if you have difficulty with any of the questions.]

Novice Transmitter Power

The FCC has issued rules explaining how transmitter power should be measured at the output of the transmitter. Novice licensees may use a maximum of 5 W **peak envelope power (PEP)** output on the 1270-MHz band, 25 W PEP on the 222-MHz band, and 200 W PEP on the 80, 40, 15 and 10-meter bands. The 200 W limitation also applies to any other licensed radio amateur who operates in the 80, 40 and 15-

Table 2-2

160 METERS

E,A,G

1800 — 1900 — 2000 kHz

Amateur stations operating at 1900–2000 kHz must not cause harmful interference to the radiolocation service and are afforded no protection from radiolocation operations.

80 METERS

3525   3675   3725   3850
              3775
3500   3750   4000 kHz

N,T †
G
A
E

5167.5 kHz (SSB only): Alaska emergency use only.

40 METERS

7025   7100   7150   7225
7000   7150   7300 kHz

N,T †
G *
A *
E *

* Phone operation is allowed on 7075–7100 kHz in Puerto Rico, US Virgin islands and areas of the Caribbean south of 20 degrees north latitude; and in Hawaii and areas near ITU Region 3, including Alaska.

30 METERS

E,A,G

10,100   10,150 kHz

Maximum power on 30 meters is 200 watts PEP output. Amateurs must avoid interference to the fixed service outside the US.

20 METERS

14,025   14,150   14,225
              14,175
14,000   14,150   14,350 kHz

G
A
E

17 METERS

18,068   18,110   18,168 kHz

E,A,G

15 METERS

21,100   21,200
21,025   21,300
              21,225
21,000   21,200   21,450 kHz

N,T †
G
A
E

12 METERS

24,890   24,930   24,990 kHz

E,A,G

10 METERS

28,100 28,500
28,000 28,300   29,700 kHz

N,T †
E,A,G

Novices and Technicians are limited to 200 watts PEP output on 10 meters.

6 METERS

50.1
50.0   54.0 MHz

E,A,G,T

2 METERS

144.1
144.0   148.0 MHz

E,A,G,T

1.25 METERS

222.0   225.0 MHz

E,A,G,T,N

Novices are limited to 25 watts PEP output from 222 to 225 MHz.

70 CENTIMETERS **

420.0   450.0 MHz

E,A,G,T

33 CENTIMETERS **

902.0   928.0 MHz

E,A,G,T

23 CENTIMETERS **

1270   1295
1240   1300 MHz

N
E,A,G,T

Novices are limited to 5 watts PEP output from 1270 to 1295 MHz.

US AMATEUR BANDS

February 1, 1994

US AMATEUR POWER LIMITS

At all times, transmitter power should be kept down to that necessary to carry out the desired communications. Power is rated in watts PEP output. Unless otherwise stated, the maximum power output is 1500 W. Power for all license classes is limited to 200 W in the 10,100–10,150 kHz band and in all Novice subbands below 28,100 kHz. Novices and Technicians are restricted to 200 W in the 28,100–28,500 kHz subbands. In addition, Novices are restricted to 25 W in the 222–225 MHz band and 5 W in the 1270–1295 MHz subband.

Operators with Technician class licenses and above may operate on all bands above 50 MHz. For more detailed information see The FCC Rule Book.

KEY

▨ = CW, RTTY and data

▢ = CW, RTTY, data, MCW, test, phone and image

■ = CW, phone and image

▨ = CW and SSB

▨ = CW, RTTY, data, phone, and image

□ = CW only

E = EXTRA CLASS
A = ADVANCED
G = GENERAL
T = TECHNICIAN
N = NOVICE

† Only Technician–class licensees who have passed a 5 – WPM code test may use these frequencies.

** Geographical and power restrictions apply to these bands. See The FCC Rule Book for more information about your area.

Above 23 Centimeters:

All licensees except Novices are authorized all modes on the following frequencies:
2300–2310 MHz
2390–2450 MHz
3300–3500 MHz
5650–5925 MHz
10.0–10.5 GHz
24.0–24.25 GHz
47.0–47.2 GHz
75.5–81.0 GHz
119.98–120.02 GHz
142–149 GHz
241–250 GHz
All above 300 GHz

For band plans and sharing arrangements, see The ARRL Operating Manual or The FCC Rule Book.

meter Novice bands. Remember that Novices may use up to 200 W PEP on any Novice frequency in the 80, 40, 15 and 10-meter bands.

In addition to these Novice bands, higher-class licensees may use a maximum of only 200 W PEP on the 30-meter band. On all other bands, the rules limit the maximum transmitter output power in the amateur service to 1500 W PEP output. Higher-class licensees may use 1500 W PEP in the Novice sections of 10 meters, 222 MHz and 1270 MHz. In practice, unless you're attempting something as challenging as "moonbounce"—bouncing a signal off the moon—amateurs rarely use more than a couple of hundred watts on the VHF and UHF bands.

So far, we have been talking about the *maximum* transmitter power the rules allow. There is another rule to consider, though. According to FCC rules, an amateur station must use the *minimum* transmitter power necessary to maintain reliable communication. What this means is simple—if you don't need 200 W to contact someone, don't use it!

[Before proceeding to the next section turn to Chapter 12 and study questions N1F01 through N1F11. Review this section if you have any difficulties.]

THE NOVICE EXAM

The FCC refers to the various exams for Amateur Radio licenses as exam *Elements*. For example, exam Elements 1A, 1B and 1C are the 5, 13 and 20-word-per-minute (WPM) code exams. Element 2 is the Novice written exam. Table 2-1 summarizes the exams and privileges that go with each license class.

The table shows that you must pass exam Element 1A and Element 2 for the Novice license. The purpose of Element 1A is to prove your ability to send and receive messages using the international Morse code at a speed of 5 words per minute. The Element 2 written exam covers basic FCC Rules and Novice operating procedures, along with basic electronics theory.

[Now turn to Chapter 12 and study question N1D03. Review this section if you have difficulty with this question.]

What Will My Novice Exam Be Like?

Three amateurs with General class licenses or higher, who are 18 or older, can give the test for a Novice license. The examiners must not be relatives of anyone taking the exam. The exams can be given at the convenience of the candidates and the examiners, at any location they agree to. (All rules of the VEC program must be followed.)

A Volunteer Examiner (VE) is someone who volunteers to test others for amateur licenses. There is probably an active Volunteer Examining Team somewhere in your area. These teams conduct tests for all Amateur Radio licenses. The Volunteer Examiners give Novice exams at all of their regular exam sessions.

If you have any trouble locating examiners, write to: Educational Activities Department, ARRL Headquarters, 225 Main Street, Newington, CT 06111. We will refer you to someone near you who can arrange your test. We will also supply a list of ARRL/VEC Volunteer Exam sessions in your area. VEC is an abbreviation for Volunteer Examiner Coordinator. The ARRL VEC oversees the thousands of ARRL Volunteer Examiners, arranges and publishes exam schedules and inspects application forms before they are submitted to the FCC.

The code test is normally given first. The examiners will usually send five minutes of 5-word-per-minute code, and then test your copy. The test usually takes one of two forms. The examiners may ask you 10 questions based on the contents of the transmission, and you must answer 7 of the 10 questions correctly. Or, the examiners may check your answer sheet for one minute of perfect ("solid") copy (25 characters in a row with no errors) out of the five-minute transmission.

The written test consists of 30 questions on general operating practices, rules and regulations, and basic radio theory. To pass, you must correctly answer 22 of the 30 questions on your exam.

Your exam must include 30 questions taken from the pool of questions printed in Chapter 12 of this book. Your examiners should choose one question from each of the 30 question-pool subsections. The questions must be used exactly as they are printed in Chapter 12. Chapter 12 has the complete pool of questions with multiple-choice answers.

[Turn to Chapter 12 and study question N1A11. If you have trouble answering this question review this section.]

THE TECHNICIAN LICENSE

With Technician license in hand, you'll be able to operate on all the VHF, UHF and microwave bands allocated to the amateur service—and there are a lot of them! Table 2-2 lists the frequency range for the most popular amateur bands. This table also shows the frequency and emission privileges for each license class on these bands. For a complete list of the amateur bands, including those in the largely experimental microwave region, see *The FCC Rule Book*, published by ARRL.

If you decide to learn the international Morse code and are able to pass a 5-word-per-minute code test, you will have earned an enhanced Technician license. A Certificate of Successful Completion of Examination from a VE session showing you passed a Morse code exam (in addition to your regular Technician class license) allows you to operate on the four HF Novice subbands that normally allow for direct worldwide communication.

THE TECHNICIAN EXAM

The Technician class exam consists of two written tests

Elements 2 and 3A. Chapters 12 and 13 of this book contain every question in the question pools from which your exam will be composed. If you study this book carefully, you will be able to pass the Technician class exam. If you wish to qualify for the enhanced Technician license, you'll also need a computer program or cassette tapes that teach international Morse code. The ARRL offers a set of two 90-minute cassettes called *Your Introduction to Morse Code*. The ARRL also offers *Morse Tutor* for IBM-PC and compatible computers.

What Will the Exam Be Like?

All US exams are administered by volunteer examiners who are certified by a Volunteer Examiner Coordinator (VEC). *The FCC Rule Book*, published by ARRL, contains details on the VE program.

To qualify for a Technician class license, you must pass FCC Exam Elements 2 and 3A. If you already hold a valid Novice license, then you have credit for passing Elements 1A and 2, and will not have to retake them. In that case when you upgrade to Technician, you will retain your Novice HF privileges.

The Element 3A exam consists of 25 questions taken from a pool of more than 250. The FCC specifies nine subelements, or divisions, for each exam element. These are the Commission's Rules, Operating Procedures, Radio-Wave Propagation, Amateur Radio Practices, Electrical Principles, Circuit Components, Practical Circuits, Signals and Emissions, and Antennas and Feed Lines.

The question pools for all amateur exams are maintained by a Question Pool Committee selected by the Volunteer Examiner Coordinators. The FCC allows Volunteer Examiners to select the questions for an amateur exam, but they must use the questions exactly as they are released by the VEC that coordinates the test session. If you attend a test session coordinated by the ARRL/VEC, your test most likely will be designed by the ARRL/VEC, and the questions and answers will be exactly as they are printed in Chapters 12 and 13.

Finding an Exam Opportunity

To determine where and when an exam will be given, contact the ARRL/VEC Office, or watch for announcements in the Hamfest Calendar and Coming Conventions columns in *QST*. Many local clubs sponsor exams, so they are another good source of information on exam opportunities. ARRL officials such as Directors, Vice Directors and Section Managers receive notices about test sessions in their area. See page 8 in the latest issue of *QST* for names and addresses.

To register for an exam, send a completed Form 610 to the VE Team responsible for the exam session if preregistration is required. Otherwise, bring the form to the session. Registration deadlines, and the time and location of the exams, are mentioned prominently in publicity releases about upcoming sessions.

FILLING OUT YOUR FCC FORM 610

What's next? More paperwork! All applications for new amateur licenses, or modifications or renewals, are made on an FCC Form 610. Only Forms with a revision date of November 1993 or later (found in the lower right corner) are acceptable. For a copy of the latest Form 610, write to: Form 610, ARRL, 225 Main Street, Newington, CT 06111-1494. Please enclose a self-addressed, stamped envelope.

You should complete the top half of the form, Section 1, as shown in Figure 2-3. Be sure to write your birth year in box 2. Probably the most common error is to write the current year in this space. Check box 4A, "Examination for a new license" if you don't already have a license. Skip lines 5 and 7, unless this is not your first license or you have another Form 610 waiting for the FCC to process it.

Unless you plan to install an antenna over 300 feet high, or your permanent station location will be in a designated wilderness area, wildlife preserve or nationally recognized scenic and recreational area, check "NO" on line 6.

Your Volunteer Examiners will fill out the Administering VE's report, Section 2 on the bottom half of the Form 610. This is where the examiners certify that you have passed the exam. They will check the appropriate box in section A and fill out sections C and D. The three Volunteer Examiners will print their names, call signs and the date of the exam, and sign the form on the lines at the bottom.

The back of the Form 610 is for a physician's certification that you have a handicap that makes it impossible for you to pass a 13 or 20 wpm Morse code exam. Detailed instructions to the physician are included on the bottom of the Form.

The 610 Form serves as the application for your license. You will use the same form to renew your license before it expires. This form also serves another very important purpose. Line 3 provides the FCC with a mailing address where they can contact you. If you move, or your address changes for any reason, use an FCC Form 610 to notify them of your new address. It's important that you receive any mail sent to you by the FCC. If you don't respond to an FCC letter concerning a rules violation, you may be fined or your license may be suspended or revoked!

Taking the Exam

By the time examination day rolls around, you should have already prepared yourself. This means getting your

Approved by OMB
3060-0003
Expires 08/31/96
See instructions for
information regarding
public burden estimate.

APPLICATION FORM 610 FOR
AMATEUR OPERATOR/PRIMARY STATION LICENSE

SECTION 1 - TO BE COMPLETED BY APPLICANT (See instructions)

1. Print or type last name **Bedlack**   Suffix   First name **Tracy**   Middle initial **A**

2. Date of birth **03-29-68**
month   day   year

3. Mailing address (Number and street) **225 Main St.**   City **Newington**   State code **CT**   ZIP code **06111**

4. I HEREBY APPLY FOR (make an X in the appropriate box(es)):

4A. ☒ EXAMINATION for a new license

4B. ☐ EXAMINATION for upgrade of my operator license class

4C. ☐ CHANGE my name on my license to my new name in Item 1. My former name was:

_____
(Last name)   (Suffix)   (First name)   (MI)

4D. ☐ CHANGE my mailing address on my license to my new address in Item 3

4E. ☐ CHANGE my station call sign systematically (See instructions)
Applicant's Initials _____

4F. ☐ RENEWAL of my license

5. Unless you are requesting a new license, attach the original or a photocopy of your license to the back of this Form 610 and complete Items 5A and 5B.

5A. Call sign shown on license

5B. Operator class shown on license

6. Would an FCC grant of your request be an action that may have a significant environmental effect?   ☒ NO   ☐ YES (Attach required statement)

7. If you have filed another Form 610 that we have not acted upon, complete Items 7A and 7B.

7A. Purpose of other form

7B. Date filed _____ _____ _____
month   day   year

WILLFUL FALSE STATEMENTS MADE ON THIS FORM ARE PUNISHABLE BY FINE AND/OR IMPRISONMENT, (U.S. CODE, TITLE 18, SECTION 1001), AND/OR REVOCATION OF ANY STATION LICENSE OR CONSTRUCTION PERMIT (U.S. CODE, TITLE 47, SECTION 312(A)(1)) AND/OR FORFEITURE (U.S. CODE, TITLE 47, SECTION 503).

I CERTIFY THAT ALL STATEMENTS AND ATTACHMENTS ARE TRUE, COMPLETE, AND CORRECT TO THE BEST OF MY KNOWLEDGE AND BELIEF AND ARE MADE IN GOOD FAITH; THAT I AM NOT A REPRESENTATIVE OF A FOREIGN GOVERNMENT; THAT I WAIVE ANY CLAIM TO THE USE OF ANY PARTICULAR FREQUENCY REGARDLESS OF PRIOR USE BY LICENSE OR OTHERWISE; AND THAT THE STATION TO BE LICENSED WILL BE INACCESSIBLE TO UNAUTHORIZED PERSONS.

8. Signature of applicant (Do not print, type, or stamp.)   (Must match name in Item 1.)   **X Tracy A. Bedlack**

9. Date signed **03-16-94**
month   day   year

SECTION 2 - TO BE COMPLETED BY ALL ADMINISTERING VE's

A. Applicant is qualified for operator license class:

☐ NOVICE   (Elements 1(A), 1(B), or 1(C) and 2)
☒ TECHNICIAN   (Elements 2 and 3(A))
☐ TECHNICIAN PLUS   (Elements 1(A), 1(B), or 1(C), 2 and 3(A))
☐ GENERAL   (Elements 1(B) or 1(C), 2, 3(A) and 3(B))
☐ ADVANCED   (Elements 1(B) or 1(C), 2, 3(A), 3(B) and 4(A))
☐ AMATEUR EXTRA   (Elements 1(C), 2, 3(A), 3(B), 4(A) and 4(B))

B. VEC receipt date:

C. Name of Volunteer-Examiner Coordinator (VEC): **ARRL**

D. Date of VEC coordinated examination session: **3/16/94**

E. Examination session location: **225 Main St. Newington, CT 06111**

I CERTIFY THAT I HAVE COMPLIED WITH THE ADMINISTERING VE REQUIREMENTS IN PART 97 OF THE COMMISSION'S RULES AND WITH THE INSTRUCTIONS PROVIDED BY THE COORDINATING VEC AND THE FCC

1st VE's name (Print First, MI, Last, Suffix)	VE's station call sign	VE's signature (must match name)	Date signed
Bart J. Jahnke	KB9NM	Bart J. Jahnke	3-16-94
2nd VE's name (Print First, MI, Last, Suffix)	VE's station call sign	VE's signature (must match name)	Date signed
ROSALIE A. WHITE	WA1STO	Rosalie A. White	3-16-94
3rd VE's name (Print First, MI, Last, Suffix)	VE's station call sign	VE's signature (must match name)	Date signed
Larry D. Wolfgang	WR1B	Larry D. Wolfgang	3-16-94

FCC Form 610
November 1993

Figure 2-3—A completed FCC Form 610. All applications for new Amateur Radio licenses must be submitted using this form.

ATTACH ORIGINAL OR A PHOTOCOPY OF YOUR LICENSE HERE:

SECTION 3 - TO BE COMPLETED BY PHYSICIAN

PHYSICIAN'S CERTIFICATION OF DISABILITY

Please see notice below

Print, type, or stamp physician's name: _____

Street address: _____

City, State, ZIP code: _____

Office telephone number: ( ) _____

I CERTIFY THAT I have read the Notice to Physician Certifying to a Disability, and that the person named in Item 1 on the reverse is severely handicapped, the duration of which will extend for more than 365 days beyond this date. Because of this severe handicap, this person is unable to pass a 13 or 20 words per minute telegraphy examination. I am licensed to practice in the United States or its Territories as a doctor of medicine (M.D.) or doctor of osteopathy (D.O.). I have considered the accommodations that could be made for this person's disability and have determined that, even with accommodations, this person would be unable to pass a 13 or 20 words per minute telegraphy examination.

WILLFUL FALSE STATEMENT IS PUNISHABLE BY FINE AND IMPRISONMENT (U.S. CODE TITLE 18, SECTION 1001)

➡ _____   _____

PHYSICIAN'S SIGNATURE (DO NOT PRINT, TYPE, OR STAMP)   M.D. or D.O.   DATE SIGNED

PATIENT'S RELEASE

Authorization is hereby given to the physician named above, who participated in my care, to release to the Federal Communications Commission any medical information deemed necessary to process my application for an amateur operator/primary station license.

➡ _____   _____

APPLICANT'S SIGNATURE (DO NOT PRINT, TYPE, OR STAMP)   DATE SIGNED

NOTICE TO PHYSICIAN CERTIFYING TO A DISABILITY

You are being asked by a person who has already passed a 5 words per minute telegraphy examination to certify that, because of a severe handicap, he/she is unable to pass a 13 or 20 words per minute telegraphy examination. If you sign the certification, the person will be exempt from the examination. Before you sign the certification, please consider the following:

THE REASON FOR THE EXAMINATION - Telegraphy is a method of electrical communication that the Amateur Radio Service community strongly desires to preserve. We support their objective by authorizing additional operating privileges to amateur operators who increase their skill to 13 and 20 words per minute. Normally, to attain these levels of skill, intense practice is required. Annually, thousands of amateur operators prove by passing examinations that they have acquired the skill. These examinations are prepared and administered by amateur operators in the local community who volunteer their time and effort.

THE EXAMINATION PROCEDURE - The volunteer examiners (VEs) send a short message in the Morse code. The examinee must decipher a series of audible dots and dashes into 43 different alphabetic, numeric and punctuation characters used in the message. To pass, the examinee must correctly answer questions about the content of the message. Usually, a fill-in-the-blanks format is used. With your certification, they will give the person credit for passing the examination, even though they do not administer it.

MUST A PERSON WITH A HANDICAP SEEK EXEMPTION?

No handicapped person is required to request exemption from the higher speed telegraphy examinations, nor is anyone denied the opportunity to take the examinations because of a handicap. There is available to all otherwise qualified persons, handicapped or not, the Technician Class operator license that does not require passing a telegraphy examination. Because of international regulations, however, any handicapped applicant requesting exemption from the 13 or 20 words per minute examination must have passed the 5 words per minute examination.

ACCOMMODATING A HANDICAPPED PERSON - Many handicapped persons accept and benefit from the personal challenge of passing the examination in spite of their hardships. For handicapped persons without an exemption who have difficulty in proving that they can decipher messages sent in the Morse code, the VEs make exceptionally accommodative arrangements. They will adjust the tone in frequency and volume to suit the examinee. They will administer the examination at a place convenient and comfortable to the examinee, even at bedside. For a deaf person, they will send the dots and dashes to a vibrating surface or flashing light. They will write the examinee's dictation. Where warranted, they will pause in sending the message after each sentence, each phrase, each word, or each character to allow the examinee additional time to absorb and interpret what was sent. They will even allow the examinee to send the message, rather than receive it.

YOUR DECISION - The VEs rely upon you to make the necessary medical determination for them using your professional judgement. You are being asked to decide if the person's handicap is so severe that he/she cannot pass the examination even when the VEs employ their accommodative procedures. The impairment, moreover, will last more than one year. This procedure is not intended to exempt a person who simply wants to avoid expending the effort necessary to acquire greater skill in telegraphy. The person requesting that you sign the certification will give you names and addresses of VEs and other amateur operators in your community who can provide you with more information on this matter.

DETAILED INSTRUCTIONS - If you decide to execute the certification, you should complete and sign the Physician's Certification of Disability on the person's FCC Form 610. You must be an M.D. or D.O. licensed to practice in the United States or its Territories. The person must sign a release permitting disclosure to the FCC of the medical information pertaining to the disability.

FCC Form 610
November 1993

schedule, supplies and mental attitude ready. Plan your schedule so you'll get to the examination site with plenty of time to spare. There's no harm in being early. In fact, you might have time to discuss hamming with another applicant, which is a great way to calm pretest nerves. Try not to discuss the material that will be on the examination as this may make you even more nervous. By this time, it's too late to study anyway!

What supplies will you need? First, be sure you bring your current *original* Amateur Radio license, if you have one. Bring along several sharpened number 2 pencils and two pens (blue or black ink). Be sure to have a good eraser. A pocket calculator may also come in handy. You may use a programmable calculator if that is the kind you have, but take it into your exam "empty" (cleared of all programs and constants in memory). Don't program equations ahead of time because you may be asked to demonstrate that there is nothing in the calculator's memory.

The VE Team is required to check two forms of identification before you enter the test room. This includes your *original* Amateur Radio license (if you have one). A photo ID of some type is best for the second form, but is not required by FCC. Other acceptable forms of ID include a driver's license, a piece of mail addressed to you, a birth certificate or some other such document.

The following description of the testing procedure applies to exams coordinated by the ARRL/VEC, although many other VECs use a similar procedure.

Code Tests

If you are interested in using Morse code and gaining the additional privileges granted to Technicians who have passed a 5-WPM code test, this information will be of interest to you.

The code tests are usually given before the written exams. The 20-WPM exam is usually given first, then the 13-WPM exam and finally the 5-WPM test. There is no harm in trying the 20- or 13-WPM exams even though you are testing for the Technician exams.

Before you take the code test, you'll be handed a piece of paper to copy the code as it's sent. The test will begin with about a minute of practice copy. Then comes the actual test: five minutes of Morse code. You are responsible for knowing the 26 letters of the alphabet, the numerals 0 through 9, the period, comma, question mark, $\overline{AR}$, $\overline{SK}$, $\overline{BT}$ and $\overline{DN}$. See Chapter 11 for explanations of what these characters mean. You may copy the entire text word for word, or just take notes on the content. At the end of the transmission, the examiner will hand you 10 questions about the text. Simply fill in the blanks with your answers. (You must spell each answer exactly as it was sent.) Some examining teams may choose to use a multiple-choice format for the code-test answers. If you get at least 7 correct, you pass! Alternatively, the exam team has the option to look at your copy sheet if you fail the 10-question exam. If you have one minute of solid copy, they can certify that you passed the test on that basis. The format

of the test transmission is similar to one side of a normal on-the-air amateur conversation.

A sending test may not be required. The Commission has decided that if applicants can demonstrate receiving ability, they most likely can also send at that speed. But be prepared for a sending test, just in case! Subpart 97.503(a) of the FCC rules says, "A telegraphy examination must be sufficient to prove that the examinee has the ability to send correctly by hand and to receive correctly by ear texts in the international Morse code at not less than the prescribed speed—"

Written Tests

The examiner will give each applicant a test booklet, an answer sheet and scratch paper. After that, you're on your own. The first thing to do is read the instructions. Be sure to sign your name every place it's called for. Do all of this at the beginning to get it out of the way.

Next, check the examination to see that all pages and questions are there. If not, report this to the examiner immediately. When filling in your answer sheet, make sure your answers are marked next to the numbers that correspond to each question.

Go through the entire exam, and answer the easy questions first. Next, go back to the beginning and try the harder questions. The really tough questions should be left for last. Guessing can only help, as there is no additional penalty for answering incorrectly.

If you have to guess, do it intelligently: At first glance, you may find that you can eliminate one or more "distracters." Of the remaining responses, more than one may seem correct; only one is the best answer, however. To the applicant who is fully prepared, incorrect distracters to each question are obvious. Nothing beats preparation!

After you've finished, check the examination thoroughly. You may have read a question wrong or goofed in your arithmetic. Don't be overconfident. There's no rush, so take your time. Think, and check your answer sheet. When you feel you've done your best and can do no more, return the test booklet, answer sheet and scratch pad to the examiner.

The Volunteer-Examiner team will grade the exam right away. The passing mark is 74%. (That means no more than 6 incorrect answers on the Element 3A exam.) You will receive a Certificate of Successful Completion of Examination (CSCE) showing all exam elements that you pass. If you are already licensed, and you pass the exam elements required to earn a higher class of license, the CSCE authorizes you to operate with your new privileges. When you use these new privileges, you must sign your call sign, followed by the slant mark ("/"; on voice, say "stroke" or "slant") and the letters KT (for Novices upgrading to Technician). There is one exception to this rule: If you previously held a Technician license and later pass the 5-WPM code test, you do not have to use the "KT" indicator when you use your new privileges or when you operate on the bands already allocated to Technicians (50 MHz and above).

If you pass only some of the exam elements required for a higher class license, you will still receive a CSCE. That certificate shows what exam elements you passed, and is valid for one year. Use it as proof that you passed those exam elements so you won't have to take them over again next time you try for the upgrade.

If you hold a CSCE for the 5-WPM code test, make sure you keep it in a safe place. Along with your actual Technician license, it serves as authorization to operate on the HF Novice bands.

NOW THAT YOU'RE A HAM

Well, the big day has finally arrived—your "ticket" came in the mail from the FCC! You have permission to operate an amateur station. Now you are ready to put your very own amateur station on the air!

As the proud owner of a new Amateur Radio ticket, you'll soon be an "on-the-air" person instead of an "off-the-air" person. You probably can't wait to make your first contact! You'll be putting all the information you had to learn for the exam to good use.

First things first. When you get your new ticket, "hot off the press," make a few photocopies of it. Then put the original license in a safe place. You must have your license (or a photocopy) available whenever you operate, so put one of those copies in your wallet. That magic piece of paper is your **operator license**, which lets you operate a station within your Novice or Technician privileges. It is also your **station license**, and lists the call sign that identifies your station.

Consider laminating your original license, as the ink sometimes lifts off the paper, even if you place the license behind glass in a frame.

CALL SIGNS

The FCC issues call signs on a systematic basis. When they process your application, you get the next call sign to come out of the FCC computer. The FCC issues call signs from four "groups," depending on the class of license.

Novice class licensees are given "Group D" call signs. These have a "two-by-three" format—two letters followed by a number, followed by three letters. The letters before the number make up the call sign *prefix*, and the letters after the number are the *suffix*. An example of a Novice call sign is KA1WWP.

Technician class licensees have "Group C" call signs.

These have a "one-by-three" format—a one-letter prefix, a number and a three-letter suffix. An example: N1NAG.

The call-sign formats of other Amateur Radio license classes follow. Group C calls are also given to General class amateurs. Advanced class amateurs get calls from Group B— the "two-by-two" group. KB9NM is an example of a call from this group. Group A calls are for Amateur Extra class operators only. These calls are of the "one-by-two" or "two-by-one" format. (K8CH is a one-by-two call; AA2Z is a two-by-one call.) In some parts of the country, all one-by-two and one-by-three call signs have been issued. If that is the case, your call sign may be assigned from a different group. For example, new Technician licensees may receive a call sign from Group D, rather than Group C.

Once you have a call sign, you may keep it as long as you want to (unless your license expires or is revoked). In other words, there's no requirement to change your call sign when you upgrade to a higher license class. The FCC gives amateurs their choice in this matter.

The number in the call sign shows the district where the call was first issued. Figure 2-4 shows the 10 US call districts. Amateurs may keep their calls when they move from one district to another. This means the number is not always an indication of where an amateur is. It tells only where he or she was living when the license was first issued. For example, WB3IOS received her license in Pennsylvania, in the third call district, but she now lives in the first district.

The first letter of a US call will always be A, N, K or W. These letters are assigned to the United States as amateur call-sign prefixes. Other countries use other prefixes— LA2UA is a Norwegian call sign, and VU2HO is from India.

[Turn to Chapter 12 and study questions N1D04, N1D07, N1D08 and N1D09. If you have any difficulty, review this section.]

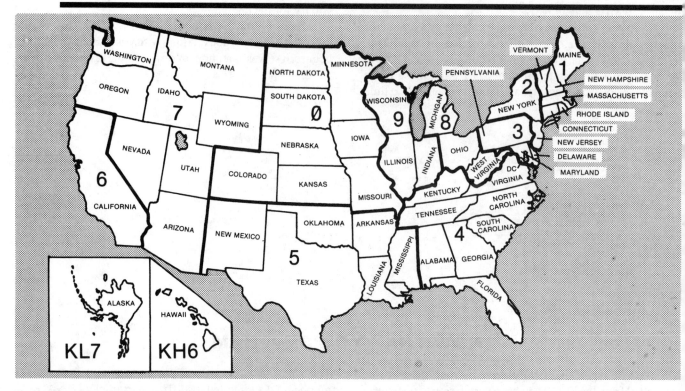

Figure 2-4—The 10 US call districts. An amateur holding the call sign WA1STO lived in the first district when the FCC assigned her that call. Alaska is part of the seventh call district, but has its own set of prefixes: AL7, KL7, NL7 and WL7. Hawaii, part of the sixth district, has the AH6, KH6, NH6 and WH6 prefixes.

OPERATING GUIDELINES

Keep your operation "above board"—set an operating example that will make you proud. You should be familiar with the basic operating and technical rules. You should also know the standard operating practices used for the various modes you operate. Chapter 9 includes detailed information about operating your station. In this section we will study the FCC Rules of operating and cover some very simple guidelines.

Logbooks are useful for recording dates, calls, names and locations of those stations you contact. When you confirm contacts by sending "QSL cards" (see Chapter 9) a logbook is a convenient way to keep track of these exchanges. Your log will provide a useful and interesting history, if you choose to keep one. Figure 2-5 shows an example of the information many amateurs keep in a logbook.

TIME OUT FOR STATION IDENTIFICATION

FCC regulations are very specific about station identification. You must identify your station every 10 minutes or less during a contact and at the end of the contact. You do not have to identify with every transmission, only at the required

intervals. Identify your station by transmitting the station call sign listed on your license.

The rules prohibit **unidentified communications or signals** (where the transmitting station's call sign is not transmitted). Be sure you understand the proper station identification procedures, so you don't violate this rule. The exception is when you are transmitting signals to control a model craft, such as a plane or boat. When controlling model craft, you don't have to transmit your call sign provided you attach a label to the transmitter with your call sign, name and address, and the transmitter power does not exceed 1 W.

You don't have to transmit your call sign at the beginning of a contact. In most cases it will help you establish the communications, though. Unless you give your call sign, the other stations won't know who you are! When you take part in a net or talk with a group of hams regularly, they may learn to recognize your voice. Then you may not have to give your call sign for them to know who you are.

You do not have to transmit both call signs when you are talking with another ham—only your own. The exception is in cases of **third-party communications** with a station in a foreign country. (You'll learn more about third-party com-

DATE	FREQ.	MODE	POWER	TIME	STATION WORKED	REPORT SENT	REPORT REC'D	TIME OFF	COMMENTS (QTH / NAME / QSL VIA)	QSL S	QSL R
16 NOV	3715	A1A	100	1800	KA1EBV	589	479	1835	MANOMET, MA – JOHN WILLIAMS	✓	
"	"	"	"	1840	WB7TPY	599	599	1855	TUFTS UNIVERSITY – DAVE GOES TO DENTAL SCHOOL		
17 NOV	28.125	A1A	100	2000	VP2ML	579	589	2003	CARIBBEAN – CHOD – NICE WX! BIG PILE UP	✓	✓
	28.140	"	"	2010	KAØHJD WØSH	599	599	2015	DESMOINES, IA – KRISTEN AND DAD, WØSH — THEY COLLECT OLD TELEGRAPH KEYS!	✓	✓
	3710	A1A	100	2033	KA1GQJ	579	589	2055	LINDA – NURSE IN A HOSPITAL	✓	
20 NOV	3715	A1A	100	1645	NA1L	559	559	1700	DALE IS A LAWYER; GOOD SIGNAL, KA1IXI, XYL – CHERYL	✓	✓
	3722	A1A	100	1702	AA2Z	599	599	1705	EAST HAMPTON, CT – WOW! MARK WORKS AT ARRL HQ!!	✓	✓
	3720	A1A	100	2315	WA1TBY			2325	NENN – QNI QTC 2		
	21.175	A1A	100	2332	N6TR	539	569	2345	OREGON – SAID HIS NAME WAS "TREE"?!		✓

Figure 2-5—Most amateurs keep a station logbook to record information about each contact.

munications later in this chapter.) Then you do have to transmit both call signs at the end of the contact.

Let's look at an example of how to identify an Amateur Radio station properly. KB4IYK and WB9RRU have been in communication for 45 minutes and are now signing off. Each operator has already transmitted his call four times (once after each 10-minute interval). Each should now transmit his call one more time as they sign off, for a total of five times during the QSO. (QSO means a two-way contact or communication with another ham.)

Suppose the QSO had lasted only 8 minutes. Each station would be *required* to transmit his call sign only once (at the end of the communication). You may identify more often than this to make it easier to communicate on a crowded band. But the rules specify only that you identify every 10 minutes, and at the end of a contact. On voice, identification is simply "from KB4IYK." In Morse code, use DE KB4IYK ("DE" means *from* in French). Note that the FCC Rules do not *require* you to transmit your call sign at the beginning of the contact. Most stations do transmit their call sign at the beginning, however.

[You should turn to Chapter 12 now, and study questions N1H06 through N1H11. Also study questions N1J06, N1J07 and N1J09. Review this section if you have any difficulty with these questions.]

POINTS OF COMMUNICATIONS

Who can you talk to with your new license? The FCC defines "points of communication" to specify what kinds of radio stations you may talk with. It's pretty simple, actually: *You may converse with all amateur stations at any time.* This includes amateurs in foreign countries, unless either amateur's government prohibits the communications. (There are a few countries in the world that do not allow Amateur Radio. There are also times when a government will not allow its amateurs to talk with people in other countries.)

The FCC must authorize any communication with any stations not licensed in the amateur service. An example of such authorization is when amateur stations communicate with military communications stations on Armed Forces Day each year.

Another example of amateurs communicating with non-amateur stations is during Radio Amateur Civil Emergency Service (RACES) operation. During an emergency, a registered RACES station may conduct civil-defense communications with stations in the Disaster Communications Service. During such emergencies, RACES stations may also communicate with other US Government stations authorized to conduct civil-defense communications.

[Turn to Chapter 12 now and study questions N1H02 and N1I02. Review this section if you have difficulty with either of these questions.]

BROADCASTING

Amateur Radio is a two-way communications service. Amateur Radio stations may not engage in **broadcasting**, the

from October 1955 QST

transmission of information intended for reception by the general public. There are also restrictions regulating one-way transmissions of information of general interest to other amateurs. Amateur stations may transmit one-way signals while in beacon operation or radio-control operation. Novice control operators don't have these privileges, however.

[Now turn to Chapter 12 and study question N1I06. If you are uncertain of the answer to this question, review this section.]

THIRD-PARTY COMMUNICATIONS

Communication on behalf of anyone other than the operators of the two stations in contact is **third-party communications**. (Many hams call it *third-party traffic*.) For example, sending a message from your mother-in-law to her relatives in Scarsdale on Valentine's Day is third-party communications. Third-party communications of a personal nature is okay, but passing messages involving business matters is not. The Amateur Service is not the place to conduct business.

You can pass third-party messages to other stations in the United States. Outside the US, FCC rules strictly limit this type of communication to those countries that have **third-party communications agreements** with the US.

In many countries, the government operates the telephone system and other communications lines. You can imagine why these governments are reluctant to allow their amateur operators to pass messages. They would be in direct competition with the government-operated communications system.

The ARRL monthly journal, *QST*, periodically publishes a list of countries with which the US has a third-party communications agreement. The list does change, and sometimes there are temporary agreements to handle special events. Check such a list, or ask someone who knows, before you try to pass a message to another country.

You may allow an unlicensed person to participate in Amateur Radio from your station. This is **third-party participation**. It is another form of third-party communications.

You (as control operator) must always be present to make sure the unlicensed person follows all the rules. You can allow your family members and friends to enjoy some of the excitement of Amateur Radio in this way. They can speak into the microphone or even send Morse code, as long as you are present to control the radio. They can also type messages on your computer keyboard to talk with someone using packet radio or radioteletype.

If you are allowing a non-amateur friend to use your station to talk to someone in the US, and a foreign station breaks in to talk to your friend, you should have your friend wait while you find out if the US has a third-party agreement with the foreign station's government.

There is another important rule to keep in mind about third-party participation. If the unlicensed person was an amateur operator whose license was suspended or revoked by the FCC, that person may not participate in any amateur communication. You can't allow that person to talk into the microphone of your transmitter or operate your Morse code key or computer keyboard.

[Now turn to Chapter 12 and study questions N1I01, N1I09, N1I10 and N1I11. Review this section as needed.]

BUSINESS COMMUNICATIONS

Amateur communication is noncommercial radio communication between amateur stations, solely with a personal aim and without pecuniary or business interest. *Pecuniary* refers to payment of any type.

This definition tells us that amateur operators should not conduct any type of business communications. You may not conduct communications for your own business, or for your employer. You can, however, use your Amateur Radio station to conduct your own personal communications. This includes using the autopatch on your local repeater for such personal calls as making an appointment or ordering food. (In 1993, the FCC amended the business communications rules in Section 97.113 to allow such communication. Prior to this rule change, such calls would be considered "illegal business communications.")

Of course you can also use the autopatch to call a garage for help when your car breaks down in the middle of a busy expressway. This is an emergency, because your property (car), and possibly your life, are in immediate danger. In an emergency, you can use Amateur Radio in any way necessary to obtain help.

There are many examples of times when personal property or lives are in immediate danger. You may use Amateur Radio to call for help in such a situation, even though you seem to be violating other rules. Just be sure you really have an emergency situation first.

No one can use an amateur station for monetary gain. You must not accept payment in any form for the use of your station at any time. This also means you may not accept payment for transmitting a message for anyone. Payment means more than just money here. It refers to any type of compensation, which would include materials or services of any type.

There are two exceptions to this rule. A club station intended primarily for transmitting Morse code practice and information bulletins of interest to all amateurs may employ a paid control operator. That person can be paid to serve as the control operator only when the station is actually transmitting code practice or bulletins. This exception allows the ARRL to pay a control operator for W1AW, the station at League Headquarters in Newington, Connecticut. The station is dedicated to Hiram Percy Maxim, the League's first president. The W1AW operator can't make general contacts with other hams after the code-practice sessions, however.

The second exception is that teachers may use Amateur Radio stations as part of their classroom instruction at an educational institution.

[Before you go on to the next section, turn to Chapter 12 and study questions N1H03, N1H04, N1H05 and N1I05. If you have any difficulty with any of them, review this section.]

OTHER ASSORTED RULES

Under FCC Rules, amateurs may not transmit music of any form. They may not use obscene, indecent or profane language. You can't use codes or ciphers to obscure the meaning of transmissions. This means you can't make up a "secret" code to send messages over the air to a friend. (Control signals transmitted for remote control of model craft are not considered codes or ciphers. Neither are telemetry signals, such as a satellite might transmit to tell listeners about its condition. A space station—satellite—control operator can use specially coded signals to control the satellite.)

Amateurs may not cause **malicious (harmful) interference** to other communications of any type, amateur or nonamateur. You may not like the other operator's practices, or you may believe he or she is violating the rules. You have no right to interfere with their communications, however.

Amateurs may not transmit **false or deceptive signals**, such as a distress call when no emergency exists. You must not, for example, start calling MAYDAY (an international distress signal) unless you are in a life-threatening situation.

Distress Calls

If you should require immediate emergency help, and you're using a voice (telephony) mode, call MAYDAY. Use whatever frequency offers the best chance of getting a useful answer. MAYDAY is from the French *m'aidez* (help me). On CW (telegraphy), use $\overline{SOS}$ to call for help. Repeat this call a few times, and pause for any station to answer. Identify the transmission with your call sign. Stations that hear the call sign will realize the $\overline{SOS}$ is legitimate. Repeat this procedure for as long as possible, or until you receive an answer. In a life- or property-threatening emergency, you may send a distress call on *any* frequency, even outside the amateur bands, if you think doing so will bring help faster.

BE ON THE FRONTIER OF TECHNOLOGICAL ADVANCES

For 90 years hams have carried on a tradition of learning by doing. From the earliest days of radio, hams have built their transmitters from scratch, wrapping strands of copper wire salvaged from Model T automobiles around oatmeal boxes. Through experimenting with building their own equipment, hams have pioneered advances in technology such as the techniques for single sideband. Hams were the first to bounce signals off the moon to extend signal range. Amateurs' practical experience has led to many technical refinements and cost reductions beneficial to the commercial radio industry. The photo at the right shows a complete transmitter constructed in a small package.

Be ready to supply the following information to the stations responding to an $\overline{SOS}$ or MAYDAY:

- *The location of the emergency*, with enough detail to permit rescuers to locate it without difficulty.
- *The nature of the distress.*
- *The type of assistance required* (medical, evacuation, food, clothing or other aid).
- *Any other information* to help locate the emergency area.

If you receive a distress signal, you are also allowed to transmit on *any* frequency to provide assistance.

Space Stations

The FCC defines a **space station** as "An amateur station located more than 50 km (about 30 miles) above the Earth's surface." This obviously includes amateur satellites, and also includes any operation from the Space Shuttle, the Russian MIR space station and any future operations by astronauts in space. Any licensed ham can be the licensee or control operator of a space station.

[Now it's time to turn to Chapter 12 again, and study a few questions. You should be able to answer questions N1I03, N1I04, N1I07 and N1I08. Also study questions N1J01 through N1J05, N1J08, N1J10 and N1J11. Review this section if you have difficulty with any of these questions.]

[**If you are studying for the Technician exam**, turn to Chapter 13 and study questions T2C01 and T2C02. Review this section, if you need to.]

Official Notices of Violation

The FCC is the agency in charge of maintaining law and order in radio operation in the US. The International Telecommunication Union (ITU) sets up international rules that the government agencies of each country must follow. Both sets of rules provide the basic structure for Amateur Radio in the United States.

Suppose you receive an official notice from the FCC informing you that you have violated a regulation. Now what should you do? Simple: Whatever the notice tells you to do.

Operating a station at your summer cottage is considered *portable operation*—operating at a place other than that shown on the station license. You may operate your amateur station anywhere the FCC has jurisdiction, at any time. Using a transceiver installed in your car is called *mobile operation*—operating a station while in motion or during halts at unspecified locations. Amateurs do *not* have to notify the FCC before placing a station in portable or mobile operation.

The FCC will mail any notices of violation to the permanent mailing address shown on your license. There may be times when you will operate your station away from home for

SOME EARLY AMATEUR INGENUITY

One young lad of seventeen, known to possess an especially efficient spark, CW and radio telephone station, was discovered to be the son of a laboring man in extremely reduced circumstances. The son had attended grammar school until he was able to work, and then he assisted in the support of his family. They were poor indeed. Yet despite this the young chap had a marvelously complete and effective station, installed in a miserable small closet in his mother's kitchen. How had he done it? The answer was that he had constructed every last detail of the station himself. Even such complicated and intricate structures as head telephones and vacuum tubes were homemade! Asked how he managed to make these products of specialists, he showed the most ingenious construction of headphones from bits of wood and wire. To build vacuum tubes he had found where a wholesale drug company dumped its broken test tubes, and where the electric light company dumped its burned-out bulbs, and had picked up enough glass to build his own tubes and enough bits of tungsten wire to make his own filaments. To exhaust the tubes he built his own mercury vacuum pump from scraps of glass. His greatest difficulty was in securing the mercury for his pump. He finally begged enough of this from another amateur. And the tubes were good ones—better than many commercially manufactured and sold. The greatest financial investment that this lad had made in building his amateur station was 25 cents for a pair of combination cutting pliers. His was the spirit that has made Amateur Radio.—*Clinton B. Desoto, in* 200 Meters and Down

a long period. Arrange to have your mail forwarded from your permanent mailing address to your temporary home in those cases.

There exists a philosophy in Amateur Radio that is deeply rooted in our history. This philosophy is as strong now as it was in the days of the radio pioneers. We are talking about the *self-policing* of our bands. Over the years, amateurs have become known for their ability to maintain high operating standards and technical skills. We do this without excessive regulation by the FCC. The Commission itself has praised the amateur service for its tradition of self-policing. Perhaps the underlying reason for this is the amateur's sense of pride, accomplishment, fellowship, loyalty and concern. Amateur Radio is far more than just a hobby to most amateurs.

As a new or prospective amateur operator, you will begin to discover the wide horizons of your new pastime. You will learn more about the rich heritage we all share. Take this sense of pride with you, and follow the Amateur's Code.

[This completes your study of the first part of the chapter. The remainder cover the rules and regulations you'll need to know to pass the Element 3A exam for the Technician license. Before moving on, turn to Chapter 12 and study question N1H01. By now you should have no difficulty with the questions in subelement N1 of the Novice question pool.]

The Amateur's Code

The Radio Amateur is:

CONSIDERATE . . . never knowingly operates in such a way as to lessen the pleasure of others.

LOYAL . . . offers loyalty, encouragement and support to other amateurs, local clubs, and the American Radio Relay League, through which Amateur Radio in the United States is represented nationally and internationally.

PROGRESSIVE . . . with knowledge abreast of science, a well-built and efficient station and operation above reproach.

FRIENDLY . . . slow and patient operating when requested; friendly advice and counsel to the beginner; kindly assistance, cooperation and consideration for the interests of others. These are the hallmarks of the amateur spirit.

BALANCED . . . radio is an avocation, never interfering with duties owed to family, job, school, or community.

PATRIOTIC . . . station and skill always ready for service to country and community.

The original Amateur's Code was written by Paul M. Segal, W9EEA, in 1928.

Additional Rules Information For The Technician Exam

KEY WORDS

Bandwidth—The width of a frequency band outside of which the mean power is attenuated at least 26 dB below the mean power of the total emission, including allowances for transmitter drift or Doppler shift.

Beacon station—An amateur station transmitting communications for the purposes of observation of propagation and reception or other related experimental activities.

Broadcasting—Transmissions intended to be received by the general public, either direct or relayed.

Control operator—An amateur operator designated by the licensee of a station to be responsible for the transmissions of an amateur station.

Control point—The locations at which the control operator function is performed.

Earth station—An amateur station located on, or within 50 km of, the Earth's surface intended for communications with space stations or with other Earth stations by means of one or more other objects in space.

Frequency coordination—Allocating repeater input and output frequencies to minimize interference between repeaters and to other users of the band.

Grace period—The time FCC allows following the expiration of an amateur license to renew that license without having to retake an examination. Those who hold an expired license may not operate an amateur station until the license is reinstated.

One-way communications—Transmissions that are not intended to be answered. The FCC strictly limits the types of one-way communications allowed on the amateur bands.

Peak envelope power (PEP)—The average power of a signal at its largest amplitude peak.

Prohibited transmissions—Certain types of transmissions, such as music and obscenities, that the FCC doesn't allow on the amateur bands.

RACES (Radio Amateur Civil Emergency Service)—A part of the amateur service that provides radio communications only for civil defense purposes.

Repeater station—An amateur station that automatically retransmits the signals of other stations.

Space station—An amateur station located more than 50 km above the Earth's surface.

Temporary state of communication emergency—When a disaster disrupts normal communications in a particular area, the FCC can declare this type of emergency. Certain rules may apply for the duration of the emergency.

Third-party communications—Messages passed from one amateur to another on behalf of a third person.

This section covers the rules and regulations you must know to pass your Technician class exam. When you have read and understood it, you're well on your way to knowing what you'll need to know to pass the Technician exam.

CONTROL OPERATOR AND CONTROL POINT

Each amateur station must have a person or persons who are responsible for the proper operation of that station. The **control operator** is that person. The control operator ensures compliance with FCC rules. The licensed amateur at the mike or key is the one who is legally in control.

The FCC assumes that you are always the control operator of your station, unless there is some written record to the contrary. (If the control operator is *not* the station licensee, responsibility for proper station operation is shared between them—they are both responsible.) The control operator function takes place at the **control point**. This control point is usually, but not always, at the radio. In the case of a remotely controlled station, the control point may be elsewhere.

[Before going on, you should turn to Chapter 13 and study questions T1A01 and T1A02.]

TECHNICIAN EMISSION PRIVILEGES

As you can see from Tables 2-1 and 2-2, Technician class operators enjoy *all frequency privileges* allocated to the amateur service *above 30 MHz*. Technician class amateurs can operate on all authorized frequencies, using all authorized modes, on the VHF, UHF and microwave amateur bands. They may use up to 1500-watts output on these bands. In addition, Technicians who pass a 5-WPM code test gain the same emission privileges that Novices have on the four HF Novice bands.

Technician class licensees may operate on the VHF and UHF bands listed below, plus several others that are higher in frequency. For a complete list of amateur frequency privileges, including the microwave bands where frequency is measured in gigahertz, see *The FCC Rule Book*, published by the ARRL.

On the 50-MHz (6-meter) band: all emission types except pulse—CW, RTTY, data, MCW, test, phone and image—are authorized on 50.1 MHz to 54.0 MHz. On 50.0 to 50.1 MHz, only CW is allowed.

On the popular 144-MHz (2-meter) band: all emission types (except pulse, but including image transmissions) from 144.1 to 148 MHz, and CW only from 144.0 to 144.1 MHz. Although the 2-meter band is best known for repeater and packet operation, hams also use it for such *weak-signal activities* as meteor scatter, moonbounce and aurora.

On the 222-MHz band (1¼-meter) band: all emission types (except pulse) from 222.0 to 225.0 MHz.

On the 420-MHz (70-cm) band: all emission types (except pulse) from 420 to 450 MHz.

On the 902-MHz (33-cm) band: all emission types from 902 to 928 MHz.

On the 1240-MHz (23-cm) band: all emission types (except pulse) from 1240 to 1300 MHz.

On the 2300-MHz (13-cm) band: all emission types from 2300 to 2310 and 2390 to 2450 MHz.

EMISSION STANDARDS

FCC Rules stipulate that no amateur station transmission shall occupy more **bandwidth** than necessary for the information rate and emission type being transmitted, in accordance with good amateur practice. Bandwidth is a measure of how much space your signal takes up. A CW signal has a smaller bandwidth than does a single-sideband phone signal, and an SSB signal in turn has a smaller bandwidth than an AM or FM voice signal.

The Rules specify certain characteristics of RTTY and other emissions, and you should be aware of them to ensure that you're not in violation of the Rules.

RTTY Sending Speed

When transmitting one of the specified digital codes, the rules impose certain sending-speed limitations. The specified codes are Baudot, AMTOR and ASCII.

In the 28- to 50-MHz range, the maximum sending speed (symbol rate) for a RTTY or packet transmission is 1200 bauds (bits per second).

Between 50 and 222 MHz, the maximum sending speed for a RTTY or packet transmission is 19.6 kilobauds.

Above 222 MHz, the maximum RTTY or packet sending speed is 56 kilobauds.

RTTY Frequency Shift

Below 50 MHz, the maximum frequency shift permitted for RTTY is 1000 Hz.

Above 50 MHz, there is no maximum frequency shift.

Authorized Bandwidth

When you are transmitting a digital code other than one of those specified in the rules, there is a bandwidth limitation imposed.

Between 50 and 222 MHz, the authorized bandwidth of a RTTY, data or multiplexed emission using an unspecified digital code is 20 kHz.

Between 222 and 450 MHz, the authorized bandwidth of a RTTY, data or multiplexed emission using an unspecified digital code is 100 kHz.

[Before continuing, take a look at questions T1A05 through T1A10, T1B08, T1B09 and T1C02 through T1C11 in Chapter 13. You may find it helpful to refer to Table 2-2 while studying these questions.]

LICENSE TERMS

All licenses issued after January 1984 have a 10-year

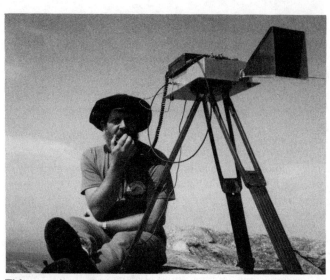

This gent brought his 10-GHz microwave station to the summit of Mount Monadnock in New Hampshire to take part in an ARRL-sponsored contest. Many hams enjoy operating away from the comforts of home.

term. Renew your license before it expires by completing an FCC Form 610 and sending it to the FCC at least 90 days before your current license expires. If you forget to renew your license, you still have a two-year **grace period** in which to send in a request for reinstatement on FCC Form 610. If you send it within the grace period, the FCC will reinstate your expired license automatically after the Form 610 is processed. You will be able to keep your old call sign under these circumstances. It is important to remember that you may not operate your station until your reinstated ticket arrives. After the two-year grace period, you must take the exams over again. If you pass, you will be assigned another call sign.

Modifying Your License

If you move or change your name, you will need to modify your Amateur Radio license. To do this, fill out a current Form 610, following the directions carefully, and mail it to the FCC in Gettysburg, Pennsylvania.

[Before proceeding, turn to Chapter 13 and review questions T1A03 and T1A04.]

TRANSMITTER POWER

In general, outside the HF Novice bands, Technician class operators may operate with a *maximum* **peak envelope power (PEP)** output of 1500 W. We've emphasized the word *maximum* because the FCC Rules also stipulate: "An amateur station must use the minimum power necessary to carry out the desired communications." Transmitter power is measured in watts of PEP output at the antenna terminals of the transmitter or amplifier.

The FCC defines **peak envelope power** as "the average power supplied to the antenna transmission line by a trans-

mitter during one RF cycle at the crest of the modulation envelope." This sounds pretty technical, but it isn't too difficult to understand the basic principle. The *modulation envelope* refers to the way the information signal varies the transmitter output. Think of it as increasing and decreasing the transmitted signal. All we have to do is find the highest point, or maximum output-signal level. Then we look at one cycle of the radio-frequency (RF) signal and measure the average power during that time. You might use a *wattmeter* to make that measurement for you, and the meter will do all the work. So you just have to read a meter scale or read the numbers off a digital display.

There are some exceptions to the maximum power rules:

- All who are authorized to operate in the 80, 40 and 15-meter Novice subbands may use a maximum of 200 watts PEP.
- Technicians who are authorized to operate on the Novice bands (those who have passed a 5-WPM code test) are also limited to 200 W PEP on the 10-meter Novice subband, 28.1 to 28.5 MHz.
- Amateurs are limited to 200 W PEP on the 30-meter band (10.1-10.15 MHz). You will be able to use this band when you upgrade to General class.
- Stations in beacon operation are limited to 100 W PEP output.
- Stations operating near certain military installations may use a maximum of 50 W PEP on the 450-MHz band.

FREQUENCY SHARING

If we are to make the best use of the limited amount of available spectrum, there must be ways to ensure that harmful interference is kept to a minimum. If two amateur stations want to use the same frequency, both stations have an equal right to do so. The FCC encourages efficient, interference-free sharing of the ham bands by limiting transmitter output power, by assigning services either primary or secondary status on a frequency band and by encouraging repeaters to be *coordinated* (operating frequencies are recommended by a regional group organized for this purpose). We've just discussed transmitter power; now we'll take a look at primary and secondary allocations, and repeater coordination.

Primary and Secondary Allocations

The basic frequency-sharing principle is straightforward: Where a band of frequencies is allocated to different services of the same category, the basic principle is the *equality of right to operate*. "Category" refers to whether a service operates on a primary or secondary basis. The amateur service is allocated many different frequency bands, but some of them are allocated on a primary basis and some are secondary. A station in a secondary service must not cause harmful interference to, and must accept interference from, stations in a primary service.

By sharing our frequencies with several other services, including the US military, hams can have the use of a greater amount of spectrum than would otherwise be the case. If you are operating on a band on which the amateur service has a secondary allocation, and a station in the primary service causes interference, you must change frequency immediately. You may also be interfering with the other station, and that is prohibited by Part 97 of the FCC Rules. *The FCC Rule Book* shows which amateur bands are primary allocations which are secondary.

Repeater Frequency Coordination

A **repeater station** is an amateur station that automatically retransmits the signals of other stations. Repeaters in the same area that use the same or similar frequencies can interfere with each other. As more and more repeater stations have been set up, there have been more and more cases of repeater interference. One effective way of dealing with the problem is **frequency coordination**. Volunteer frequency coordinators have been appointed to ensure that new repeaters use frequencies that will tend not to interfere with existing repeaters in the same area. The FCC encourages frequency coordination, but the process is organized and run by hams and groups of hams who use repeaters.

The FCC has ruled in favor of coordinated repeaters if there is harmful interference between two repeaters. In such a case, if a frequency coordinator has coordinated one but not the other, the licensee of the *uncoordinated* repeater is responsible for solving the interference problem. If both repeaters are coordinated, or if neither is, then both licensees are equally responsible for resolving the interference.

EARTH STATIONS

Earth stations are stations located on the Earth, or within 50 km of it, intended for communications with **space stations** or with other Earth stations by means of one or more objects in space. Any amateur can be the control operator of an Earth station, subject to the limitations of his license class. Some specific transmitting frequencies are authorized for Earth stations. These are located in the 40, 20, 17, 15, 12, 10 and 2-meter bands, in the 23- and 70-cm bands, and in all other amateur bands that are higher in frequency. A complete list appears in *The FCC Rule Book*.

[Before moving along, turn to Chapter 13 and review questions T1B01, T1B02, T1B04, T1B05, T1B06, T1B10, T1B11, T1D06, T1D07 and T1D08.]

EMERGENCY COMMUNICATIONS

The FCC not only permits, but encourages, licensed hams to assist in emergencies, notwithstanding the FCC Rules that apply at all other times. The Rules state: "When normal communication systems are overloaded, damaged or disrupted because a disaster has occurred, or is likely to occur . . . an amateur station may make transmissions necessary to meet essential communication needs and facilitate relief actions."

The FCC recognizes that "amateurs may provide essential communications in connection with the immediate safety of human life and immediate protection of property when normal communication systems are not available." This is the one exception to the prohibition on business communications on the ham bands.

If you're in the middle of a hurricane, forest fire or blizzard, and you offer your communications services to the local authorities, you can do whatever you need to do to help deal with the emergency. This includes allowing a physician to operate your radio or helping the Red Cross to assess damages.

Other cases aren't as clear cut. If you find yourself stuck on a highway, and there's no telephone nearby, can you call a towing service by using a repeater autopatch or by asking a ham who is listening on the repeater to do so? Sure. Stranded motorists have been killed on the highway while waiting for assistance.

In the wake of a major disaster, the FCC may suspend or change its Rules to help deal with the immediate problem. Part 97 stipulates that when a disaster disrupts normal communications systems in a particular area, the FCC may declare a **temporary state of communication emergency**. The declaration will set forth any special conditions or rules to be observed during the emergency. Amateurs who want to request that such a declaration be made should contact the FCC Engineer in Charge in the area concerned.

[Before going on, review these questions from Chapter 13: T1E10 and T1E11.]

RACES

RACES (Radio Amateur Civil Emergency Service) is a part of the amateur service that provides radio communications only for civil defense purposes. It is active *only* during periods of local, regional or national civil emergencies.

You must be registered with the responsible civil defense organization to operate as a RACES station. RACES stations may not communicate with amateurs not operating in a RACES capacity. Restrictions do not apply when stations are operating in a non-RACES amateur capacity, such as the ARES, the Amateur Radio Emergency Service.

Only civil-preparedness communications can be transmitted during RACES operation. These are defined in Section 97.407 of the FCC regulations. Rules permit tests and drills for a maximum of one hour per week. All test and drill messages must be clearly identified as such.

[Now study questions T2C04, T2C05 and T2C06 in Chapter 13. Review as needed.]

Tactical Communications

Tactical communications is first-response communications in an emergency involving a few people in a small area.

This type of communications is unformatted and seldom written. It may be urgent instructions or requests such as "Send an ambulance," or "Who has the medical supplies?"

Tactical communications usually use 2-meter repeater net frequencies or the 146.52-MHz simplex calling frequency. Compatible mobile, portable and fixed-station equipment is plentiful and popular for these frequencies. Tactical communications is particularly important when working with local government and law-enforcement agencies. Use the 12-hour local-time system for times and dates when working with relief agencies. Most may not understand the 24-hour system or Coordinated Universal Time (UTC).

Another way to make tactical communications efficient is to use tactical call signs, which describe a function, location or agency. Their use promotes coordination with individuals or agencies who are monitoring. When operators change shifts or locations, the set of tactical call signs remains the same. Amateurs may use such tactical call signs as *parade headquarters*, *finish line*, *Red Cross* or *Net Control*. This procedure promotes efficiency and coordination in public-service communication activities. Tactical call signs do not fulfill the identification requirements of Section 97.119 of the FCC Rules, however. Amateurs must also identify their station operation with their FCC-assigned call sign. Identify at the end of operation and at intervals not to exceed 10 minutes during operation.

Health-and-Welfare Traffic

There can be a large amount of radio traffic to handle during a disaster. Phone lines still in working order are often overloaded. They should be reserved for emergency use by those people in peril. Shortly after a major disaster, *Emergency* messages leave the disaster area. These have life-and-death urgency or are for medical help and critical supplies. Handle them first. Next is *Priority* traffic. These are emergency-related messages, but not as important as Emergency messages. Then, handle *health-and-welfare* traffic, which pertains to the well being of evacuees or the injured. This results in timely advisories to those waiting outside the disaster area.

Hams inside disaster areas cannot immediately find out about someone's relative when their own lives may be threatened—they are busy handling emergency messages. After the immediate emergency subsides, concerned friends and relatives of possible victims can send health-and-welfare inquiries into the disaster area.

Emergency Equipment

When cities, towns, counties, states or the federal government call for emergency communications, amateurs answer. They press their mobile and portable radio equipment into service, using alternatives to commercial power. Generators, car batteries, windmills or solar energy can provide power for equipment during an emergency, when nor-

mal ac power is often out of service. A dipole antenna is the best choice in emergencies—it can be installed easily, and wire is light and portable. Carry an ample supply of wire and you'll be ready to go on the air at any time and any place. If you're planning to use a battery powered hand-held transceiver, bring along at least one spare, *charged* battery pack.

[Before going on, turn to Chapter 13 and study questions T2C07 through T2C12. Review this section if you have difficulty with any of these questions.]

ONE-WAY TRANSMISSIONS

Certain types of transmissions are designated as one-way. Only certain kinds of **one-way communications** or transmissions are permitted on the amateur bands.

Emergencies

Since normal restrictions are suspended when life or property is in immediate danger, as discussed above, one-way transmissions are not considered broadcasting (which is not allowed) under emergency conditions.

Beacon Stations

A **beacon station** is simply a transmitter that alerts listeners to its presence. In the amateur service, beacons are used primarily for the study of radio-wave propagation—to allow amateurs to tell when a band is open to different parts of the country or world. The FCC defines a beacon station as an amateur station transmitting communications for the purposes of observation of propagation and reception or other related experimental activities.

The FCC Rules address beacon operation this way:

- automatically controlled beacon stations are limited to certain parts of the 28, 50, 144, 222 and 432-MHz amateur bands, and all amateur bands above 450 MHz.

- the transmitter power of a beacon must not exceed 100 W.

- any license class, except Novice, can operate a beacon station.

Remote Control of Model Craft

Amateurs are permitted to use radio links to control model craft. This is called "telecommand (control)" of model craft. Most model control activity takes place on the 6-meter band. Remote control operation is permitted with these restrictions:

- station identification is not required for transmission directed only to the model craft. The control transmitter must have a label indicating the station's call sign and the licensee's name and address.

- control signals are not considered codes and ciphers.

• transmitter power cannot exceed 1 watt.

Other Permitted One-Way Transmissions

Aside from the examples listed above (emergency communications, remote control of a device and beacon operation), Part 97 allows the following kinds of one-way amateur communications:

• brief transmissions necessary to make adjustments to a station;
• brief transmissions necessary to establishing two-way communications with other stations;
• transmissions necessary to assisting persons learning, or improving proficiency in, the international Morse code; and
• transmissions necessary to disseminate information bulletins (these are to be directed only to amateurs and must consist solely of subject matter of direct interest to the amateur service).

[Before continuing, review questions T1B03, T1D01, T1D05, T1D09, T1D10 and T1D11 in Chapter 13.]

THOU SHALT NOT . . . PROHIBITED TRANSMISSIONS

There are some specific things you cannot do on the amateur bands—the FCC calls them **prohibited transmissions**. These include:

• accepting direct and indirect payment for operating an amateur station
• broadcasting
• news gathering
• transmitting music
• aiding criminal activities
• transmitting codes and ciphers
• engaging in obscenity, profanity and indecency.

Payment

You must never accept any money, or other type of material compensation, for operating your station. The amateur service is a voluntary, noncommercial communication service under Part 97. Note that this does not prohibit an organization from providing food during your participation in emergency communications, for example.

There is one exception to the prohibition against receiving payment for operating an amateur station. Payment is allowed to the control operators of a club station if that station is used primarily to transmit Morse code practice and information bulletins of interest to Amateur Radio operators. Payment can be made only for the time the station is transmitting code practice and bulletins, and the code-practice sessions must last at least 40 hours per week. In addition, the station must transmit on at least six medium- and high-frequency bands, and the schedule of transmissions must be

Shuttle astronaut Kathy Sullivan holds the amateur call N5YYV. She put it to good use when she flew aboard shuttle mission STS-45 in March and April 1992, making 2-meter voice and packet contacts with hams and schoolchildren across the country. One FCC rule applies specifically to shuttle communications: Amateurs can retransmit US Government shuttle communications to a ham audience—if they get NASA's permission beforehand. (*photo courtesy NASA*)

published 30 days in advance. This exception allows the control operators at the ARRL Headquarters station, W1AW, to be paid employees of the League.

Broadcasting

Broadcasting refers to transmissions intended to be received by the general public. These may be either direct or relayed. As such, broadcasting must be left to those services that are directly authorized to do so by the FCC—commercial television and radio, for example. Broadcasting is specifically prohibited in the amateur service.

News Gathering

You must not allow your amateur station to be used for any activity directly related to program production or news gathering for broadcast purposes. There is only one exception: You can transmit news information about an event if the following requirements are met: (1) the news information involves the immediate safety of life of individuals or the immediate protection of property; (2) the news information is directly related to the event; (3) the information can't be transmitted by any other means because normal communications systems have been disrupted or because there are no other communications systems available at the place where the information is originated; and (4) other means of communication could not be reasonably provided before or during the event.

Music

The transmission of music is strictly prohibited in the amateur bands. This includes accidental transmission of music playing in the background while talking on the microphone.

There is an exception to the rule against transmitting music. A station may retransmit signals from the US Government space shuttle, with prior approval from the National Aeronautics and Space Administration (NASA). Such stations are allowed to transmit the incidental music that may be included in those shuttle transmissions.

Criminal Activity

Amateur Radio may not be used for any purpose, or in connection with any activity, that is contrary to Federal, state or local law.

Codes and Ciphers

You may not hide the meaning of your communications by putting them into codes and ciphers, except for a few specific exceptions cited in Part 97. Codes and ciphers refer to text that has been transformed to conceal its meaning. This restriction is consistent with the FCC's obligations under the international regulations that amateur communications be made in plain language.

Obscenity, Profanity and Indecency

The rules say, quite simply, that you cannot transmit obscene, indecent or profane language or meaning. The FCC has developed standards of obscenity and indecency, but the best rule of thumb is to avoid making any transmissions that could be interpreted as being offensive.

[Before going on, check these questions in Chapter 13: T1E01 through T1E06.]

RETRANSMITTING RADIO SIGNALS

Radio amateurs may retransmit only signals originating in the amateur service, except for emergency communications. Some amateurs like to record W1AW news bulletins and transmit them over a local repeater as a service. No problem, as long as this is done manually (with a control operator present). You must never record an individual's transmissions and play them back over the air without permission. That's poor amateur practice, and transmission of false signals.

You may retransmit US Government space shuttle communications for the exclusive use of other amateurs, provided you get prior approval from the National Aeronautics and Space Administration (NASA). You may also retransmit propagation and weather-forecast information intended for use by the general public and originated from US Government stations and communications.

Automatic retransmission of amateur signals may be done only by repeater, auxiliary, digipeater and space stations.

THIRD-PARTY COMMUNICATIONS

Third-party communications is amateur communication by or under the supervision of the control operator at an amateur station to another amateur station on behalf of anyone other than the control operator. A third-party message is one the control operator (first party) of your station sends to another station (second party) for anyone else (third party). Third-party messages include those that are spoken, written, keystroked, keyed, photographed or otherwise originated by or for a third party, and transmitted by your amateur station live or delayed. A third party may also be a person permitted by the control operator to participate in Amateur Radio communications.

International third-party traffic is prohibited, except when:

- communicating with a person in a country with which the US shares a third-party agreement, or
- in cases of emergency where there is an immediate threat to lives or property, or
- the third party is eligible to be a control operator of the station.

If a third party is ineligible to be a control operator, he or she may participate in radio communications if the control operator is present at the control point and continuously monitors and supervises the third party's participation. In international third-party communications, your station identification must include the other station's call sign and your call sign at the end of the communication.

STATION IDENTIFICATION

Under FCC Rules, you must clearly make known the source of your transmissions to anyone receiving them. No station may transmit unidentified communications or signals, or transmit as the station call any call not authorized to the station.

You must identify your station at the end of each contact, and every 10 minutes during the contact. When you identify by phone, you must use English, and the FCC recommends that you use a phonetic alphabet to make it easier to understand the letters. There is information about a standard phonetic alphabet you should use in Chapter 9.

Many repeaters now feature voice identification, but it's okay to identify a repeater with Morse code, too. If a repeater identifies with Morse code, the code speed must be 20 WPM or less.

If you're a Novice class operator who has passed the Technician exam and holds a Certificate of Successful Completion of Examination, special rules apply. To identify when operating on a voice mode with your new Technician privileges, give your call sign followed by a word that describes the slant mark and the identifier "KT." An example: KA9GND slant KT. Another: KA2FCC stroke KT. When operating CW or on digital modes, use the fraction bar (/) followed by KT, as in KA1WWP/KT. You only need to use the /KT identifier until your new license arrives from the FCC.

Technicians who pass the 5-WPM code test and thus qualify for the enhanced Technician license (including HF Novice privileges) need not use the "KT" identifier when using their new privileges or when operating on the Technician bands (50 MHz and above). You must retain your Certificate of Successful Completion of Exam (CSCE) that shows you passed the code test. This is all the authorization you will receive to operate on HF.

You should know one more thing about station identification: There is only one type of emission that can always be used when you identify your station, regardless of frequency or operating mode: Morse code (CW).

[Congratulations! Be sure to review questions in Chapter 13 numbered T1A11, T1B07, T1C01, T1D02, T1D03, T1D04 and T1E07 through T1E09. Review the material in this chapter until you feel confident, and then move on to Chapter 3.]

KEY WORDS

Alternating current (ac)—Electrical current that flows first in one direction in a wire and then in the other. The applied voltage is also changing polarity. This direction reversal continues at a rate that depends on the frequency of the ac.

Ampere (A)—The basic unit of electrical current. Current is a measure of the electron flow through a circuit. If we could count electrons, we would find that if there are 6.24×10^{18} electrons moving past a point in one second, we have a current of one ampere.[1] We abbreviate amperes as amps.

Audio frequency (AF)—The range of frequencies that the human ear can detect. Audio frequencies are usually listed as 20 Hz to 20,000 Hz.

Battery—A device that stores electrical energy. It provides excess electrons to produce a current and the voltage or EMF to push those electrons through a circuit.

Centi—The metric prefix for 10^{-2}, or divide by 100.

Closed, or complete circuit—An electrical circuit with an uninterrupted path for the current to follow. Turning a switch on, for example, closes or completes the circuit, allowing current to flow.

Conductor—A material that has a loose grip on its electrons, so an electrical current can pass through it.

Current—A flow of electrons in an electrical circuit.

Deci—The metric prefix for 10^{-1}, or divide by 10.

Direct current (dc)—Electrical current that flows in one direction only.

Electromotive force (EMF)—The force or pressure that pushes a current through a circuit.

Electron—A tiny, negatively charged particle, normally found in an area surrounding the nucleus of an atom. Moving electrons make up an electrical current.

Energy—The ability to do work; the ability to exert a force to move some object.

Frequency—The number of complete cycles of an alternating current that occur per second.

Giga—The metric prefix for 10^{9}, or times 1,000,000,000.

Hertz (Hz)—An alternating-current frequency of one cycle per second. The basic unit of frequency.

Insulator—A material that maintains a tight grip on its electrons, so that an electric current cannot pass through it (within voltage limits).

Kilo—The metric prefix for 10^{3}, or times 1000.

Mega—The metric prefix for 10^{6}, or times 1,000,000.

Metric prefixes—A series of terms used in the metric system of measurement. We use metric prefixes to describe a quantity as compared to a basic unit. The metric prefixes indicate multiples of 10.

Metric system—A system of measurement developed by scientists and used in most countries of the world. This system uses a set of prefixes that are multiples of 10 to indicate quantities larger or smaller than the basic unit.

Micro—The metric prefix for 10^{-6}, or divide by 1,000,000.

Milli—The metric prefix for 10^{-3}, or divide by 1000.

Ohm—The basic unit of electrical resistance, used to describe the amount of opposition to current.

Ohm's Law—A basic law of electronics. Ohm's Law gives a relationship between voltage (E), current (I) and resistance (R). The voltage applied to a circuit is equal to the current through the circuit times the resistance of the circuit (E = IR).

Open circuit—An electrical circuit that does not have a complete path, so current can't flow through the circuit.

Parallel circuit—An electrical circuit where electrons follow more than one path in going from the negative supply terminal to the positive terminal.

Pico—The metric prefix for 10^{-12}, or divide by 1,000,000,000,000.

Power—The rate of energy consumption. We calculate power in an electrical circuit by multiplying the voltage applied to the circuit times the current through the circuit (P = IE).

Power supply—An electrical circuit that provides excess electrons to flow into another circuit. The power supply also supplies the voltage or EMF to push the electrons along. Power supplies convert a power source (such as the ac mains) to a form useful for various circuits.

Radio frequency (RF)—The range of frequencies that can be radiated through space in the form of electromagnetic radiation. We usually consider RF to be those frequencies higher than the audio frequencies, or above 20 kilohertz.

Resistance—The ability to oppose an electric current.

Resistor—Any material that opposes a current in an electrical circuit. An electronic component especially designed to oppose current.

Series circuit—An electrical circuit in which the electrons must all flow through every part of the circuit. There is only one path for the current to follow.

Short circuit—An electrical circuit in which the current does not take the desired path, but finds a shortcut instead. Often the current goes directly from the negative power-supply terminal to the positive one, bypassing the rest of the circuit.

Volt (V)—The basic unit of electrical pressure or EMF.

Voltage—The EMF or pressure that causes electrons to move through an electrical circuit.

Voltage source—Any source of excess electrons. A voltage source produces a current and the force to push the electrons through an electrical circuit.

Watt (W)—The unit of power in the metric system. The watt describes how fast a circuit uses electrical energy.

Wavelength—The distance an ac signal will travel during the time it takes the signal to go through one complete cycle.

---

[1]Numbers written as a multiple of some power are expressed in exponential notation. This notation is explained in detail on page 3-2.

Chapter 3

Understanding Basic Theory

This chapter introduces you to basic electronics and radio theory. First we present the theory that you'll need to pass your Novice exam. This basic theory will be a foundation for you to build on. Later in the chapter is a section devoted to the theory needed to pass the Technician exam. Even if you don't plan to try for your Technician license right away, you'll probably want to study this section, too. The theory you learn in this chapter will also help you to assemble and operate your amateur station.

There is much for you to learn, and most of the material presented here will be new to you. To make your study as easy as possible, Chapter 3 is divided into four main sections. We start with basic electrical principles. Here you'll learn the basics like voltage, current, conductors and insulators. Then we move on to electronics fundamentals, where you learn about resistance, Ohm's Law and power. You'll also learn about direct and alternating currents. The Technician sections of this chapter explain more about alternating current, as it applies to inductance and capacitance.

To get the most from this chapter, you should take it one section at a time. Study the material in each section and really

know it before you go on to the next section. The sections build on each other, so you may find yourself referring to sections you've already studied from time to time.

We use many technical terms in electronics. We have provided definitions that are as simple and to-the-point as possible. You will want to refer often to the **Key Words** at the beginning of this chapter. Don't be afraid to turn back to any section you have already studied. This review is helpful if you come across a term that you are not sure about. You will probably not remember every bit of this theory just by reading the chapter once.

This chapter includes many drawings and illustrations to help you learn the material. Pay attention to these graphics, and you'll find it easier to understand the text. We'll direct you to the Question Pools (Chapters 12 and 13) at appropriate points in the text. Use these directions to help you study the Novice and Technician Question Pools. When you think you've learned the section, you are ready to move on.

If you have trouble understanding parts of this chapter, ask your instructor or another experienced ham for help. Many other books can help, too. ARRL's *Understanding Basic Electronics* is written for students with no previous electronics background. The booklet *First Steps in Radio* is a collection of articles from a *QST* series by Doug DeMaw. *First Steps* covers a wide range of technical topics written especially for a beginner. To study more advanced theory, you may want to purchase a copy of *The ARRL Handbook*.

These publications are available from your local ham dealer or from ARRL Headquarters.

Remember! Take it slowly, section by section. Before you know it, you'll have learned what you need to know to pass your test and get on the air. Good luck!

BASIC ELECTRICAL PRINCIPLES

In this section, you will learn what electricity is and how it works. We'll introduce you to the atom and the electron, the basic elements of electricity. There are no questions about atoms and electrons on the Novice or Technician exams, but understanding them will help you understand the rest of this chapter.

THE METRIC SYSTEM

We'll be talking about the units used to describe several electrical quantities later in this chapter. Before we do that, let's take a few minutes to become familiar with the **metric system**. This simple system is a standard system of measurement used all over the world. All the units used to describe electrical quantities are part of the metric system.

In the US, we use a measuring system known as the US Customary system. In this system there is no logical progression between the various units of length, weight, volume or other quantities. For example, we have 12 inches in 1 foot, 3 feet in 1 yard and 1760 yards in 1 mile. For measuring the volume of liquids we have 2 cups in 1 pint, 2 pints in 1 quart and 4 quarts in 1 gallon. To make things even more difficult, we use some of these same names for different volumes when we measure dry materials! As you can see, this system of measurements can be very confusing. Even those who are very familiar with the system do not know all the units used for different types of measurements.

It is exactly this confusion that led scientists to develop the orderly system we know today as the metric system. This system uses a basic unit for each different type of measurement. For example, the basic unit of length is the meter. (This unit is spelled metre nearly everywhere in the world except the US!) The basic unit of volume is the liter (or litre). The unit for mass (or quantity of matter) is the gram. The newton is the metric unit of force, or weight, but we often use the gram to indicate how "heavy" something is. We can express larger or smaller quantities by multiplying or dividing the basic unit by factors of 10 (10, 100, 1000, 10,000 and so on). These multiples result in a standard set of prefixes, which can be used with all the basic units. Table 3-1 summarizes the most-used **metric prefixes**. These same prefixes can be applied to any basic unit in the metric system. Even if you come across some terms you are unfamiliar with, you will be able to recognize the prefixes.

We can write these prefixes as powers of 10, as shown

in Table 3-1. The power of 10 (called the *exponent*) shows how many times you must multiply (or divide) the basic unit by 10. For example, we can see from the table that **kilo** means 10^3. Let's use the meter as an example. If you multiply a meter by 10 three times, you will have a *kilo*meter. (1 meter $\times 10^3 = 1$ m $\times 10 \times 10 \times 10 = 1000$ meters, or 1 kilometer.) If you multiply 1 meter by 10 six times, you have a **mega**meter. (1 meter $\times 10^6 = 1$ m $\times 10 \times 10 \times 10 \times 10 \times 10 \times 10 = 1,000,000$ meters or 1 megameter.)

Notice that the exponent for some of the prefixes is a negative number. This indicates that you must *divide* the basic unit by 10 that number of times. If you divide a meter by 10, you will have a **deci**meter. (1 meter $\times 10^{-1} = 1$ m $\div 10 = 0.1$ meter, or 1 decimeter.) When we write 10^{-6}, it means you must divide by 10 six times. (1 meter $\times 10^{-6} = 1$ m $\div 10 \div 10 \div 10 \div 10 \div 10 \div 10 = 0.000001$ meter, or 1 **micro**meter.)

We can easily write very large or very small numbers with this system. We can use the metric prefixes with the basic units, or we can use powers of 10. Many of the quantities used in basic electronics are either very large or very small numbers, so we use these prefixes quite a bit. You should be sure you are familiar at least with the following prefixes and their associated powers of 10: **giga** (10^9), **mega** (10^6), **kilo** (10^3), **centi** (10^{-2}), **milli** (10^{-3}), **micro** (10^{-6}) and **pico** (10^{-12}).

Let's try an example. We have a receiver dial calibrated

Table 3-1

International System of Units (SI)—Metric Units

Prefix	Symbol	Multiplication Factor	
tera	T	10^{12}	= 1,000,000,000,000
giga	G	10^9	= 1,000,000,000
mega	M	10^6	= 1,000,000
kilo	k	10^3	= 1,000
hecto	h	10^2	= 100
deca	da	10^1	= 10
(unit)		10^0	= 1
deci	d	10^{-1}	= 0.1
centi	c	10^{-2}	= 0.01
milli	m	10^{-3}	= 0.001
micro	μ	10^{-6}	= 0.000001
nano	n	10^{-9}	= 0.000000001
pico	p	10^{-12}	= 0.000000000001

in megahertz (MHz), and it shows a signal at a frequency of 3.725 MHz. Where would a dial calibrated in kilohertz show the signal? From Table 3-1 we see that kilo means times 1000, and mega means times 1,000,000. That means that our signal is at 3.725 MHz × 1,000,000 = 3,725,000 hertz. There are 1000 hertz in a kilohertz, so 3,725,000 divided by 1000 gives us 3725 kHz.

How about another one? If we have a current of 3000 milliamps, how many amps is this? From Table 3-1 we see that milli means multiply by 0.001 or divide by 1000. Dividing 3000 milliamps by 1000 gives us 3 amps. The metric prefixes make it easy to use numbers that are a convenient size simply by changing the units. 3.725 MHz is certainly easier to work with than 3,725,000 hertz!

Notice that it doesn't matter what the units are or what they represent. Meters, hertz, amperes, volts, farads or watts make no difference in how we use the prefixes. Each prefix represents a certain multiplication factor, and that value never changes.

By now you should begin to understand how to change prefixes in the metric system. First write the number and find the proper power of ten (from memory or Table 3-1), and then move the decimal point to change to the basic unit. Then divide by the multiplication factor for the new prefix you want to use. With a little more practice you'll be changing prefixes with ease.

There is another method you can use to convert between metric prefixes, but it involves a little trick. Learn to write the chart shown in Figure 3-1 on a piece of paper when you are going to make a conversion. Always start with the large prefixes on the left and go toward the right with the smaller ones. Sometimes you can make an abbreviated list, using only the units from kilo to milli. If you need the units larger than kilo or smaller than milli, be sure to include the dots as shown in Figure 3-1. (They mark the extra decimal places between the larger and smaller prefixes, which go in steps of 1000 instead of every 10.) Once you learn to write the chart correctly, it will be very easy to change prefixes.

Let's work through an example to show how to use this chart. For this example, we'll use a term that you will run into quite often in your study of electronics: **hertz**. Hertz (abbreviated Hz) is a unit that refers to the frequency of a radio or television wave.

Change 3700 kilohertz to hertz. Since we are starting with kilohertz (kilo), begin at the k on the chart. Now count each symbol to the right, until you come to the basic unit (U). Did you count three places? Well that's how many places you must move the decimal point to change from kilohertz (kHz) to hertz (Hz). Which way do you move the deci-

mal point? Notice that you counted to the right on the chart. Move the decimal point in the same direction. Now you can write the answer: 3700 kHz = 3,700,000 Hz!

Suppose a meter indicates a current of 5,400 milliamperes (abbreviated mA) through a circuit. How many amperes (abbreviated A) is that? First, write the list of metric prefixes. Since you won't need those smaller than milli or larger than kilo, you can write an abbreviated list. You don't have to write the powers of ten, if you remember what the prefixes represent. To change from milli to the unit, we count 3 decimal places toward the left. This tells us to move the decimal point in our number three places to the left.

5,400 mA = 5.4 A

Let's try one more example, for some extra practice changing metric prefixes by moving the decimal point in a number. What if someone told you to tune your radio receiver to 3,726,000 Hz? You probably won't find any radio receiver with a dial marking like this! To make the number more practical, we'll write the frequency with a prefix that's more likely to appear on a receiver dial.

Our first step is to select a new prefix to express the number. We can write the number with only one or two digits to the left of the decimal point. It looks like we'll need the entire prefix chart for this one, so write it down as described earlier. (You can look at the chart in Figure 3-1, but you should practice writing it for those times when you don't have the book—like your exam!)

The next job is to count how many places you can move the decimal point. The number you end up with should have one or two digits to the left of the decimal point. Remember that metric prefixes larger than kilo represent multiples of 1000, or 10^3. Did you count six places to move the decimal point in our example, 3,726,000 Hz? That would leave us with 3.726×10^6 Hz.

Now go back to the chart and count six places to the left. (This is the same number of places and the same direction as we moved the decimal point.) The new spot on the chart indicates our new metric prefix, mega, abbreviated M. Replacing the power of ten with this prefix, we can write our frequency as 3.726 MHz.

[Before you go on to the next section, turn to Chapter 12. Be sure you can answer all of the questions in the Novice Question Pool with numbers between N5A01 and N5A11. Review this section if you have any difficulty.]

10^{18}	10^{15}	10^{12}	10^{9}	10^{6}	10^{3}	10^{2}	10^{1}	10^{0}	10^{-1}	10^{-2}	10^{-3}	10^{-6}	10^{-9}	10^{-12}	10^{-15}	10^{-18}
E	· · P	· · T	· · G	· · M	· · k	h	da	U	d	c	m	· · μ	· · n	· · p	· · f	· · a

Figure 3-1—This chart shows the symbols for all metric prefixes, with the power of ten that each represents. Write the abbreviations in decreasing order from left to right. The dots between certain prefixes indicate there are two decimal places between those prefixes. You must be sure to count a decimal place for each of these dots when converting from one prefix to another. When you change from a larger to a smaller prefix, you are moving to the right on the chart. The decimal point in the number you are changing also moves to the right. Likewise, when you change from a smaller to a larger prefix, you are moving to the left, and the decimal point also moves to the left.

ELECTRICITY

The word is a spine-tingling mystery. It's the force behind our space-age civilization. It's one of nature's greatest powers. We love it; we fear it. We use it in our work and play. But what is it?

It's a mystery only in our minds. Actually, electricity is the marvelous stuff which, when untamed, we call lightning. One lightning bolt produces enough electricity to supply your needs for a lifetime. In another form, electricity is the power in a battery that cranks the engine to start your car. Electricity also ignites the gasoline in the engine. Yet, with all its power, electricity is the careful messenger carrying information from your brain to your muscles, enabling you to move your arms and legs. You can buy a small container of electricity no bigger than a dime (a battery). Electric utilities generate and transmit huge amounts of electricity every day. From lightning bolts to brain waves, it's all the same stuff: electrons.

INSIDE ATOMS

Everything you can see and touch is made up of *atoms*. Atoms are the building blocks of nature. Atoms are too small to see, but the *subatomic particles* inside atoms are even smaller.

Each atom has a *nucleus* in its center. Other particles orbit around this central core. Think of the familiar maps of our solar system: planets orbit the sun. In an atom, charged particles orbit the central core (the nucleus). Other charged particles make up the nucleus. Figure 3-2 is a simplified illustration of an atom's structure.

Some particles have *negative charges* while others have *positive charges*. The core of an atom contains positively charged particles. Negative particles, called **electrons**, orbit around the nucleus. Scientists have identified more than 100 different kinds of atoms. The number of positively and negatively charged particles in an atom determines what type of element that atom is. Different kinds of atoms combine to form various materials. For example, a hydrogen atom has one positively charged particle in its nucleus and one electron around the outside. An oxygen atom has eight positive particles in the nucleus and eight electrons around the outside. When two hydrogen atoms combine with one oxygen atom, we have water.

Have you ever tried to push the north poles of two magnets together? Remember that soft, but firm, pressure holding them apart? Similar poles in magnets repel each other; opposite poles attract each other. You can see this if you experiment with a pair of small magnets.

Charged particles behave in a way similar to the two magnets. A positively charged particle and a negatively charged particle attract each other. Two positive or two negative particles repel each other. Like repels like; opposites attract.

Electrons stay near the central core, or nucleus, of the atom. The positive charge on the nucleus attracts the negative electrons. Meanwhile, since the electrons are all negatively

A phenomenon that has fascinated mankind for ages, lightning is simply a natural source of electricity.

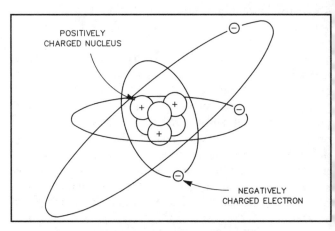

Figure 3-2—Each atom is a microscopic particle composed of a central, dense, positively charged nucleus, surrounded by tiny negatively charged electrons. There are the same number of positive particles in the nucleus and negative electrons outside the nucleus.

charged, they repel each other. This makes the electrons move apart and fill the space around the nucleus. (Scientists sometimes refer to this area as an electron "cloud.") An atom has an equal balance of negative and positive charges and shows no electrical effect to the outside world. We say the atom is *neutral*.

ELECTRON FLOW

In many materials, especially metals, it's easy to dislodge an electron from an atom. When the atom loses an

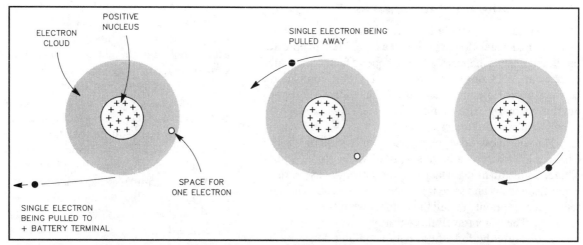

Figure 3-3—An ion is an electrically charged particle. When an electron (a negative ion) moves from one atom to another, the atom losing the electron becomes a positive ion. This electron movement represents an electric current. The electrons are moving from right to left in this drawing.

electron, it upsets the stability and electrical balance of the atom. With one particle of negative charge gone, the atom has an excess of positive charges. The free electron has a negative charge. We call the atom that lost the electron a positive *ion* because of its net positive charge. (An ion is a charged particle.) If there are billions of similar ions in one place, the quantity of charge becomes large enough to cause a noticeable effect.

Positively charged ions can pull negative particles (electrons) from neutral atoms. These electrons can move across the space between the atom and the ion and orbit the positively charged ion. Now the positively charged ion has become a neutral atom again, and another atom has become a positively charged ion! If this process seems confusing, take

a look at Figure 3-3. Here we show a series of atoms and positive ions, with electrons moving from one atom to the next. The electrons are moving from right to left in this diagram. We call this flow of electrons *electricity*. Electricity is nothing more than the flow of electrons.

How difficult is it for a positive ion to rip an electron away from an atom? That depends on the individual atoms making up a particular material. Some atoms hold firmly to their electrons and won't let them flow away easily. Other atoms keep only a loose grip on electrons and let them slip away easily. This means that some materials carry electricity better than others. **Conductors** are materials that keep only a loose grip on their electrons. We call those materials that try to hold on to their electrons **insulators**.

DEMONSTRATING ELECTRON FLOW

Here's an easy way to see the power of electrical charges. On a dry day hang some metal foil from dry thread. Rub one end of a plastic comb on wool (or run it through your hair). Then bring the end of the comb near the metal foil.

Rubbing the comb on wool detaches electrons from the wool fibers, depositing them on the surface of the comb. When you bring the charged end of the comb near the foil, free electrons in the foil will be repelled by the negative charge on the comb. The free electrons will move as far from the comb as they can—on the edge farthest from the comb. This leaves the near edge of the foil with a shortage of electrons (in other words, with a positive charge). The near edge of the foil is then attracted to the negatively charged comb.

As soon as the foil touches the comb, some of the excess electrons on the comb flow onto the foil. (Electricity!) The foil will then have a net negative charge, as the comb does, and the foil will be repelled by the comb. This simple experiment shows the attraction, repulsion and flow of electrons—the heart of electricity.

Why Do Electrons Flow?

There are many similarities between electricity flowing in a wire and water flowing through a pipe. Most people are familiar with what happens when you open a faucet and water comes out. We can use this to make a useful comparison between water flow and electron flow (electricity). Throughout this chapter we use examples of water flowing in a pipe to help explain electronics.

Do you know how your town's water system works? Chances are, there is a large supply of water stored somewhere. Some towns use a lake or river. Other towns get their water from wells and store it in a tank or reservoir. The system then uses gravity to pull the water down from the tank or reservoir. The water travels through a system of pipes to your house. Because the force of gravity is pulling down on the water in the tank, it exerts a pressure on the water in the pipes. This makes the water flow out of the faucets in your house with some force. If you have a well, you probably have a storage tank. The storage tank uses air pressure to push the water up to the top floor of your house.

In these water systems, a pump takes water from the large supply and puts it into a storage tank. Then the system uses air pressure or the force of gravity to push the water through pipes to the faucets in your home.

We can compare electrons flowing through wire to water flowing through a pipe. We need some force to make water flow through a pipe; what force exerts pressure to make electrons flow through a wire?

VOLTAGE

The amount of pressure that it takes to push water to your house depends on how far you live from the reservoir. If the water has to travel over hills along the way, even more pressure will be required. The pressure required to make electrons flow in an electrical circuit also depends on the opposition that the electrons must overcome. The pressure that forces the electrons through the circuit is known as **electromotive force** or, simply, **EMF**.

EMF is similar to water pressure. More pressure moves more water. Similarly, more EMF moves more electrons. We measure EMF in a unit called the **volt (V)**, so we sometimes refer to the EMF as a **voltage**. If more voltage is applied to a circuit, more electrons will flow. We measure voltage with a device called a *voltmeter*.

An electric wall outlet in your home usually supplies about 120 volts. If you have an outlet for an electric stove or an electric clothes dryer in your home, that outlet probably provides 240 V. A car **battery** is normally rated at 12 volts. A single D-cell battery supplies 1.5 V. **Voltage sources** come in a wide variety of ratings, depending on their intended use. With our system of metric prefixes, we can express a thousand volts as 1 kilovolt or "1 kV."

Another way to think about this electrical voltage pushing electrons through a circuit is to remember that like-charged objects repel. If we have a large group of electrons, the negative charge of these electrons will act to repel, or push, other electrons through the circuit. In a similar way, a large group of positively charged ions attract, or pull, electrons through the circuit.

Because there are two types of electric charge (positive and negative), there are also two polarities associated with a voltage. A **voltage source** always has two terminals, or poles: the positive terminal and the negative terminal. The negative terminal repels electrons (negatively charged particles) and the positive terminal attracts electrons. If we connect a piece of wire between the two terminals of a voltage source, electrons will flow through the wire. We call this flow an electrical **current**.

Batteries

In our water system, a pump supplies pressure to pull water from the source and force it into the pipes. Similarly, electrical circuits require an electron source and a "pump" to move the electrons along.

A **battery** is one example of a **power supply**. We use a battery as both the source of electrons and the pump that moves them along. The battery is like the storage tank in our water system. A battery provides pressure to keep the electrons moving.

There is an excess of electrons at the negative terminal of a power supply. At the positive terminal there is an excess of positive ions. With a conducting wire connected between the two terminals, the electrical pressure (voltage) generated by the power supply will cause electrons to move through the conductor.

Batteries come in all shapes and sizes. Some batteries are tiny, like those used in hearing aids and cameras. Other batteries are larger than the one in your car. A battery is one kind of voltage source.

CURRENT

You have probably heard the term "current" used to describe the flow of water in a stream or river. Similarly, we call the flow of electrons an electric **current**. Each electron is extremely small. It takes quintillions and quintillions of electrons to make your toaster heat bread or your TV draw pictures. (A quintillion is a one with 18 zeros after it—1,000,000,000,000,000,000. Using powers of 10, as described earlier, we could also write this as 1×10^{18}.)

The basic unit of electromotive force (EMF) is the volt. The volt was named in honor of Allessandro Giuseppe Antonio Anastasio Volta (1745-1827). This Italian physicist invented the electric battery.

The action of electric current on a magnet was first applied to telegraphy by André Marie Ampère (1775-1836) in 1820. An ampere is the basic unit of electrical current.

When water flows from your home faucet, you don't try to count every drop. The numbers would be very large and unmanageable, and the drops are coming out much too fast to count! To measure water flow, you count larger quantities such as gallons and describe the flow in terms of gallons per minute. Similarly, we can't deal easily with large numbers of individual electrons, nor can we count them conveniently. We need a shorthand way to measure the number of electrons. So, as with gallons per minute of water, we have amperes of electric current. We measure current with a device called an *ammeter*.

Suppose you are looking through a "window" into a wire, and can count electrons as they move past you. (See Figure 3-4.) If you count 6,240,000,000,000,000,000 (6.24 × 10^{18}) electrons moving past your "window" each second, the circuit has a current of one **ampere**. (Don't worry! You won't have to remember this number.) So when you express a circuit's current in amperes, remember that it is a measure of the number of electrons flowing through the circuit. A circuit with a current of 2 amperes has twice as many electrons flowing out of the supply as a circuit with a current of 1 ampere.

Write "2 A" for two amperes or "100 mA" (milliamps) for 0.1 ampere (sometimes also abbreviated amp or amps). You can use all the metric prefixes with the ampere. Most of the time you will see currents expressed in amps, milliamps and microamps. See Table 3-1 to review the list of metric prefixes.

[Now study questions N5B01 through N5B06 in Chapter 12. Review this section if you have trouble answering these questions.]

Figure 3-4—A window like this into a wire might allow you to count electrons as they flow past, to measure the current. Even so, you would have to count 6.24 × 10^{18} electrons per second for a current of only 1 ampere!

CONDUCTORS

As we pointed out earlier, some atoms have a firm grasp on their electrons and other atoms don't. More current can flow in materials made of atoms that have only a weak hold on their electrons. Some materials, then, conduct electricity better than others.

Silver is an excellent **conductor**. The loosely attached electrons in silver atoms require very little voltage (pressure) to produce an electric current. Copper is much less expensive than silver and conducts almost as well. We can use copper to make wire needed in houses, and in radios and other electronic devices. Steel also conducts, but not as well as copper. In fact, most metals are fairly good conductors, so aluminum, mercury, zinc, tin and gold are all conductors.

INSULATORS

Other materials keep a very firm grip on their electrons.

These materials do not conduct electricity very well. Materials such as glass, rubber, plastic, ceramic, mica, wood and even air are poor conductors. Pure distilled water is a fairly good **insulator**. Most tap water is a good conductor, however, because it has minerals and other impurities dissolved in it. Figure 3-5 lists some common insulators and conductors.

The electric power company supplies 120 V on the wires into your home. That voltage is available for your use at electrical outlets or sockets. Why don't the electrons spill out of the sockets? The insulation between the two sides of the outlet prevents the electrons from flowing from one side to the other. The air around the socket acts as an insulator to stop them from flowing into the room.

Because insulators are *poor* conductors rather than *non*conductors, every insulator has a *breakdown voltage*. A voltage higher than the breakdown voltage will force electrons to move through the insulator. The insulator will start to conduct electricity. Depending on the material, it may be

Figure 3-5—Here's one way to show how an insulator differs from a conductor. Wood, on the left, holds onto its electrons pretty tightly, keeping them from flowing between its atoms. Metals, on the other hand, are more generous with their electrons. Electrons are more easily pulled away from the metal atoms, and the metal atoms are then left with a positive charge. If the metal atoms attract extra electrons from neighboring atoms, they become negatively charged.

WOOD

METAL

damaged if you exceed the breakdown voltage. Better insulators have higher breakdown voltages. Breakdown voltage also depends on the thickness of the insulating material. A thin layer of one insulating material (like Teflon or mica) may be just as good as a much thicker layer of another material (like paper or air).

A good example of an insulator that will conduct at very high voltage is air. Air is a fine insulator at the voltages normally found in homes and industry. When a force of millions of volts builds up, however, there's enough pressure to send a bolt of electrons through the air—lightning. You can produce a voltage large enough to make a spark jump through a thin air layer by shuffling your feet across the carpet on a dry day. When you reach for a metal object like a doorknob, you can often feel the spark jump from your finger. If the room is darkened, you can also see the spark.

When you are insulating wires or components, always be sure to use the right insulating material. Make sure you use enough insulation for the voltages you're likely to encounter. Heat-shrinkable tubing or other insulating tubing is often convenient for covering a bare wire or a solder connection. You can also wrap the wire with electrical tape. Several layers of tape, wrapped so it overlaps itself, will provide enough insulation for up to a few hundred volts.

[Before you go on to the next section, turn to Chapter 12 and study questions N5B07, N5B08 and N5B09. If you have any difficulty, review this section.]

ELECTRONICS FUNDAMENTALS

In this section, you'll learn about resistance. You'll see how resistance fits into one of the most fundamental laws of electronics, Ohm's Law. We will also briefly discuss series and parallel circuits. While there are no questions about these circuits on the Novice exam, understanding them will make it much easier for you to understand some of the other concepts you *will* be tested on! If you are preparing for the Technician exam, be sure you understand these circuits.

This section shows you how to do some basic circuit calculations. We have tried to keep the arithmetic simple, and to explain all of the steps in the solution. Associating numbers with a concept often makes it easier to understand. Read these examples and then try the calculations yourself. Be sure

The basic unit of resistance is the ohm, named in honor of Georg Simon Ohm (1787-1854).

you can work the problems in the question pool when you study those questions.

RESISTANCE

What if you partially blocked a water pipe with a sponge? Eventually, the water would get through the sponge, but it would have less pressure than before. It takes pressure to overcome the resistance of the sponge.

Figure 3-6—This photograph shows some of the many types of resistors. Large power resistors are at the top of the photo. The small resistors are used in low-power transistor circuits.

Similarly, materials that conduct electrical current also present some opposition, or **resistance**, to that flow. **Resistors** are devices that are especially designed to make use of this opposition. Resistors limit the amount of current that flows through a circuit. Figure 3-6 shows some common resistors.

In a water pipe, increasing the pressure forces more water through the sponge (the resistance). In an electrical circuit, increasing the voltage forces more current through the resistor. The relationship between voltage, current and resistance is predictable. We call this relationship **Ohm's Law** and it is a basic electronics principle.

The **ohm** is the basic unit used to measure resistance. The abbreviation for ohms is Ω, the Greek capital letter omega. This unit is named for Georg Simon Ohm, a German physics teacher and mathematician. Of course we also use the metric prefixes with ohms, when appropriate. So you will often see resistors specified as having 47 kilohms or 1.2 megohms of resistance. Written with abbreviations, these resistors would be 47 kΩ and 1.2 MΩ.

[Turn to Chapter 12 and study questions N5B10 and N5B11 now. Review this section if you can't answer them.]

[**If you are preparing for the Technician exam**, also study question T5A01 in Chapter 13.]

OHM'S LAW

The amount of water flowing through a pipe increases as we increase the pressure and decreases as we increase the resistance. If we replace "pressure" with "voltage," we can write a mathematical relationship for an electric circuit:

$$\text{current} = \frac{\text{voltage}}{\text{resistance}} \qquad \text{(Eq 3-1)}$$

This equation tells us the current through a circuit equals the voltage applied to the circuit divided by the circuit resistance.

THERE REALLY WAS AN OHM (1787-1854)

Although we take Ohm's Law for granted, it wasn't always so widely accepted. In 1827, Georg Simon Ohm finished his renowned work, *The Galvanic Circuit Mathematically Treated*. Scientists at first resisted Ohm's techniques. The majority of his colleagues were still holding to a non-mathematical approach. Finally, the younger physicists in Germany accepted Ohm's information in the early 1830s. The turning point for this basic electrical law came in 1841 when the Royal Society of London awarded Ohm the Copley Medal.

As the oldest son of a master locksmith, Georg received a solid education in philosophy and the sciences from his father. At the age of 16 he entered the University of Erlangen (Bavaria) and studied for three semesters. His father then forced him to withdraw because of alleged overindulgences in dancing, billiards and ice skating. After 4½ years, his brilliance undaunted, Ohm returned to Erlangen to earn his PhD in mathematics.

Enthusiasm ran high at that time for scientific solutions to all problems. The first book Ohm wrote reflected his highly intellectualized views about the role of mathematics in education. These opinions changed, however, during a series of teaching positions. Oersted's discovery of electromagnetism in 1820 spurred Ohm to avid experiments, since he taught Physics and had a well-equipped lab. Ohm based his work on direct scientific observation and analyses, rather than on abstract theories.

One of Ohm's goals was to be appointed to a major university. In 1825 he started doing research with the thought of publishing his results. The following year he took a leave of absence from teaching and went to Berlin.

In Berlin, Ohm made his now-famous experiments. He ran wires between a zinc-copper battery and mercury-filled cups. While a Coulomb torsion balance (voltmeter) was across one leg of the series circuit, a "variable conductor" completed the loop. By measuring the loss in electromagnetic force for various lengths and sizes of wire, he had the basis for his formulas.

Because *The Galvanic Circuit* produced near-hostility, Ohm withdrew from the academic world for nearly six years before accepting a post in Nuremberg. English and French physicists do not seem to have been aware of the profound implications of Ohm's work until the late 1830s and early 1840s.

Following his belated recognition, Ohm became a corresponding member of the Berlin Academy. Late in 1849 he went to the University of Munich and in 1852, only two years before his death, he achieved his lifelong dream of a full professorship at a major university.

If the voltage stays constant but more current flows in the circuit, we know there must be less resistance. The relationship between current and voltage is a measure of the resistance:

$$\text{resistance} = \frac{\text{voltage}}{\text{current}} \qquad \text{(Eq 3-2)}$$

We can state this equation in words as: The circuit resistance is equal to the voltage applied to the circuit divided by the current through the circuit.

Finally, we can determine the voltage if we know how much current is flowing and the resistance in the circuit:

$$\text{voltage} = \text{current} \times \text{resistance} \qquad \text{(Eq 3-3)}$$

The voltage applied to a circuit is equal to the current through the circuit times the circuit resistance.

Scientists are always looking for shorthand ways of writing these relationships. They use symbols to replace the words: E represents voltage (remember EMF?), current is I (from the French word *intensité*) and resistance is R. We can now express **Ohm's Law** in a couple of letters:

$$E = IR \text{ (volts = amperes} \times \text{ohms)} \qquad \text{(Eq 3-4)}$$

This is the most common way to express Ohm's Law, but we can also write it as:

$$I = \frac{E}{R} \text{ (amperes = volts divided by ohms)} \qquad \text{(Eq 3-5)}$$

and

$$R = \frac{E}{I} \text{ (ohms = volts divided by amperes)} \qquad \text{(Eq 3-6)}$$

E is EMF in volts, I is the current in amperes and R is the resistance in ohms. If you know two of the numbers, you can calculate the third. Figure 3-7 shows a diagram to help you solve Ohm's Law problems. Simply cover the symbol of the quantity that you do not know. If the remaining two are side-by-side, you must multiply them. If one symbol is above the other, then you must divide the quantity on top by the one on the bottom.

If you know current and resistance in a circuit, Ohm's Law will give you the voltage (Eq 3-4). For example, what is the voltage applied to the circuit if 5 amperes of current flows through 20 ohms of resistance? From Eq 3-4 or Figure 3-7, we see that we must multiply 5 amperes times 20 ohms to get the answer, 100 volts. The EMF in this circuit is 100 volts.

$$E = I R \qquad \text{(Eq 3-4)}$$

$$E = 5 \text{ amperes} \times 20 \text{ ohms}$$

$$E = 100 \text{ volts}$$

Figure 3-7—A simple diagram to help you remember the Ohm's Law relationships. To find any quantity if you know the other two, simply cover the unknown quantity with your hand or a piece of paper. The positions of the remaining two symbols show if you have to multiply (if they are side by side) or divide (if they appear one over the other as a fraction).

Figure 3-8— Some Ohm's Law problems and solutions.

E = ? RESISTOR R = 10 Ω
BATTERY
AMMETER
I = 2 A

GIVEN: I = 2 AMPERES
R = 10 OHMS

FIND: E (VOLTAGE)

E = I × R = 2 × 10 = 20

VOLTAGE EQUALS 20 VOLTS

E = 12 V RESISTOR R = 6 Ω
BATTERY
AMMETER
I = ?

GIVEN: E = 12 VOLTS
R = 6 OHMS

FIND: I (CURRENT)

$$I = \frac{E}{R} = \frac{12}{6} = 2$$

CURRENT EQUALS TWO AMPERES

E = 6 V RESISTOR R = ?
BATTERY
AMMETER
I = 2 A

GIVEN: E = 6 VOLTS
I = 2 AMPERES

FIND: R (RESISITANCE)

$$R = \frac{E}{I} = \frac{6}{2} = 3$$

RESISTANCE EQUALS THREE OHMS

Suppose you know voltage and resistance—100 volts in the circuit to push electrons against 20 ohms of resistance? Eq 3-5 gives the correct equation, or you can use the diagram in Figure 3-7. You must divide 100 volts by 20 ohms to find that 5 amperes of current is flowing.

$$I = \frac{E}{R} \qquad \text{(Eq 3-5)}$$

$$I = \frac{100 \text{ volts}}{20 \text{ ohms}}$$

$$I = 5 \text{ amperes}$$

If you know voltage and current in a circuit, you can calculate resistance. Using our example, 100 volts is pushing 5 amperes through the circuit. This time Eq 3-6 is the one to use, and you can also find this from Figure 3-7. 100 divided by 5 equals 20, so the resistance is 20 ohms.

$$R = \frac{E}{I} \qquad \text{(Eq 3-6)}$$

$$R = \frac{100 \text{ volts}}{5 \text{ amperes}}$$

$$R = 20 \text{ ohms}$$

If you know E and I, you can find R. If you know I and R, you can calculate E. If E and R are known, you can find I.

Put another way, if you know volts and amperes, you can calculate ohms. If amperes and ohms are known, volts can be found. Or if volts and ohms are known, amperes can be calculated. Figure 3-8 illustrates some simple circuits and how Ohm's Law can be used to find an unknown quantity in the circuit. Make up a few problems of your own and test how well you understand this basic law of electricity. You'll soon find that this predictable relationship, symbolized by the repeatable equation of Ohm's Law, makes calculating values of components in electrical circuits easy!

[Before you read further, turn to Chapter 12 and study questions N5C01 through N5C04.]

[**If you are preparing for the Technician exam**, also study questions T5B01 through T5B11 in Chapter 13. Review the material in this section if you have difficulty with any of these questions.]

SERIES AND PARALLEL CIRCUITS

There are two basic ways that you can connect the parts in an electric circuit. If we hook several resistors together in a string, we call it a **series circuit**. If we connect several resistors side-by-side to the same voltage source, we call it a **parallel circuit**.

In our water-pipe example, what would happen to the current through the pipe if we placed another sponge in it? You're right: The second sponge would further reduce the flow. We could do the same thing with a single, larger sponge. The total resistance in a series circuit is the sum of all the resistances in the circuit.

In a series circuit, the same current, I, flows through

Figure 3-9—Resistance limits the amount of current that can flow in a circuit. Adding a second resistance reduces the current because the total resistance is larger. The total resistance of a string of series-connected resistors is the sum of all the individual resistances.

Figure 3-10—Resistors in parallel. In this circuit, the electrical current splits into three separate paths through the resistors. The full battery voltage is applied to each resistor. Current through each individual resistor is independent of the other resistors. For example, if R_2 and R_3 are changed to a value different than the original value, the current in R_1 would remain unchanged. Even if R_2 and R_3 are removed, the current in R_1 remains unchanged.

each resistor, since it has no other path to follow. When a voltage source (like a battery) is hooked to our string of resistors, we can calculate the voltage across each resistor. How? Using Ohm's Law, of course.

Remember E = IR? The voltage across any resistor in the circuit will be its value in ohms multiplied by the current in amperes. We call the voltage across the resistor the voltage drop. If we calculate the voltage drop across each resistor in the circuit and add them together, we will get the battery voltage. Figure 3-9 shows that with more resistors in a series circuit, there will be less current.

Now let's look at the case of a parallel circuit. This is similar to having two water pipes running side by side. More water can flow through two parallel pipes of the same size than through a single one. With two pipes, the current is greater for a given pressure.

If these pipes had sponges in them, the flow would be reduced in each pipe. There are still two paths for the water to take. More water will flow than if there was a single pipe with a sponge in it. Now let's go back to electrical resistors and voltage. Adding a resistor in parallel with another one provides two paths for the electrical current to follow. This reduces the total resistance. You can connect more than two resistors in parallel, providing even more paths for the electrons. This will reduce the resistance still more. Figure 3-10 shows that as we add resistors in parallel, we provide more current paths. The result is less total resistance in the circuit and more current.

OPEN AND SHORT CIRCUITS

You've probably heard the term **short circuit** before. A short circuit happens when the current flowing through the components doesn't follow the course we expect it to. Instead, the current finds another path, a shorter one, between the terminals of the power source. This is why we call this path a short circuit. Because there is less opposition to the flow of electrons, there is a larger current. Often the current through the new (short) path is so large that the wires or components can't handle it. When this happens, the wires and components can be damaged. Figure 3-11 shows a bare wire causing a short circuit.

Some people think that a short circuit occurs when there is no resistance in the connection between the positive and negative terminals of the power supply. Actually there will always be *some* resistance but it may be such a small amount that it can be ignored.

In the extreme case, if a short circuit develops in our house wiring, the wire may overheat and can even start a fire. This is why it is important to have a properly rated fuse connected in series with a circuit. We will describe fuses in more detail in Chapter 4.

Question N5C09 in the Novice Question Pool asks, "Which electrical circuit uses too much current?" This question refers to a short circuit, but the terminology may be confusing. A circuit doesn't *use* current, in the sense of something being consumed, or used up. Current flows *through* the circuit, but the same current that flows out of a battery or other power supply will flow back into the supply at the other terminal. You can imagine electrons leaving the negative battery terminal and flowing through the circuit. All the electrons that leave the battery eventually return to the positive terminal. The circuit doesn't *use*, or *consume*, any of the electrons.

Too much current is also a relative term. A short circuit draws more current than is intended. If you connect two or three electric space heaters and a few lights to one 15-amp household circuit, they may draw too much current. This will blow a fuse or trip a circuit breaker, but it does not mean you have a short circuit. Each heater is drawing the current it is designed to draw.

The opposite of a short circuit is an **open circuit**. In an open circuit the current is interrupted, just as it is when you

Figure 3-11—Part A shows a light bulb in a working circuit. In B, the insulation covering the wires has broken, and the two wires are touching, so we have a short circuit.

Figure 3-12—Part A shows a light bulb in a working circuit. In B, one wire has broken, preventing the current from flowing through the bulb. This is an example of an open circuit.

turn a light switch off. There is no current through an open circuit. The switch breaks (opens) the circuit, putting a layer of insulating air in the way so no current can flow. This break in the current path presents an extremely high resistance. An open circuit can be good, as when you throw the on/off switch to off. An open circuit can be bad if it's an unwanted condition caused by a broken wire or a bad component. Figure 3-12 illustrates an open circuit. Sometimes the resistance in the circuit path is so large that it is impractical to measure it. When this is true, we say that there is an infinite resistance in the path, creating an open circuit.

A **closed**, or **complete circuit** has an uninterrupted path for the current to follow. Turning a switch on, for example, closes or completes the circuit, allowing current to flow.

[At this time you should turn to Chapter 12 and study the questions N5C08 and N5C09. If you have any difficulty with these questions, review this section.]

ENERGY AND POWER

We define **energy** as the ability to do work. An object can have energy because of its position (like a rock ready to fall off the edge of a cliff). An object in motion also has energy (like the same rock as it falls to the bottom of the cliff). In electronics, a power supply or battery is the source of electrical energy. We can make use of that energy by connecting the supply to a light bulb, a radio or other circuit.

We discussed voltage drops in the SERIES AND PARALLEL CIRCUITS section. As electrons move through a circuit and go through resistances, there is a voltage drop. These voltage drops occur because *energy* is "used up" or "consumed." Actually, we can't lose the electrical energy; it is just changed to some other form. The resistor heats up because of the current through it; the resistance converts electrical energy to heat energy. More current produces still more heat, and the resistor becomes warmer. If too much current flows, the resistor might even catch fire!

As electrons flow through a light bulb, the resistance of the bulb converts some electrical energy to heat. The filament in the bulb gets so hot that it converts some of the electrical energy to light energy. Again, more current produces more light and heat.

You should get the idea that we can "use up" a certain amount of energy by having a small current go through a resistance for a long time or by having a larger current go through it for a shorter time. When you buy electricity from a power company, you pay for the electrical energy that you use each month. You might use all of the energy in one day, or use a small amount every day. The electric meter on your house just measures how much energy you use and the power company sends someone around to read the meter each month.

Sometimes it is important to know how fast a circuit can use energy. You might want to compare how bright two different light bulbs will be. If you're buying a new freezer, you might want to know how much electricity it will use in a month. You will have to know how fast the freezer or the light bulbs use electrical energy. We use the term **power** to define the rate of energy consumption. The basic unit for measuring power in the metric system is the **watt**. You have probably seen this term used to rate electrical appliances. You know that a light bulb rated at 75 watts will be brighter than one rated at 40 watts. (Sometimes we abbreviate watts with a capital W.)

In electronics, there is a very simple way to calculate power. Just multiply amperes times volts. The equation is:

$$P = IE \qquad\qquad\text{(Eq 3-7)}$$

where:

  P is power, measured in watts
  I is the current in amperes
  E is the EMF in volts

For example, suppose a 12-V battery is pushing 3 A of current through a light bulb that is operating normally. Using Eq 3-7, we can find the power rating for this light bulb.

$$P = IE = 3\text{ A} \times 12\text{ V} = 36\text{ W}$$

If you know the power in a system and the voltage applied to the circuit, you can compute the current. Or, if you know power and current, you can find voltage. In other words, if you know any two parts you can find the third. Figure 3-13 shows a diagram to help you with calculations like these, or we can write the equations:

$$I = \frac{P}{E} \qquad\qquad\text{(Eq 3-8)}$$

$$E = \frac{P}{I} \qquad\qquad\text{(Eq 3-9)}$$

If you apply a 10-V EMF to a 20-W circuit, divide 20 by 10 to find that there is a 2-A current.

$$I = \frac{P}{E} = \frac{20\text{W}}{10\text{ V}} = 2\text{ A}$$

What voltage produces a 5-A current through a 60-W circuit? To find voltage you simply divide power by current.

$$E = \frac{P}{I} = \frac{60\text{ W}}{5\text{ A}} = 12\text{ V}$$

Suppose you turn on a single light. One-half ampere of current flows through the bulb with 120 volts applied. What's the power rating of the light bulb? Eq 3-7 shows us how to find this power. Multiply 120 volts times ½ (0.5) ampere to calculate that the light is a 60-watt bulb.

Figure 3-13—A simple diagram to help you remember the power equation relationships. To find any quantity given the other two, simply cover the unknown quantity with your hand or a piece of paper. The remaining two symbols will indicate if you have to multiply (if they are side by side) or divide (if they appear one over the other as a fraction).

Understanding Basic Theory   **3-13**

$$P = IE = 0.5 \ A \times 120 \ V = 60 \ W$$

By the way, the E and I in the power equation are the same E and I in Ohm's Law. So if we know any two of the four quantities voltage, current, resistance and power, we can calculate the other two.

Remember that a light bulb converts electrical energy into heat and light. The bulb has resistance; it opposes the flow of electrons. Is it possible to find the resistance of the wire inside the light bulb?

Yes. We know the voltage applied to the wire and the amount of current through it. This means that we can use Ohm's Law to calculate the resistance of the wire in the light bulb. Eq 3-6 gives the form of Ohm's Law that we need to find this answer. To find the resistance inside the bulb, divide 120 V by 0.5 A.

$$R = \frac{E}{I} = \frac{120 \ V}{0.5 \ A} = 240 \ \Omega$$

The bulb is the equivalent, then, of a 240-ohm resistor in the circuit.

The resistance of the bulb depends on whether the filament is hot or cold. In this example, the answer of 240 ohms is correct only when the bulb is lit. If you use an ohmmeter (a meter for measuring resistance—only when the bulb is *not* connected to the voltage source!), you will find a different value because the bulb is cold.

[Before you go on to the last Novice section in this chapter, turn to Chapter 12 and study questions N5C05, N5C06 and N5C07. Review this section if you have any problems.]

DIRECT AND ALTERNATING CURRENT

In this section, you will learn what we mean by **direct current** and **alternating current**. You will also learn the meaning of some important terms that go along with alternating current, such as **frequency** and **wavelength**.

Two Types of Current

Until now, we have been talking about **direct current** electricity, known as **dc** for short. In direct current, the electrons flow in one direction only—from negative to positive. In our water-flow analogy, this is like water that can flow in only one direction. We know that water can flow in more than one direction, however. The tides in the ocean are a good example.

There is a second kind of electricity called **alternating current**, or **ac**. In ac, the terminals of the power supply change from positive to negative to positive and so on. Because the

---

WHY USE ALTERNATING CURRENT IN OUR HOMES?

Why do power companies use alternating current in the power lines that run to your home? One important reason is so they can use transformers to change the voltage. This allows the company to use an appropriate voltage for each part of their distribution system. In this way the power company can minimize the power losses in the transmission lines. The generator at the power station produces ac by moving a wire (actually many turns of wire) through a magnetic field in an alternator. The resulting output has a relatively low voltage. Why don't the power companies send this directly through the power lines to your house? At first, this seems like a good idea. It would eliminate the many transformers and power

stations that often clutter our landscape.

The answer can be found in Ohm's Law. Even a very good conductor, such as the copper used in the power company's high-voltage lines, has a certain amount of resistance. This factor becomes very important when we consider the very long distances the generated electricity must travel.

Remember that the voltage drop across a resistance is given by the formula E = IR, where I is the value of current and R is the value of resistance. If we can reduce either the resistance of the wire or the value of the current through the wire, we can reduce the voltage drop. The resistance of the wire is relatively constant, although we can reduce it somewhat by using a very large diameter wire. If we increase the voltage, a smaller current will be required for the same power transfer from the generating station to your home.

Using a very high voltage also provides more "overhead." If the power company starts with 750,000 volts, and the voltage has dropped to 740,000 volts by the time it reaches the first substation, they just use a transformer rated for 740,000-V input to give the desired output. If they send 50,000 volts on to the next substation, there is still plenty of overhead. By the time it gets to the power lines outside your house, the voltage has dropped to around 3000 volts. A pole transformer then steps it down to 240 volts to supply power to your house. This voltage is normally split in half to provide two 120-V circuits to your house.

The basic unit of frequency is the hertz. This unit is named in honor of Heinrich Rudolf Hertz (1857-1894). This German physicist was the first person to demonstrate the generation and reception of radio waves.

poles change and electrons always flow from negative to positive, ac flows first in one direction, then the other. The current alternates in direction.

We call one complete round trip a *cycle*. The **frequency** of the ac is the number of complete cycles, or alternations, that occur in one second. We measure frequency in *hertz* (abbreviated Hz). Frequency is a measure of the number of times in one second the alternating current flows back and forth. One cycle per second is 1 Hz. 150 cycles per second is 150 Hz. One thousand cycles per second is one kilohertz (1 kHz). One million cycles per second is one megahertz (1 MHz).

[Now study the following questions in Chapter 12: N5C10, N5C11, N5D01 and N5D02. Review this section if you have difficulty answering any of these questions.]

More Ac Terminology

Batteries provide direct current. To make an alternating current from this direct-current source, you would have to switch the polarity of the voltage source rapidly. Imagine trying to turn the battery around so the plus and minus terminals changed position very rapidly. This would not be a very practical way to produce ac! You must have a power supply where the polarity is constantly changing. The terminals must be positive and negative one moment, and then negative and positive, constantly switching back and forth.

The power company has a more practical way to create ac: they use a large machine called an *alternator*. The ac supplied to your home goes through 60 complete cycles each second. Thus, the electricity from the power company has a frequency of 60 Hz.

This 60-hertz ac electricity builds slowly to a peak current or voltage in one direction, then decreases to zero and reverses to build to a peak in the opposite direction. If you plot these changes on a graph, you get a gentle up-and-down curve. We call this curve a *sine wave*. Figure 3-14 shows several cycles of a sine-wave ac signal.

Alternating current can do things direct current can't. For instance, a 120-V ac source can be increased to a 1000-V source with a *transformer*. Transformers can change the value of an ac voltage, but not a dc voltage. The power company supplies 120-V ac to your house.

Frequency and Wavelength

We discussed the frequency of an ac signal earlier. From this discussion, you must realize that alternating currents and voltages can change direction at almost any rate imaginable. Some signals have low frequencies, like the 60-Hz-ac electricity the power company supplies to your house. Other signals have higher frequencies; for example, radio signals can alternate at more than several million hertz.

When we talk about such a wide range of frequencies, it is common to break the wide range into several smaller ranges. One common breaking point is the difference between audio- and radio-frequency signals. If you connected an ac signal having a frequency anywhere between 20 Hz and 20,000 Hz (20 kHz) to a loudspeaker, you would hear a sound. Because these signals can produce sounds, they are called **audio-frequency (AF)** signals. The higher the frequency of the signal, the higher the pitch of the sound you would hear. (Caution—DO NOT connect the 60-Hz power from a household receptacle to a speaker even though 60 Hz is in the audio-frequency range! You may be seriously injured or KILLED by the voltage of this signal!)

Not all people can hear this full range of signals from 20 Hz to 20 kHz. Some people hear the low frequencies better than the high frequencies, and others hear the high frequencies better. This is the general range of frequencies that humans can expect to hear, however. (Dogs can hear signals at a much higher frequency, which is why dog whistles, used for training, don't produce a sound you can hear.)

If a signal has a frequency above the audio-frequency range (20 kHz), we call it a **radio-frequency (RF)** signal. Signals in the RF range can also be broken into smaller groups, such as very-low frequency (VLF), high frequency (HF), very-high frequency (VHF), ultra-high frequency (UHF) and so on. Don't worry about the names of these ranges

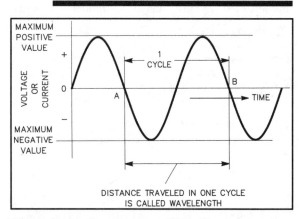

Figure 3-14—The sine wave is one way to show alternating current. Let's follow one cycle starting on line "0" at point A, indicated near the center of the graph. The wave goes in a negative direction to its most negative point, then heads back up to zero. After the wave goes through zero, it becomes more and more positive, reaches the positive peak, then goes back to zero again. This is one full cycle of alternating current.

for your exam. You will probably hear the terms as you listen in on other hams' discussions, though. The Novice bands are in the HF, VHF and UHF ranges.

If we know the frequency of an ac signal, we can use that frequency to describe the signal. We can talk about 60-Hz power or a 3745-kHz radio signal. **Wavelength** is another quality that can be associated with every ac signal. As its name implies, wavelength refers to the distance that the wave will travel through space in a single cycle. All such signals (sometimes called electromagnetic waves) travel though space at the speed of light, 300,000,000 meters per second $(3.00 \times 10^8 \text{ m/s})$. We use the lower-case Greek letter lambda (λ) to represent wavelength.

The faster a signal alternates, the less distance the signal will be able to travel during one cycle. There is an equation that relates the frequency and the wavelength of a signal to the speed of the wave:

$$c = f \lambda \qquad \text{(Eq 3-10)}$$

where:
 c is the speed of light, 3.00×10^8 meters per second
 f is the frequency of the wave in hertz
 λ is the wavelength of the wave in meters

We can solve this equation for either frequency or wavelength, depending on which quantity we want to find.

$$f = \frac{c}{\lambda} \qquad \text{(Eq 3-11)}$$

and

$$\lambda = \frac{c}{f} \qquad \text{(Eq 3-12)}$$

From these equations you may realize that as the frequency increases the wavelength gets shorter. As the frequency decreases the wavelength gets longer. Suppose you are transmitting a radio signal on 7.125 MHz. What is the wavelength of this signal? We can use Eq 3-12 to find the answer. First we must change the frequency to hertz:

7.125 MHz = 7,125,000 Hz.

Then we use this value in Eq 3-12.

$$\lambda = \frac{c}{f} = \frac{3.00 \times 10^8 \text{ m/s}}{7.125 \times 10^6 \text{ Hz}}$$

$$\lambda = \frac{300,000,000 \text{ m/s}}{7,125,000 \text{ Hz}} = 42 \text{ meters}$$

Of course, you already knew that this frequency was in the 40-meter Novice band, so this answer should not surprise you.

As another example, what is the wavelength of a signal that has a frequency of 3.725 MHz?

$$\lambda = \frac{c}{f} = \frac{3.00 \times 10^8 \text{ m/s}}{3.725 \times 10^6 \text{ Hz}}$$

$$\lambda = \frac{300,000,000 \text{ m/s}}{3,725,000 \text{ Hz}} = 80.5 \text{ meters}$$

Even if you have trouble with this arithmetic, you should be able to learn the frequency and wavelength relationships for the six Novice bands, as shown in Table 3-2.

Table 3-2

Novice-Band Frequencies and Wavelengths

Frequency Range (megahertz)		Approximate Wavelength (meters)
3.675	— 3.725	80
7.1	— 7.15	40
21.1	— 21.2	15
28.1	— 28.5	10
222	— 225	1.25
1270	— 1295	0.23 (23 centimeters)

Notice that higher-frequency signals have shorter wavelengths. Lower-frequency signals have longer wavelengths. As you increase a signal frequency the wavelength gets shorter. As a signal's wavelength increases, the frequency goes down.

[Congratulations! You are now well on your way to knowing all the electronics you will need to pass your Novice exam. Turn to Chapter 12 and study questions N5D03 through N5D11. Don't hesitate to come back to this chapter to review any sections that you are still a little uncertain about.]

[**If you are preparing for the Technician exam**, you should continue with this chapter.]

Electrical Theory for the Technician Exam

KEY WORDS

Capacitor—An electrical component usually formed by separating two conductive plates with an insulating material. A capacitor stores energy in an *electric field*.

Capacitance—The ability of a capacitor to store energy in an *electric field*.

Farad—The basic unit of capacitance.

Henry—The basic unit of inductance.

Induced EMF—A voltage produced by a change in magnetic lines of force around a conductor. When a magnetic field is formed by current in the conductor, the induced voltage always opposes change in that current.

Inductance—The ability of a coil to store energy in a *magnetic field*.

Inductor—An electrical component usually composed of a coil of wire wound on a central core. An inductor stores energy in a *magnetic field*.

Parallel circuit—An electrical circuit in which the electrons follow more than one path in going from the negative supply terminal to the positive terminal.

Series circuit—An electrical circuit in which all the electrons must flow through every part of the circuit. There is only one path for the electrons to follow.

CALCULATING SERIES AND PARALLEL RESISTANCE

Sometimes it's necessary to calculate the total resistance of resistors connected in a **series circuit**. Resistors in series are connected end to end like a string of sausages. At other times you must calculate total resistance in a parallel circuit. In a **parallel circuit**, resistors are side by side like a picket fence. There will be times when you need a certain amount of resistance somewhere in a circuit. There may be no standard resistor value that will give the necessary resistance. Sometimes you may not have a certain value on hand. By combining resistors in parallel or series you can obtain the desired value.

When you connect resistors in series, as in Figure 3-15, the total resistance is simply the sum of all the resistances. Let's use the analogy of a sponge in a water pipe again. Connecting resistors in series is like putting several sponges in the same pipe. The total resistance to the water would be the sum of all the individual resistances. Resistors in series add.

$$R_{TOTAL} = R_1 + R_2 + R_3 + \ldots + R_n \qquad \text{(Eq 3-13)}$$

where n is the total number of resistors.

The total resistance of a string of resistors in series will always be greater than any individual resistance in the string. All the circuit current flows through each resistor in a series circuit. A series circuit with resistor values of 2 ohms, 3 ohms and 5 ohms would have a total resistance of 10 ohms. If a circuit has two equal-value resistors connected in series, the total resistance will be twice the value of either resistor alone.

In a parallel circuit, things are a bit different. When we connect two or more resistors in parallel, more than one path for current exists in the circuit. See Figure 3-16. This is like connecting another pipe into our water-pipe circuit. When there is more than one path, more water can flow during a given time. With more than one resistor, more electrons can flow. This means there is less resistance and a greater current.

The formula for calculating the total resistance of resistors connected in parallel is:

$$R_{TOTAL} = \cfrac{1}{\cfrac{1}{R_1} + \cfrac{1}{R_2} + \cfrac{1}{R_3} + \ldots + \cfrac{1}{R_n}} \qquad \text{(Eq 3-14)}$$

where n is the total number of resistors. For example, if we connect three 100-ohm resistors in parallel, their total resistance is:

$$R_{TOTAL} = \cfrac{1}{\cfrac{1}{100} + \cfrac{1}{100} + \cfrac{1}{100}}$$

$$R_{TOTAL} = \cfrac{1}{\cfrac{3}{100}} = \cfrac{1}{0.03} = 33.33 \text{ ohms}$$

If we connect a 25-ohm resistor, a 100-ohm resistor and

Figure 3-15—The total resistance of a string of series-connected resistors is the sum of all the individual resistances.

Figure 3-16—In this circuit, the electrical current splits into three separate paths through the resistors. Each resistor is exposed to the full battery voltage. Current through each individual resistor is independent of the other resistors. For example, if R$_2$ and R$_3$ are changed to a value different than the original value, the current in R$_1$ remains unchanged. Even if R$_2$ and R$_3$ are removed, the current in R$_1$ remains unchanged.

a 10-ohm resistor in parallel, our equation becomes:

$$R_{TOTAL} = \cfrac{1}{\cfrac{1}{25} + \cfrac{1}{100} + \cfrac{1}{10}}$$

$$R_{TOTAL} = \cfrac{1}{0.04 + 0.01 + 0.1} = \cfrac{1}{0.15} = 6.67 \text{ ohms}$$

To calculate the total resistance of two resistors in parallel, Equation 3-14 reduces to the "product over sum" formula.

Divide the product of the two resistances by their sum:

$$R_{TOTAL} = \frac{R_1 \times R_2}{R_1 + R_2} \qquad \text{(Eq 3-15)}$$

If we connect two 50-ohm resistors in parallel, the total resistance would be:

$$R_{TOTAL} = \frac{50 \times 50}{50 + 50} = \frac{2500}{100} = 25 \text{ ohms}$$

The total resistance of two equal-value resistors connected in parallel is half the value of one of the resistors.

Now let's look at any parallel combination of resistors. The total resistance is always less than the smallest value of the parallel combination. You can use this fact to make a quick check of your calculations. The result you calculate should be smaller than the smallest value in the parallel combination. If it isn't, you've made a mistake somewhere!

INDUCTANCE

The motion of electrons produces magnetism. Every electric current creates a *magnetic field* around the wire in which it flows. (A magnetic field represents the invisible magnetic force, such as the attraction and repulsion between magnets.) Like an invisible tube, the magnetic field is positioned in concentric circles around the conductor. See Figure 3-17A. The field is established when the current flows, and collapses back into the conductor when the current stops. The field increases in strength when the current increases and decreases in strength as the current decreases. The force produced around a straight piece of wire by this magnetic field is usually negligible. When the same wire is formed into a coil, the force is much greater. In coils, the magnetic field around each turn also affects the other turns. Together, the combined forces produce one large magnetic field, as shown in Figure 3-17B. Much of the energy in the magnetic field concentrates in the material in the center of the coil (the *core*). Most practical **inductors** consist of a length of wire wound on a core.

An inductor stores energy in a magnetic field. This property of a coil to store energy in a magnetic field is called **inductance**. Magnetic fields can also set electrons in motion. When a magnetic field increases in strength, the voltage on a conductor within that field increases. When the field strength decreases, so does the voltage.

Let's apply a dc voltage to an inductor. As there is no current to start with, the current begins to increase when the voltage is applied. This will establish an electrical current in the inductor. The voltage induced by the magnetic field of the inductor opposes the applied voltage. Therefore, the inductor will oppose the increase in current. This is a basic property of inductors. Any changes in current through the inductor, whether increasing or decreasing, are opposed. A voltage is induced in the coil that opposes the applied voltage, and tries to prevent the current from changing. This is called **induced EMF** (voltage) or back EMF. Gradually a current will be

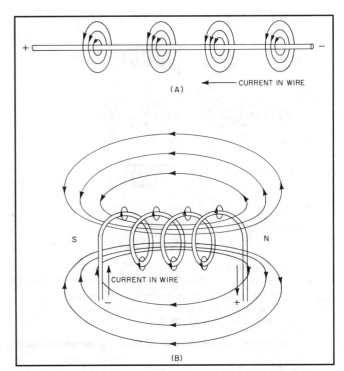

Figure 3-17—A magnetic field surrounds a wire with a current in it. If the wire is formed into a coil, the magnetic field becomes much stronger as the lines of force reinforce each other.

produced by the applied voltage. (The term "gradually" is relative. In radio circuits, the time needed to produce the current in the circuit is often measured in microseconds.)

The "final" current that flows through the inductor is limited only by any resistance that might be in the circuit. There is very little resistance in the wire of most coils. The current will be quite large if there is no other resistance.

In the process of getting this current to flow, energy is stored. This energy is in the form of a magnetic field around the coil. When the applied current is shut off, the magnetic field collapses. The collapsing field returns energy to the circuit as a momentary current that continues to flow in the same direction as the original current. This current can be quite large for large values of inductance. The induced voltage can rise to many times the applied voltage. This can cause a spark to jump across switch or relay contacts when the circuit is broken to turn off the current. This effect is called "inductive kickback."

When an ac voltage is applied to an inductor, the current through the inductor will reverse direction every half cycle. This means the current will be constantly changing. The inductor will oppose this change. Energy is stored in the magnetic field while the current is increasing during the first half cycle. This energy will be returned to the circuit as the current starts to decrease. A new magnetic field will be produced during the second half cycle. The north and south poles of the field will be the reverse of the first half cycle. The energy stored in that field will be returned to the circuit as the current

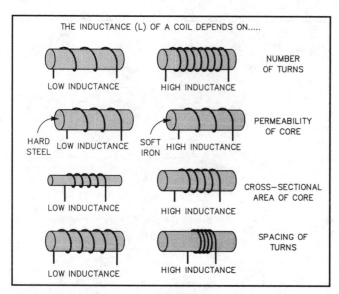

Figure 3-18—The value of an inductor depends on the material used in the core, the number of windings, and the length and diameter of the coil.

2) the number of turns used to wind the coil;
3) the length of the coil (spacing of turns);
4) the diameter of the coil (cross-sectional area).

Changing any of these factors changes the inductance. See Figure 3-18.

The basic unit of inductance is the **henry**, named for the American physicist Joseph Henry. The henry is often too large for practical use in measurements. We use the millihenry (10^{-3}) abbreviated mH, or microhenry (10^{-6}) abbreviated µH.

INDUCTORS IN SERIES AND PARALLEL

In circuits, inductors combine like resistors. The total inductance of several inductors connected in series is the sum of all the inductances:

$$L_{TOTAL} = L_1 + L_2 + L_3 + ... + L_n \qquad \text{(Eq 3-16)}$$

where n is the total number of inductors.

For two equal-value inductors connected in series, the total inductance will be twice the value of one of the inductors.

For parallel-connected inductors:

$$L_{TOTAL} = \cfrac{1}{\cfrac{1}{L_1} + \cfrac{1}{L_2} + \cfrac{1}{L_3} + ... + \cfrac{1}{L_n}} \qquad \text{(Eq 3-17)}$$

You should recognize this equation from our study of combining resistors in parallel, and realize that for two parallel-connected inductors, Equation 3-17 reduces to:

$$L_{TOTAL} = \frac{L_1 \times L_2}{L_1 + L_2} \qquad \text{(Eq 3-18)}$$

For two equal-value inductors connected in parallel, the total inductance will be half the value of one of the components.

[Turn to Chapter 13 and study questions T5A02, T5A03, T5A04, T5A08 and T5A09. Review this section as needed.]

again starts to decrease. Then a new magnetic field will be produced on the next half cycle. This process keeps repeating, as long as the ac voltage is applied to the inductor.

FACTORS THAT DETERMINE INDUCTANCE

The inductance of a coil determines several circuit conditions. One is the amount of opposition to changes in current. Another is the amount of energy stored in the magnetic field. The back EMF induced in the coil also depends on the inductance. In turn, the inductance of a coil, usually represented by the symbol L, depends on four things:

1) the type of material used for the core (permeability of the material), and its size and location in the coil;

CAPACITANCE

A simple **capacitor** is formed by separating two conductive plates with an insulating material. Connect one plate to the positive terminal of a voltage source. Connect the other plate to the negative terminal. We can build up a surplus of electrons on one plate, as shown in Figure 3-19. At some point, the voltage across the capacitor will equal the applied voltage, and the capacitor is said to be charged. This stored electric charge produces an *electric field*, which is an invisible electric force of attraction or repulsion acting between

charged objects. (In this case the electric field is between the capacitor plates.) The capacitor stores energy as an electric field between the capacitor plates.

If we connect a load to a charged capacitor, it will discharge through the load, releasing stored energy. The basic property of a capacitor (called **capacitance**) is this ability to store a charge in an electric field.

The basic unit of capacitance is the **farad**, named for Michael Faraday. Like the henry, the farad is usually too

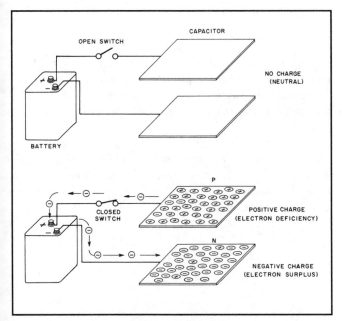

Figure 3-19—When a voltage is applied to a capacitor, an electron surplus (negative charge) builds up on one plate, while an electron deficiency forms on the other plate to produce a positive charge.

Figure 3-20—The capacitance of a capacitor depends on the area of the plates, the distance between the plates and the type of dielectric material used.

large a unit for practical measurements. For convenience, we use microfarads (10^{-6}), abbreviated μF, or picofarads (10^{-12}), abbreviated pF.

FACTORS THAT DETERMINE CAPACITANCE

The capacitance value of a capacitor is determined by three factors. Increased plate-surface area will increase capacitance. Increased spacing between plates reduces capacitance. The type of insulating material (dielectric) used between the plates will affect capacitance. See Figure 3-20.

Reducing the spacing between plates or using a better insulator will increase the capacitance. For a given plate size, using multiple plates in the construction of a capacitor increases capacitance (by giving the effect of increased plate size). The effects of these factors are covered in more detail in Chapter 4.

CAPACITORS IN SERIES AND PARALLEL

We can increase the value of capacitance by increasing the total plate area. We can effectively increase the total plate area by connecting two capacitors in parallel. The two parallel-connected capacitors act like one larger capacitor, as shown in Figure 3-21A and B. The total capacitance of several parallel-connected capacitors is the sum of all the values.

$$C_{TOTAL} = C_1 + C_2 + C_3 + ... + C_n \qquad \text{(Eq 3-19)}$$

where n is the total number of capacitors.

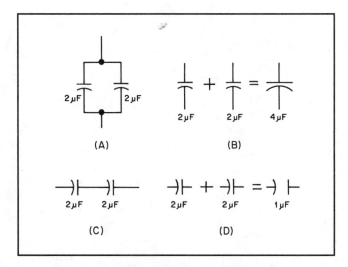

Figure 3-21—Parallel-connected capacitors are shown at A. This connection has the effect of increasing the total plate area, as shown at B. This increases the capacitance. Series connection, shown at C, has the effect of increasing the spacing between the plates, as shown at D. This decreases the capacitance.

If you connect two equal-value capacitors in parallel, the total capacitance of this combination will be twice the value of either capacitor alone.

Connecting capacitors in series has the effect of increasing the distance between the plates, thereby reducing the total capacitance, as shown in Figure 3-21C and D. For capacitors in series, we use the familiar reciprocal formula:

$$C_{TOTAL} = \cfrac{1}{\cfrac{1}{C_1} + \cfrac{1}{C_2} + \cfrac{1}{C_3} + \ldots + \cfrac{1}{C_n}} \qquad \text{(Eq 3-20)}$$

When there are only two capacitors in series, Equation 3-20 reduces to the product over sum formula:

$$C_{TOTAL} = \frac{C_1 \times C_2}{C_1 + C_2} \qquad \text{(Eq 3-21)}$$

Using this formula for two equal-value resistors or two equal-value inductors in *parallel*, we discovered the total value is half the value of either one. Two equal-value capacitors connected in *series* total one-half the value of either single capacitor.

[This completes your study of Chapter 3. Now turn to Chapter 13 and study exam questions T5A05, T5A06, T5A07, T5A10 and T5A11. Before proceeding to the next chapter, review the material in this section if you have difficulty with any of these questions.]

KEY WORDS

Battery—A device that converts chemical energy into electrical energy.

Chassis ground—The common connection for all parts of a circuit that connect to the negative side of the power supply.

Double-pole, double-throw (DPDT) switch—A switch that has six contacts. The DPDT switch has two center contacts. The two center contacts can each be connected to one of two other contacts.

Double-pole, single-throw (DPST) switch—A switch that connects two contacts to another set of contacts. A DPST switch turns two circuits on or off at the same time.

Earth ground—A circuit connection to a ground rod driven into the earth or to a cold-water pipe made of copper that goes into the ground.

Fixed resistor—A resistor whose value cannot be adjusted by the user.

Fuse—A thin metal strip mounted in a holder. When too much current passes through the fuse, the metal strip melts and opens the circuit.

NPN transistor—A transistor that has a layer of P-type semiconductor material sandwiched between layers of N-type semiconductor material.

PNP transistor—A transistor that has a layer of N-type semiconductor material sandwiched between layers of P-type semiconductor material.

Potentiometer—Another name for a **variable resistor**. The value of a potentiometer can be changed without removing it from a circuit.

Resistor—A circuit component that controls current through a circuit.

Schematic symbol—A drawing used to represent a circuit component on a wiring diagram.

Single-pole, double-throw (SPDT) switch—A switch that connects one center contact to one of two other contacts.

Single-pole, single-throw (SPST) switch—A switch that only connects one center contact to another contact.

Switch—A device used to connect or disconnect electrical contacts.

Transistor—A solid-state device comprising three layers of semiconductor material. An NPN bipolar transistor has a layer of P-type semiconductor material sandwiched between two layers of N-type material. A PNP bipolar transistor has a layer of N-type semiconductor material sandwiched between two layers of P-type material.

Triode—A vacuum tube with three active elements: cathode, plate and control grid.

Variable resistor—A resistor whose value you can change without removing it from a circuit.

Chapter 4

Circuit Components

B efore we look at the operation of electronic circuits, let's discuss some basic information about the parts that make up those circuits. This chapter presents the information about circuit components that you need to know for your Novice or Technician exam. You will find descriptions of several types of fuses, switches, resistors and semiconductor devices. We combine these components with other devices to build practical electronic circuits.

Every circuit component has a **schematic symbol**. A schematic symbol is nothing more than a drawing used to represent a component. We use these symbols when we are making a circuit diagram, or wiring diagram, to show how the components connect for a specific purpose. You will learn the schematic symbols for the circuit components discussed in this chapter. As you discover more about electronics, you will learn how these symbols can be used to illustrate practical circuit connections.

RESISTORS

Resistors are important components in electronic circuits. We talked about the concept of resistance in Chapter 3. A resistor opposes the flow of electrons. We can control the electron flow (the current) by varying the resistance in a circuit.

Most resistors have standard fixed values, so they can be called **fixed resistors**. **Variable resistors**, also called **potentiometers**, allow us to change the value of the resistance without removing and changing the component. Potentiometers are used as the volume and tone controls in most stereo amplifiers. Figure 4-1 shows two types of fixed resistors, a potentiometer and their schematic symbols.

[Now turn to Chapter 12 and study questions N6A04, N6A05 and N6A06. Review this section if you have any problems.]

Figure 4-1—Fixed resistors come in many standard values. Most of them look something like the ones shown at A. Variable resistors (also called potentiometers) are used wherever the value of resistance must be adjusted after the circuit is complete. Part B shows a potentiometer.

Figure 4-2—A single-pole, single-throw (SPST) switch can connect or disconnect one circuit. A single-pole, double-throw (SPDT) switch can connect one center contact to one of two other contacts. A double-pole, single-throw switch connects or disconnects two circuits at the same time. A double-pole, double-throw (DPDT) switch is like two SPDT switches in one package. Each half of the DPDT switch can connect one contact to two other contacts.

SWITCHES

How do you control the lights in your house? What turns on your car radio? A **switch**, of course.

The simplest kind of switch just connects or disconnects a single electrical contact. Two wires connect to the switch; when you turn the switch on, the two wires are connected. When you turn the switch off, the wires are disconnected. This is called a **single-pole, single-throw switch**. It connects a single pair of wires (single pole) and has only two positions, on or off (single throw). Sometimes we abbreviate single pole, single throw as **SPST**. Figure 4-2A shows a simple SPST switch and the symbol we use to represent it on schematic diagrams.

The switches shown in Figure 4-2 are called *knife switches* because a metal blade pivots at one end to make or break the contact. These switches show the basic operation, and physically look like the schematic symbols. They can be dangerous if used with high-voltage circuits, however. Do not use a knife switch in a circuit that carries 120-volt house current unless it is enclosed and operated with an insulated handle designed for this purpose. Knife switches are avail-

able, and can be used in low-voltage circuits.

Most of the time you'll want to use switches with contacts sealed inside a protective case. When mounted in the wall or on a control panel, the switch lever is safely insulated from the circuit.

If we want to control more devices with a single switch, we need more contacts. If we add a second contact to a single-pole, single-throw switch we can select between two devices. This kind of switch is called a **single-pole, double-throw switch**. Sometimes we abbreviate single pole, double throw as **SPDT**. An SPDT switch connects a single wire (single pole) to one of two other contacts (double throw). The switch connects a center wire to one contact when the switch is in one position. When you flip the switch to the other position, the switch connects the center wire to the other contact. Electric circuits that allow you to turn lights on or off from either of two locations use SPDT switches. Figure 4-2B shows an SPDT knife switch and the schematic symbol for any SPDT switch.

Suppose you want to switch two circuits on and off at the same time. Then you use a **double-pole, single-throw (DPST) switch**. This type of switch connects two input lines to their respective output lines at the same time. See Figure 4-2C for an example of this switch and its schematic-diagram symbol.

We can add even more contacts to the switch. A **double-pole, double-throw switch** has two sets of three contacts. We use the abbreviation **DPDT** for double pole, double throw. You can think of a DPDT switch as two SPDT switches in the same box with their handles connected. A DPDT switch has two center contacts. The switch connects each of these two center contacts (double pole) to one of two other contacts (double throw). Figure 4-2D shows a DPDT switch and its schematic symbol.

All these switches are very useful, but they only connect a center contact to one or two other contacts. What if we want to connect a single contact to *several* other contacts? We might want to use one switch to connect our transmitter to several different antennas. We can do this with a *rotary switch*. As its name implies, a rotary switch turns around a central shaft to connect one center contact to several outer contacts.

Switches like this can have many contacts. They can also have more than one center contact, or pole. We specify the particular kind of switch by the number of contacts (positions) it has around the outside, and by the number of center contacts and switch arms (poles) it has. Figure 4-3 shows a photograph of a two-pole, five-position rotary switch; it has five separate contacts and two rotary arms. The schematic symbol in Figure 4-4 is a single-pole, six-position rotary switch.

[You should now turn to Chapter 12. Study questions N6A01, N6A02, N6A09, N6A10, N6A11 and N6A12. Review this section if you have any problems.]

FUSES

What would happen if part of your receiver suddenly developed a short circuit? The current, flowing without opposition through portions of the circuit, could easily damage components not built to withstand such high current.

To protect against unexpected short circuits and other problems, most electronic equipment includes one or more

Figure 4-3—A rotary switch can connect one wire to several contacts. Most antenna switches use a rotary switch to connect a transceiver to several antennas. This photograph shows a two-pole, five-position switch; it can connect each center contact to five outside contacts.

Figure 4-4—The schematic symbol for a single-pole, six-position rotary switch.

fuses. A **fuse** is simply a device made of metal that will heat up and melt when a certain amount of current flows through it. The amount of current that causes each fuse to melt (or "blow") is determined by the manufacturer. When the fuse (usually placed in the main power line to the equipment) blows, it creates an open circuit, stopping the current.

Fuses come in many shapes and sizes. Figure 4-5 shows some of the more common fuse types. This figure also shows the three common symbols used to represent a fuse on schematic diagrams.

A fuse in a transistor radio using little power may be designed to blow at 500 mA. The fuses for your home's 120-V circuits may be designed to blow at 15 or 20 A. The principle is the same: When excessive current flows through the fuse it melts, creating an open circuit to protect your equipment. Remember that fuses are designed to protect against too much current, not too much voltage.

[To check your understanding of this section, study exam question N6A07. Review this section if you have any problems.]

BATTERIES

We talked about batteries in Chapter 3. Simply put, a **battery** changes chemical energy into electrical energy.

FUSE
SCHEMATIC SYMBOL

OR

OR

Figure 4-5—Some common fuses. Fuses protect circuits from excessive current. The inset drawing shows common fuse schematic symbols.

Figure 4-6—Batteries come in all shapes and sizes. A battery changes chemical energy into electrical energy.

When we connect a wire between the terminals of a battery, a chemical reaction takes place inside the battery. This reaction produces free electrons, and these electrons flow through the wire from the negative terminal to the positive terminal. Batteries may be small or large, round or square. Hearing aid and calculator batteries are tiny. The battery that starts your car is very large. Figure 4-6 shows several different batteries.

Batteries are made up of *cells*. The cells produce a small voltage. The voltage a cell produces depends on the chemical process taking place inside the cell. Rechargeable nickel-cadmium cells produce about 1.2 volts per cell. Common carbon-zinc and alkaline flashlight cells produce about 1.5 volts per cell. The lead-acid cells in a car battery each produce about 2 volts.

The number of cells in a battery depends on the voltage we want to get out of the battery. If we only need a low voltage the battery may contain only one cell. Small hearing-aid batteries usually contain only one cell. Part A of Figure 4-7 shows the schematic symbol for a single-cell battery. The two vertical lines in the schematic symbol represent the two electrodes in the cell. The long line represents the positive terminal and the short line represents the negative terminal.

Figure 4-7—Some small batteries contain only one cell. We use the symbol at A for a single-cell battery. Manufacturers add several cells in series to produce more voltage. At B, the schematic symbol for a multiple-cell battery.

SINGLE CELL (A)

MULTIPLE—CELL BATTERY (B)

To produce a battery with a higher voltage, several cells must be connected in series so that their outputs add. Each cell produces a small voltage. The battery manufacturer connects several cells in series to produce the desired battery voltage.

Part B of Figure 4-7 shows the schematic symbol for a multiple-cell battery. We use two sets of long and short lines to show the many cells in the battery. Again, the long line at one end represents the positive terminal, and the short line at the other end represents the negative terminal. The schematic symbol does not indicate the number of cells in the battery.

[Make a quick trip to Chapter 12 now and look at questions N6A03 and N6A08. Review this section if those questions confuse you.]

ANTENNAS AND GROUNDS

When you use your receiver or transmitter, you must connect it to an antenna. You should also connect all the equipment in your station to a good earth ground. We'll go over these connections later, in Chapter 8. There are really two kinds of ground connections: **chassis ground** and **earth ground**. The metal box that your radio is built on is called a chassis. Most manufacturers use the chassis as a common connection for all the places in the circuit that connect to the negative side of the power supply. This common connection is called the chassis ground. Figure 4-8A shows the schematic symbol for a chassis ground.

Figure 4-8—The metal chassis of a radio is sometimes used to make a common ground connection for all the circuit points that connect to the negative side of the power supply. We use the symbol at A to show those connections on a schematic diagram. Your radio equipment should be connected to a ground rod, or a copper cold-water pipe that goes into the ground, for safety. We use the symbol at B to show an earth-ground connection.

CHASSIS GROUND (A)   EARTH GROUND (B)

To keep your station safe, you should also connect the chassis ground to an earth ground. You can provide this earth ground by driving an 8-foot or longer copper-coated steel rod into the earth. (Such *ground rods* are available at electrical-supply stores.) Then connect a heavy copper strap or wire braid between the ground rod and the equipment chassis.

Another way to provide an earth ground is to connect the copper strap or wire braid to a copper cold-water pipe that goes into the earth. You have to be certain that the copper pipe continues into the earth, however. If the pipe changes to plastic pipe it will not provide a ground connection.

This connection is called an earth ground because it goes into the earth. An earth ground has a different schematic symbol than a chassis ground. Figure 4-8B shows the schematic symbol for an earth ground.

Your receiver won't hear signals and no one will hear your transmitter without an antenna. The antenna is a very important part of any radio installation. We use the symbol shown in Figure 4-9 to represent the antenna on a schematic diagram.

An antenna radiates the radio-frequency energy from a transmitter. This means the radio energy leaves the antenna and travels off into the air around it. We call this energy a *radio wave*. When the radio wave travels past another antenna it produces a small current in the wire, creating a signal for a receiver. Antennas radiate and receive radio energy.

There are many types of antennas. You will learn more about the various kinds

Figure 4-9—Most receivers and transmitters are useless without an antenna. This schematic symbol shows where the antenna connects in a circuit.

ANTENNA

of antennas later in this book. You will also learn how to choose antennas for various kinds of radio equipment.

[Now turn to Chapter 12 and study questions numbered N6B03, N6B04, N6B05 and N6B06. If you have any trouble answering the questions, review this section.]

TRANSISTORS

Many of the great technological advances of recent times—men and women in space, computers in homes, ham radio stations tiny enough to be carried in a pocket—all have been made possible by *semiconductor* electronics. Not simply a partial conductor as the name implies, a semiconductor has some of the properties of a conductor and some properties of an insulator.

Diodes and **transistors** are two types of semiconductors, the *solid-state devices* that have replaced vacuum tubes in most uses. Most semiconductor devices are much smaller than comparable tubes, and they produce less heat. Semiconductors are also usually less expensive than tubes.

Portable broadcast-band radios, which weighed several

Figure 4-10—Transistors are packaged in many different cases.

pounds and were an armful shortly after World War II, weigh ounces and can be carried in your pocket today. A complete ham radio station, which would have filled a room 50 years ago, now can be built into a container the size of a shoebox, or smaller. Solid-state technology has made all this possible.

Transistors come in many shapes and sizes. Figure 4-10 shows some of the more common case styles for transistors. The most common type of transistor is the *bipolar* **transistor**. Bipolar transistors are made of two different kinds of material (*bi* means two, as in *bi*cycle). The two kinds of semiconductor material are known as *N-type material* and *P-type material*. The N refers to the way electrons, or negative electric charges, move through the material. The P refers to the way positive charges move through this material. (Don't worry about the details of how this works now. You can learn about how transistors actually work after you pass your license exam.)

There also are two kinds of bipolar transistors. Each kind of bipolar transistor has a separate schematic symbol.

Figure 4-11—This is the schematic symbol for an NPN transistor. Remember that the arrow is "*not pointing in.*"

Figure 4-11 shows the schematic symbol for an **NPN transistor**. You can remember this symbol by remembering that the arrow is "*not pointing in.*" An NPN transistor has a layer of P-type semiconductor material sandwiched between two layers of N-type material.

Figure 4-12 shows the symbol for the other kind of bipolar transistor, the **PNP transistor**. Remember this sym-

Figure 4-12—This is the schematic symbol for a PNP transistor. Remember that the arrow "*points in proudly.*"

Figure 4-13—These diagrams show simple transistor circuits for NPN and PNP transistors. Batteries B₁ and B₂ provide the proper operating voltages and polarities to produce a current through the collector/emitter part of each circuit.

bol by saying that the arrow "*points in proudly.*" A PNP transistor has a layer of N-type semiconductor material sandwiched between two layers of P-type material.

The schematic symbols show that transistors have three leads, or electrodes. Each of the electrodes connects to a different part of the transistor. Transistors can amplify small signals; this is what makes them so useful. Usually, we use a low-voltage signal applied to the *base* of the transistor to control the current through the *collector* and *emitter*.

Figure 4-13A shows a simple NPN transistor circuit. Battery B₁ applies a positive voltage to the base of the transistor and battery B₂ applies a positive voltage to the collec-

tor. A small current flows through the emitter and base portion of the circuit. This small current controls a (usually larger) current through the collector and emitter portion of the circuit. If the base current decreases the collector current also decreases. If the base current increases, so will the collector current. If you remove B₁ or otherwise open the base circuit, there will be no collector current.

Figure 4-13B is a similar circuit for a PNP transistor. Notice this time B₁ applies a negative voltage to the transistor base lead and B₂ applies a negative voltage to the collector. Once again a small current through the base and emitter portion of the circuit controls the current through the collector and emitter part of the circuit.

[Time for another look at the question pool. Study the questions in Chapter 12 with numbers N6B01, N6B02, N6B07 and N6B08. Review this section if you have problems.]

VACUUM TUBES

The development of the vacuum tube was an important milestone in the history of radio. The vacuum tube was the first *active* electronic device—that is, the vacuum tube can amplify, or produce an enlarged version of the input signal.

From the outside, a tube looks like a glass bulb with pins sticking out of the bottom. Sometimes there are also leads coming out of the top or sides. Some tubes have a metal collar or band around the base, and some tubes have ceramic or metal envelopes (outer shell). Tubes are quite fragile, and will break easily if mishandled. They usually plug into a socket wired into a circuit. Figure 4-14 shows some common tubes.

Figure 4-14—Here are some common vacuum tubes. This figure shows tiny receiving tubes and large transmitting tubes.

Figure 4-15—A triode vacuum tube has three elements. This is the schematic symbol for a triode.

Tubes are named for the number of elements they have inside them. All tubes have at least two elements, the *plate* and the *cathode*. A tube with only two elements has a limited usefulness, however. Tubes became much more useful when inventors added a third element, the *control grid*. This grid controls the flow of electrons through the tube. By control-ling the flow of electrons through the tube, you can use it as an amplifier.

A tube with three elements is called a **triode** (*tri* means three, as in *tri*cycle). Figure 4-15 shows the schematic symbol for a triode vacuum tube.

While transistors have replaced tubes in many modern applications, tubes are still used in many types of electronic circuits. Radio and television sets using tubes are still in operation. The *picture tube* or *cathode-ray tube* (*CRT*) is a vacuum tube. Tubes can handle high power and high voltage. This makes them ideal for use in the output stage of an amateur transmitter or as high-power amplifiers.

[Turn to Chapter 12 and study questions N6B09, N6B10 and N6B11. If you're preparing for the Novice exam now, that's it for this chapter! By now, you should have a basic understanding of how all of the circuit components included on the Novice exam work. If you had trouble with any of the related questions in the question pool, review those sections before you proceed to the next chapter.]

[**If you're preparing for the Technician exam**, continue with this chapter after you review what you've learned so far.]

Additional Component Information for the Technician Exam

KEY WORDS

Capacitor—An electronic component composed of two or more conductive plates separated by an insulating material.

Color code—A system in which numerical values are assigned to various colors. Colored stripes are painted on the body of resistors and sometimes other components to show their value.

Core—The material used in the center of an inductor coil.

Electrolytic capacitor—A polarized capacitor that relies on a chemical reaction to form a thin insulating layer between the plates.

Electric field—An invisible force of nature. An electric field exists in a region of space if an electrically charged object placed in the region is subjected to an electrical force.

Film resistor—A resistor made by depositing a thin layer of resistive material on a ceramic form.

Inductor—An electrical component usually composed of a coil of wire wound on a central core.

Potentiometer—A resistor whose resistance can be varied continuously over a range of values.

Reactance—The property of an inductor or capacitor (measured in ohms) that impedes current in an ac circuit without converting power to heat.

Resistor—Any material that opposes a current in an electrical circuit. An electronic component specifically designed to oppose current.

Toroidal core—A donut-shaped form used to hold the turns of an inductor. Toroidal cores are usually made from ferrite or powdered-iron materials.

Variable capacitor—A capacitor that can have its value changed within a certain range.

Variable resistor—A resistor whose value can be adjusted over a certain range.

Wire-wound resistor—A resistor made by winding a length of wire on an insulating form.

This section presents the additional information about circuit components that you will need to know to pass your Technician class written exam. You will find descriptions of resistors, capacitors and inductors. You can combine these components with other devices to build practical electronic circuits. We describe some of these circuits in Chapter 5.

Turn to the Element 3A questions in Chapter 13 when directed to do so. How well you understand the questions will show you where you need to do some extra studying. You should thoroughly understand how these components work. When you do, you will have no problem learning to connect them to make a circuit perform a specific task.

Figure 4-16—Part A shows the schematic symbol for a resistor. Part B shows a typical resistor, including the colored stripes that identify the resistor value. Resistors made in this way range in power rating from ⅛ watt to 2 watts.

RESISTORS

When electrons flow from one point in a circuit to another point, there is *current* in that circuit. A perfect insulator would allow no electrons to flow through it (zero current), while a perfect conductor would allow infinite current (unlimited electron flow). In practice, however, there is no such thing as a perfect conductor nor a perfect insulator. Partial opposition to electron flow occurs when the electrons collide with other electrons or atoms in the conductor. The result is a reduction in current, and the conductor produces heat.

Resistors allow us to control the current in a circuit by controlling the opposition to electron flow. As they oppose the flow of electrons they dissipate electrical energy in the form of heat. The more energy a resistor dissipates, the hotter it will become. Figure 4-16A shows the schematic symbol for all nonadjustable resistors. Part B shows a common resistor. You will learn how to interpret the colored stripes on a resistor later in this chapter.

All conductors exhibit some resistance. Thick (heavy) conductors have only a little resistance. Thinner conductors have more resistance. This is just what we would expect. Think of electron flow in a conductor as being like water flow in a pipe. If we reduce the size of the pipe, less water can flow through it. Different conductor materials also have different amounts of resistance. Copper is a good conductor, and so has a small resistance. A nickel wire has more resistance than a copper wire, if both have the same diameter.

We use numbers in the *American Wire Gauge (AWG)* scale to measure wire sizes. Smaller numbers in this scale indicate larger diameter wires. A number 12 wire (a common size for house wiring) has a larger diameter than a number 30 wire (which compares to the thickness of a hair).

One simple way of producing a resistor is to use a length of wire. For example, the resistance of 1000 feet of number 28 nickel wire is 337 ohms. Therefore, if we need a resistor of 3.4 ohms, we can use 10 feet of this size nickel wire. You might think 10 feet of wire makes a pretty inconvenient resistor, but we can wind it over a form of some sort. See Figure 4-17A. This makes a unit of a much more convenient size. This is precisely how **wire-wound resistors** are constructed.

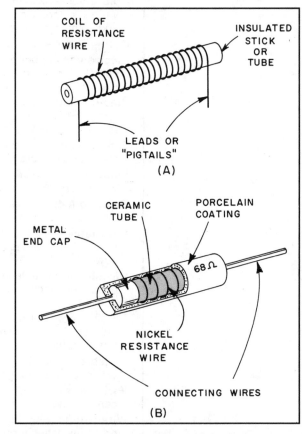

Figure 4-17—A simple wire-wound resistor is shown at A. At B is a commercially produced wire-wound resistor. These resistors are produced by winding resistive wire around a nonconductive form. Most wire-wound resistors are then covered with a protective coating.

Figure 4-18—Composition resistors are composed of a mixture of carbon and clay. The resistance is controlled by varying the amount of carbon in the material when the resistor is manufactured.

Figure 4-19—The construction of a typical metal-film resistor is shown at A. Film resistors are formed by depositing a thin film of material on a ceramic form. Material is then trimmed off in a spiral to produce the specified resistance for that particular resistor. The excess material may either be trimmed with a mechanical lathe or a laser. A lathe produces a rather rough spiral, like that shown in part B. The laser produces a finer cut, shown at C.

See Figure 4-17B.

There is a problem with this type of construction. At radio frequencies, the wire-wound construction causes the resistor to act as an **inductor**. (You will learn more about inductors later in this chapter.) It does not behave as a pure resistance. For RF circuits, resistors must be noninductive. Wire-wound resistors are common in dc circuits that carry large currents, because they can get rid of large amounts of heat.

There is another way to form a resistor. This is to connect leads to both ends of a block or cylinder of material that has a high resistance. Figure 4-18 shows a resistor made in this way. It uses a mixture of carbon and clay as the resistive element. This type of resistor is a *carbon-composition resistor*. The proportions of carbon and clay determine the value of the resistor.

An advantage of carbon-composition resistors is the wide range of available values. Other advantages are low inductance and capacitance plus good surge-handling capability. The ability to withstand small power overloads without being completely destroyed is another advantage. The main disadvantage is that the resistance of the composition resistor will vary widely. Variations are caused by operating temperature changes and resistance changes as the resistor gets older.

There are two types of **film resistors**. These are made by depositing a thin layer of resistive material on a ceramic form. The ceramic form is usually shaped as a cylinder. Metal end caps provide a connection between the resistive film and the resistor leads used to connect the part in a circuit. The entire assembly then gets a protective epoxy or plastic coating.

You can easily recognize the "dog bone" shape of film resistors. They are thinner in the middle than at the ends. This is because of the metal end caps used to connect the leads to the resistive element.

Carbon-film resistors are manufactured by using high temperatures to break down certain gaseous hydrocarbons. The resulting carbon is then deposited in a thin layer or film on the ceramic form. The thickness of the deposited film provides a means to control the final resistance.

The major advantages of carbon-film resistors are low cost and improved stability with age and temperature changes. These resistors cannot withstand electrical overloads or surges. They can be used as fuses for some applications.

The *metal-film resistor* has replaced the carbon-composition type in many low-power applications today. Metal-film resistors are formed by depositing a thin layer of resistive alloy on a cylindrical ceramic form. Nichrome (an alloy made of nickel and chromium) or other materials are used. See Figure 4-19. This film is then trimmed away in a spiral fashion to form the resistance path. The trimming can be done on a mechanical lathe or by using a laser. The resistor is then

Figure 4-20—The typical construction of a wire-wound variable resistor. As the shaft is rotated, the moving contact is electrically connected to different parts of the wire winding. This effectively changes the length of wire between the end terminal and moving-contact terminal. Increasing or decreasing wire length between terminals thus increases or decreases resistance.

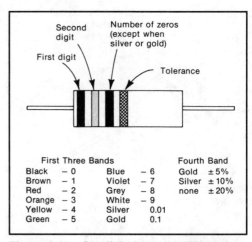

Figure 4-21—Small resistors are labeled with a color code to show their value. For example, proceeding from left to right, a resistor with color bands of yellow/violet/brown/gold is a 470-Ω resistor with a 5% tolerance.

covered with an insulating material to protect it.

Because of the spiral shape of the resistance path, metal-film resistors will exhibit some inductance. The effects of this inductance increase at higher frequencies. The higher the resistance of the unit, the greater the inductance will be. This is caused by the increased path length. Metal-film resistors provide much better temperature stability than other types.

Most resistors have standard fixed values. They are called *fixed resistors*. **Variable resistors** are sometimes called **potentiometers**. They can be used to adjust the voltage, or potential, in a circuit. Potentiometers are also used a great deal in electronics. The construction of a wire-wound variable resistor and the schematic symbol for all variable resistors is shown in Figure 4-20. Variable resistors are also made with a ring of carbon compound in place of the wire windings. A connection is made to each end of the resistance ring. A third contact is attached to a movable arm, or wiper. The wiper can be moved across the ring. As the wiper moves from one end of the ring to the other, the resistance varies from minimum to maximum.

Color Codes

It is not always practical to print the resistance values on the side of a small resistor. A **color code** shows the value of the resistor. As shown in Figure 4-21, the color of the first three bands shows the resistor value. The color of the fourth

band indicates the resistor tolerance.

Standard fixed resistors are usually found in values ranging from 2.7 Ω to 22 MΩ (22 megohms, or 22,000,000 ohms). Resistance tolerances on these standard values can be ±20%, ±10%, ±5% and ±1%. A tolerance of 10% on a 200-Ω resistor means that the actual resistance of a particular unit may be anywhere from 180 Ω to 220 Ω. (Ten percent of 200 is 20, so the resistance can be 200 + 20 or 200 – 20 ohms.)

Here is another example to help you understand this calculation. Find the range of possible values for a 100-Ω resistor that has a 10% tolerance rating.

First calculate the range of possible variation:

$100 \ \Omega \times 10\% = 100 \ \Omega \times 0.10 = 10 \ \Omega$

Now use this value to calculate the possible range of resistances:

$100 \ \Omega - 10 \ \Omega = 90 \ \Omega$

$100 \ \Omega + 10 \ \Omega = 110 \ \Omega$

A 100-Ω resistor that has a 10% tolerance rating will have a value somewhere between 90 and 110 Ω.

Suppose you are building a sensitive circuit that requires a resistor that is no more than 105 Ω. Would you select a 10%, 100-Ω resistor for this circuit? Probably not, because when you go to your local electronics parts store to buy the resistor, you may get a unit that has a resistance as high as 110 Ω. What tolerance component should you choose? A 5% tolerance resistor will suit your purpose in this circuit. You might also select a 1% resistor, because the value will only vary between 99 and 101 ohms for those parts. Resistors with the smaller tolerance rating are more expensive, however. As a general rule you should select the highest tolerance rating that will work properly in your circuit. (Higher tolerance ratings mean there is a wider range of possible values for a standard stock part.)

You can also buy resistors with even smaller tolerance ratings. *Precision resistors* are specified with 0.5%, 0.25%, 0.1% and 0.05% tolerance ratings. For a price, you could probably buy resistors with even smaller tolerances. If you have an ohmmeter that is accurate enough, you also can measure the values of standard parts to select those within the limits you require.

The tolerance rating of a resistor is not necessarily an indication of its manufactured quality. Questions T6A10 and T6A11 refer to the tolerance ratings of high-quality and low-quality resistors, however. The intent of these questions is to refer to a resistor with a small tolerance as a high-quality resistor and one with a larger tolerance as a low-quality resistor.

Power Ratings

Resistors are also rated according to the amount of power they can safely handle. Power ratings for wire-wound resistors typically start at 1 watt and range to 10 W or larger. Resistors with higher power ratings are also physically large. Greater surface area is required to dissipate the increased heat generated by large currents. Carbon-composition and film resistors are low-power components. You will find these resistors in ⅛-watt, ¼-W, ½-W, 1-W and 2-W power ratings. A resistor that is being pushed to the limit of its power rating will feel very hot to the touch. Sometimes a resistor gets very hot during normal operation. You should probably replace it with a unit having a higher power-dissipation rating.

[Now turn to Chapter 13 and study all of the questions in the T6A series. This includes questions T6A01 through T6A11. Review this section as needed.]

INDUCTORS

In Chapter 3 we learned that an **inductor** stores energy in a magnetic field. When a voltage is first applied to an inductor, with no current flowing through the circuit, the inductor will oppose the current. Remember from Chapter 3 that inductors oppose any change in current. A voltage that opposes the applied voltage is induced in the coil, and this voltage tries to prevent a current. Gradually, though, the current will build up. Only the resistance in the circuit will limit the final current value. The resistance in the wire of the coil will be very small.

In the process of getting this current to flow, the coil stores energy. The energy is in the form of a magnetic field around the wire. The increasing field strength produces a voltage that opposes the current. When you shut off the applied current, the magnetic field collapses. The collapsing field returns the stored energy to the circuit. The energy will have the form of a momentary current in the same direction as the original current. This current is produced by the EMF generated by the collapsing magnetic field. Again, this volt-age opposes any change in the current already flowing in the coil.

We can summarize the actions of an inductor by considering two factors. Inductors store energy in magnetic fields, and they oppose any change in current through the inductor. The answer to question T6B02 refers to storing a charge electromagnetically. Electromagnetic here means an *electric current* produces the *magnetic field*. An inductor **does not** store electric *charge*, however. It stores *energy*.

The amount of opposition of an inductor to changes in current is called *inductive* **reactance**. The reactance of the inductor, the amount of energy stored in the magnetic field and the back EMF induced in the coil all depend on the amount of inductance.

We use a capital L to represent the amount of inductance that a coil exhibits. Inductance depends on four things:

1) the type of material used in the **core**, and its size and location in the coil
2) the coil diameter
3) the length of the coil and
4) the number of turns of wire used to wind the coil

Changing any of these factors changes the inductance.

An inductor's **core** refers to the area inside the inductor, where the magnetic field is concentrated. If we add an *iron* or *ferrite core* to the coil, the inductance increases. We can also use brass as a core material in radio-frequency variable inductors. It has the opposite effect of iron or ferrite: The inductance decreases as the brass core enters the coil. A movable core can be placed in a coil, as shown in Figure 4-22. The amount of inductance can be varied as the core moves in and

Figure 4-22—The inductance of a coil increases as an iron core is inserted. Many practical inductors of this type use screw threads to allow moving the core in precise amounts when making small changes in inductance.

Figure 4-23—Various types of coils, and their schematic symbols. The adjustable inductor uses a movable powdered-iron core or slug. They are sometimes called slug-tuned or permeability-tuned inductors.

AIR–CORE

ADJUSTABLE

IRON–CORE

TAP

TAPPED

out of the coil. The common schematic symbols for various inductors are shown in Figure 4-23.

The adjustable coil shown in Figure 4-23 has a threaded core so you can adjust it into and out of the coil. Such inductors usually serve to adjust the operating frequency of a circuit. These adjustments often require you to turn the core material into or out of the coil while the circuit is operating. You must use a plastic tool to make these adjustments rather than a metal tool like a screwdriver. Bringing a metal tool close to the coil would change the inductance. This would result in an incorrect circuit adjustment

Ferrite comes from the Latin word for iron. So what is the difference between an iron and a ferrite core? Well, iron *cores* can be made from iron sheet metal. The layers of material that make up an ac power transformer core are made that way. Cores can also be made from powdered iron mixed with a bonding material. This holds the powdered iron together so it can be molded into the desired shape. Manufacturers mix other metal alloys with the iron "ferrous" material when they make some cores. We call these *ferrite cores*. They can select the proper alloy to provide the desired characteristics, producing cores designed to operate best over a specific frequency range. We use iron cores most often at low frequencies, such as audio frequencies. Audio and ac-power circuits are good applications for these cores. Ferrite cores are manufactured in a wide range of compositions. This optimizes their operation for specific radio-frequency ranges.

It takes some amount of energy to magnetize the core material for any inductor. This energy represents a loss in the coil. Winding the coil on an iron material increases the inductance of the coil. The energy needed to magnetize the core also increases. Some coils are wound on thin plastic forms, or made self supporting. Then the only material inside the coil is air. These are air-wound coils. They are among the lowest loss types of inductors. Air-wound coils are normally used in high-power RF circuits where energy loss must be kept to a minimum.

Toroid Cores

A *toroidal inductor* is made by winding a coil of wire on a donut-shaped core called a toroid. **Toroidal core** materials are usually powdered-iron or ferrite compounds. The toroidal inductor is highly efficient. This is because there is no break in the circular core. All the magnetic lines of force remain inside the core. This means there is very little *mutual coupling* between two toroids mounted close to each other in a circuit.

Figure 4-24 shows a toroidal inductor and the schematic symbol used to represent them. Notice that the schematic symbol is the same one used for any inductor that has an iron or ferrite core material.

Mutual coupling occurs when the magnetic flux of one coil passes through the windings of the other coil. When coils have mutual coupling, a current in one coil will induce a voltage (and a current) in the other. Mutual coupling can be a problem, especially in an RF circuit. There, signals from one stage could be coupled into another stage without follow-

SCHEMATIC SYMBOL

Figure 4-24—A toroidal coil. This type of coil has self-shielding properties. Some inductors in some circuits will interact with the magnetic fields of other inductors. To prevent this a metal shield is used to enclose them. Toroidal inductors confine magnetic fields so well that shields are usually not necessary.

Figure 4-25—Capacitors store energy in an electric field. The final amount of energy stored in a fully charged capacitor depends on the capacitor value and the applied voltage. If this circuit is used to fully charge the capacitor from a 6-volt battery, the amount of energy stored in the capacitor could be increased by using a 12-volt battery. Another way to increase the amount of stored energy is to use a larger-value capacitor.

ing the proper signal path. Toroidal cores can often help eliminate the mutual coupling that would occur between other types of coils.

In theory, a toroidal inductor requires no shield to prevent its magnetic field from spreading out and interfering with outside circuits. This self-shielding property also helps keep outside forces from interfering with the toroidal coil. Toroids can be mounted so close to each other that they can almost touch, but because of the way they are made, there will be almost no inductive coupling between them. Toroidal coils wound with only a small amount of wire on ferrite or iron cores can have a very high inductance value.

[Turn to Chapter 13 and study exam questions T6B01 through T6B08. Review any material you have difficulty with before going on.]

CAPACITORS

A **capacitor** is composed of two or more conductive plates with an insulating material between them. In Chapter 3 we learned that the basic property of a capacitor is the ability to store an electric charge. In an uncharged capacitor, the potential difference between the two plates is zero. As we charge the capacitor, this potential difference increases until it reaches the full applied voltage. The difference in potential creates an **electric field** between the two plates. See Figure 4-25.

A field is an invisible force of nature. We put energy into the capacitor by charging it. Until we discharge it, or the charge leaks away somehow, the energy is stored in the electric field. When the field is not moving, we sometimes call it an *electrostatic field*.

When we apply a dc voltage to a capacitor, current will flow in the circuit until the capacitor is fully charged to the applied voltage. After the capacitor has charged to the full applied voltage, no more current will flow in the circuit. For this reason, capacitors can be used to block dc in a circuit. Because ac voltages are constantly reversing polarity, we can block dc with a capacitor while permitting ac to pass.

When the voltage across a capacitor increases, the capacitor stores more charge in reaction to the increased voltage. When the voltage decreases, the capacitor returns some of its stored charge. These actions tend to oppose any change in the voltage applied to a capacitor. This opposition to changes in voltage (and current) is called *capacitive* **reactance**.

We can summarize the operation of a capacitor by considering two factors. A capacitor stores electric charge in the form of an electrostatic field and it opposes any change in the applied voltage.

Factors That Determine Capacitance

Remember from Chapter 3 that the value of a capacitor is determined by three factors:

1) the type of insulating material or *dielectric* used between the plates
2) the area of the plate surfaces
3) the distance between the plates

There are two factors involved with the surface area of the plates. The first factor is the number of plates. By adding additional sets of plates, each separated by insulating material, you can increase the effective plate area. Figure 4-26 shows the construction of a capacitor with two plates for each side. Additional layers of foil and insulating material can be added to increase the capacitance. Alternate foil layers connect to opposite sides of the capacitor in this construction.

The second surface-area factor involves knowing the surface area of one side of one plate. If you measure this surface area, and know how many plates are used to build a certain capacitor, you could calculate its capacitance. (You would also have to know the *dielectric constant* of the insulator and the distance between the plates.)

As the surface area increases, the capacitance increases. For a given plate area, as the spacing between the plates decreases, the capacitance increases.

Practical Capacitors

Practical capacitors are described by the material used for their dielectric. Mica, ceramic, plastic-film, polystyrene, paper and electrolytic capacitors are in common use today. They each have properties that make them more or less suitable for a particular application. Figure 4-27A shows the symbol used to represent capacitors on schematic diagrams.

Electrolytic capacitors have a dielectric that is formed after the capacitor is manufactured. Two common types of electrolytic capacitors are aluminum-electrolytic capacitors and tantalum-dielectric capacitors.

Electrolytic capacitors can be made with a very high capacitance value in a small package. Electrolytic capacitors are polarized—dc voltages must be connected to the positive and negative capacitor terminals with the correct polarity. The positive and negative electrodes in an electrolytic capacitor are clearly marked. Connecting an electrolytic capacitor incorrectly causes gas to form inside the capacitor and the capacitor may actually explode. This can be very dangerous. At the very least the capacitor will be destroyed by connecting it incorrectly.

When we draw electrolytic capacitors on schematic diagrams, we must indicate the proper polarity for the capacitor. We do this by adding a + sign to the capacitor symbol, as Figure 4-27B shows.

In some circuits it is necessary or desirable to vary the capacitance at some point. An example is in the tuning circuit

Figure 4-26—This drawing shows how the total plate surface area can be increased by stacking several plates, with insulating material between them. This increases the capacitance of the unit, compared to the capacitance if only two plates of the same size are used. The leads connect to alternate plates.

Figure 4-27—A shows the schematic symbol representing fixed capacitors. This symbol has the appearance of two plates separated by an insulator. B shows the symbol representing electrolytic capacitors, which must be connected to dc voltages with the proper polarity.

of a receiver or transmitter VFO. We could use a rotary switch to select one of several different fixed-value capacitors. It is much more convenient to use a **variable capacitor**, however. A basic air-dielectric variable capacitor is shown in Figure 4-28A. This figure also shows the schematic symbol for a variable capacitor. One set of plates is fixed in one position to the capacitor frame. These plates form the *stator*. The other set of plates rotates by turning the control shaft. The movable plates form the *rotor*. The capacitance can be varied smoothly between the maximum and minimum values.

Another type of variable capacitor is the compression variable shown in Figure 4-28B. In this type of variable capacitor, the spacing between the two plates is varied. This type of variable capacitor is usually used as a trimmer capaci-

Figure 4-28—Variable capacitors can be made with air as the dielectric, or with mica or ceramic dielectrics.

tor. A trimmer capacitor is a control used to peak or fine tune a part of a circuit. It is usually left alone once adjusted to the correct value. You should use a plastic screwdriver-like adjusting tool when tuning a trimmer capacitor. The metal of a normal screwdriver can affect the capacitance and cause false readings as you try to adjust it.

[This completes your study of Chapter 4. Now turn to Chapter 13 and study questions T6B09 through T6B14. Review any material you have difficulty with before going on.]

KEY WORDS

Antenna-matching network—A device that matches the antenna system input impedance to the transmitter, receiver or transceiver output impedance. Also called an **antenna tuner** or *Transmatch*.

Antenna tuner—See **antenna-matching network**.

Antenna switch—A switch used to connect one transmitter, receiver or transceiver to several different antennas.

Block diagram—A drawing using boxes to represent sections of a complicated device or process. The block diagram shows the connections between sections.

Electronic keyer—A device that generates Morse code dots and dashes electronically.

Feed line—The wires or cable used to connect a transmitter, receiver or transceiver to an antenna.

Hand key—A simple switch used to send Morse code. Also called a **telegraph key**.

Impedance-matching network—A device that matches the impedance of an antenna system to the impedance of a transmitter or receiver. Also called an **antenna-matching network** or *Transmatch*.

Modem—Short for *mo*dulator/*dem*odulator. A modem modulates a radio signal to transmit data and demodulates a received signal to recover transmitted data.

Microphone—A device that converts sound waves into electrical energy.

Packet radio—A communications system in which information is broken into short bursts. The bursts (packets) also contain addressing and error-detection information.

Power supply—A circuit that provides a direct-current output at some desired voltage from an ac input voltage.

Radioteletype (RTTY)—Radio signals sent from one teleprinter machine to another machine. Anything that one operator types on his teleprinter will be printed on the other machine.

Receiver—A device that converts radio signals into audio signals.

SWR meter—A measuring instrument that can indicate when an antenna system is working well.

Telegraph key—See **hand key**.

Teleprinter—A machine that can convert keystrokes (typing) into electrical impulses. The teleprinter can also convert the proper electrical impulses back into text. Computers have largely replaced teleprinters for amateur radioteletype work.

Terminal node controller (TNC)—A TNC accepts information from a computer and converts the information into packets. The TNC also receives packets and extracts information to be displayed by a computer.

Transceiver—A radio transmitter and receiver combined in one unit.

Transmitter—A device that produces radio-frequency signals.

Chapter 5

Practical Circuits

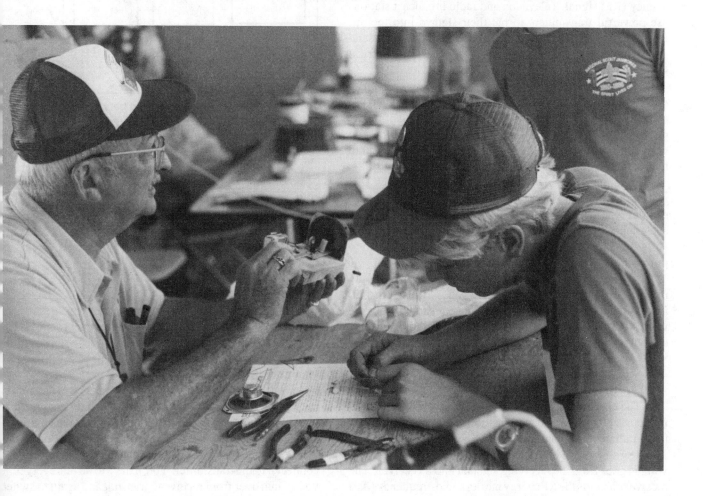

I n the last chapter, you learned about some of the components that make up electronic circuits. In this chapter, we'll introduce some of the basic equipment that goes into an Amateur Radio station. We will briefly cover receivers, transmitters, transceivers, antenna switches and other equipment. We'll show you how to connect the equipment to make a fully functional ham radio station.

Some of the terms may be confusing at first. Don't worry if you don't understand everything. Most of the ideas presented in this chapter are reinforced later in the book. We'll talk more about connecting your station equipment in Chapters 8 and 9.

Throughout this chapter we use **block diagrams**. In a block diagram, each part of a station is shown as a box. The diagram shows how all the boxes connect to each other. Study the block diagrams carefully and remember to turn to Chapters 12 and 13 when instructed to study exam questions.

BASIC STATION LAYOUT

Figure 5-1 shows a block diagram of a very simple Amateur Radio station. Let's discuss the blocks one by one.

TRANSMITTERS

A **transmitter** is a device that produces a radio-frequency (RF) signal. Television and radio broadcast stations use powerful transmitters to put their signals into the air. Radio amateurs use lower-powered transmitters to send signals to each other. A transmitter produces an electrical signal that can be sent to a distant receiver.

We call the signal from a transmitter the *radio-frequency carrier* or *RF carrier*. To transmit Morse code, you could use a telegraph key as a switch to turn the carrier on and off in the proper code pattern. If you want to transmit voice signals, you need extra circuitry in the transmitter to add voice content to the carrier. We call this extra circuitry a *modulator*.

Many modern radios require 12 V dc to operate. This makes them ideal for use in a car as part of a *mobile* radio station. You will need a separate **power supply** to operate such radios in your house. The power supply (usually) converts the 120 V ac from your wall sockets into 12 V dc to power the radios.

Transceivers usually require a *heavy-duty* power supply to provide the current needed to operate the radio in transmit. If you find that your radio works fine in your car, but does not work when you move it into your home, you should suspect a problem with the power supply.

RECEIVERS

The transmitter is a sending device. It sends a radio-frequency (RF) signal to a transmitting antenna, and the antenna radiates the signal into the air. Some distance away, the signal produces a voltage in a receiving antenna. That ac voltage goes from the receiving antenna into a **receiver**. The receiver converts the RF energy into an audio-frequency (AF) signal. You hear this AF signal in headphones or from a loudspeaker.

Just about everyone is familiar with receivers. Receivers take electronic signals out of the air and convert them into signals that we can see or hear. Your clock radio is a receiver and so is your television set. If you look around the room you're in right now, you'll probably see at least one receiver. The receiver is a very important part of an Amateur Radio station.

TRANSCEIVERS

In many modern Amateur Radio stations, the transmitter and receiver are combined into one box. We call this combination a **transceiver**. It's really more than just a transmitter and receiver in one box, though. Some of the circuits in a transceiver are used for both transmitting and receiving.

Figure 5-1—A TR switch connects between the transmitter, receiver and antenna.

Transceivers generally take up less space than a separate transmitter and receiver.

SWITCHES

Now you know the basic parts of an amateur station. The folks at the radio club gave you an old receiver, transmitter and power supply. You bring them home and set them on your desk. Now what?

Well, you know that you need to connect an antenna to the receiver if you want to hear signals. You also have to connect an antenna to the transmitter when you send out a signal. But you only have one antenna. What should you do?

You could disconnect the antenna from the receiver and connect it to the transmitter. You'd have to do this every time you switch over from receive to transmit and again to switch from transmit to receive. Most radios have their antenna connectors on the rear panel, however. Besides, these connectors aren't designed to make it easy to remove and reconnect cables rapidly.

What you need is a *transmit-receive switch* (also known as a *TR switch*). We use a TR switch to connect one antenna to a receiver and a transmitter. Remember our discussion of switches from Chapter 4? The simplest TR switch is a *single-pole, double-throw (SPDT) switch*, like the one in Figure 5-2. The antenna connects to the center arm, and the receiver and transmitter connect to the outside contacts. If you throw the switch one way, the transmitter connects to the antenna. The other switch position connects the receiver to the antenna.

Many TR switches use relays to switch the antenna. A relay uses a magnetic coil to make or break contacts. You can think of a relay as a remotely operated switch. When you

Figure 5-2—The simplest TR switch is an SPDT switch with the center arm connected to the antenna. One position connects the antenna to the receiver; the other position connects the antenna to the transmitter.

throw a switch at your operating position, current passes through the relay coil and the relay arm switches from one set of contacts to another. A transmit-receive relay may have several contacts. Extra contacts can be used to quiet the receiver in the transmit mode or to switch accessory devices.

Transceivers have a TR switch built into the radio, so only one antenna connection is needed. The transceiver switches the antenna between the receiver and transmitter sections. Some amateurs may still want to use a separate receiver with their transceiver. In this case they often use an external TR switch. A receiver should **never** connect to the output from a transmitter or transceiver. The transmitted signal would most likely destroy the receiver.

THE SIMPLEST STATION

The block diagram in Figure 5-1 shows how we connect all this equipment. The power supply connects to the transmitter. The receiver and transmitter both connect to the TR switch. The TR switch connects the transmitter and receiver to the antenna, one at a time. This is just about all you need for the most basic Amateur Radio station.

What kind of wire should you use to make all these interconnections? The power-supply wiring can be just about any insulated wire with a diameter large enough to carry the maximum current the equipment needs. Of course it needs to be insulated to prevent short circuits and shock hazards.

The wires between the transmitter and receiver or transceiver and the antenna are called the **feed line**. Feed lines, or *transmission lines,* are specially constructed cables. There are several common feed-line types. Coaxial cable, or coax (pronounced ko'-aks) may be the most common feed line type. You will learn more about coaxial cable and other feed lines in Chapter 7.

CONNECTING MANY ANTENNAS

What if you have more than one antenna? Again, you could disconnect the antenna from your transmitter or receiver and reconnect another feed line. This can be very

inconvenient. A simpler technique is to use an **antenna switch**. We mentioned antenna switches in Chapter 4. An antenna switch connects one transmitter, receiver or transceiver to several antennas. You can switch from one antenna to another with a simple flick of the switch.

The antenna switch connects at the point where the feed lines from all the antennas come into the station. See Figure 5-3. An antenna switch connects one receiver, transmitter or transceiver to one of several antennas. Even if you only have one antenna you may find an antenna switch useful. Many amateurs use an antenna switch to select between the station antenna and a *dummy antenna*, used for transmitter tuning and testing.

IMPEDANCE-MATCHING NETWORKS

Another useful accessory that you will see in many ham shacks is an **impedance-matching network**. (Impedance is similar to resistance.) This device may let you use one antenna on several bands. The matching network may also allow you to use your antenna on a band it is not designed for. Sometimes we call the impedance-matching network an **antenna tuner** or *Transmatch*. These names indicate the main function of the impedance-matching network. The network matches (tunes) the impedance of the load (the antenna and feed line) to the impedance of your transmitter. We usually connect the impedance-matching network right where the antenna comes into the station. See Figure 5-4.

Figure 5-3—An antenna switch can connect a transmitter to one of several antennas.

Figure 5-4—An antenna tuner (or impedance-matching network) connects directly to the antenna feed line. Placing an SWR meter between the tuner and the transmitter lets you see when the tuner is adjusted properly.

MONITORING THE SYSTEM

Another thing you may want to add to your station is an **SWR meter**. This device is also called an SWR bridge. The SWR meter measures something called the *standing-wave ratio*. You don't need to know too much about SWR right now. We will go over it in more detail later in the book.

Standing-wave ratio is a good indicator of how well your antenna system is working. If you install an SWR meter in your station, you can keep an eye out for problems with your antenna. If you spot the problems early you can head them off before they damage your equipment.

The SWR meter can be connected at several points in your station. One good place to connect the meter is between the antenna switch and transceiver, as Figure 5-3 shows. If you use a separate receiver and transmitter, you can connect the SWR meter between the TR switch and the rest of the antenna system. If you are using an impedance-matching network to tune your antenna system, you will need an SWR meter. Connect the SWR meter between the impedance-matching network and the transmitter or transceiver. The SWR meter then indicates when the matching network is adjusted properly. See Figure 5-4. Most amateurs connect their SWR meter as close to the transmitter output as possible.

[Now turn to Chapter 12. Study questions N7A01 through N7A13. Review this section if you have any problems.]

STATION ACCESSORIES

So far, we have been talking about very basic station layout. We showed you how to connect a transmitter, receiver and antenna switch to make a simple station. To communicate effectively, you also will need a few accessories. Let's look at what you need.

MORSE CODE KEYING

Morse code is transmitted by switching the output of a transmitter on and off. Inventive radio operators have developed many devices over the years to make this switching easier.

The simplest kind of code-sending device is one you're probably already familiar with: the **hand key**, *straight key* or **telegraph key**. These are all different names for the same piece of equipment. See Figure 5-5. A hand key is a simple switch. When you press down on the key, the contacts meet and the transmitter produces a signal.

The code you make with a hand key is only as good as your "fist," or your ability to send well-timed code. An **electronic keyer**, like the one in Figure 5-6, makes it easier to send well-timed code. You must connect a *paddle* to the keyer. The paddle has two switches, one on each side. When you press one side of the paddle, one of the switches closes and the keyer sends a continuous string of dots. When you press the other side of the paddle, the keyer sends dashes. With a little practice and some rhythm, you can send perfectly timed code with a keyer. You may want to start out with a hand key, however. Using a hand key can help you develop the rhythm you need to send good code. When you

Figure 5-5—A hand key (straight key or telegraph key) is the simplest type of code-sending device.

Figure 5-6—You can produce perfect code characters with an electronic keyer and a little practice.

can send good code with a hand key, you're ready to try a keyer.

Both the hand key and the electronic keyer connect directly to the transmitter. The key in the block diagram in Figure 5-7 connects to the transmit section of the transceiver.

MICROPHONES

If you want to transmit voice, you'll need a **microphone**. A microphone converts sound waves into electrical signals that can be used by a transmitter. All voice transmitters require a microphone of some kind. Like a code key, the microphone connects directly to the transmitter. The microphone in Figure 5-8 connects to the transmit section of the transceiver.

[Time for a trip to Chapter 12. Study questions N7B01 through N7B05. Review this section if you have problems.]

Figure 5-7—The keying device connects directly to the transmitting section of the transceiver.

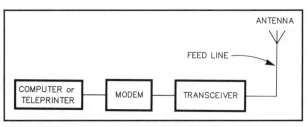

Figure 5-8—A microphone also connects directly to the transmitter. You will need a microphone to transmit voice.

RADIOTELETYPE AND DATA COMMUNICATIONS

Radioteletype (RTTY) and data communications are Amateur Radio transmissions that are designed to be received and printed automatically. Sometimes called *digital communications*, they often involve direct transfer of information (data) between computers. When you type information into your computer, the computer (with the help of some accessory equipment) processes the information. The computer then sends the information to your transmitter and the transmitter sends it out over the air. The station on the other end receives the signal, processes it and prints it out on a computer screen or printer. RTTY and data communications have become a popular form of ham-radio communication. Here we will talk about setting up a station for **radioteletype** and **packet radio** communications.

RADIOTELETYPE

Radioteletype (RTTY) communications have been around for a long time. You may have seen big noisy **teleprinter** machines in old movies. A teleprinter is something like an electric typewriter. When you type on the teleprinter keys, however, the teleprinter sends out electrical codes that represent the letters you are typing. If we send these codes to another teleprinter machine, the second machine reproduces everything you type. Hams have been converting this equipment and using it on the air for years. You can also send and receive radioteletype with a computer. These days computers are so cheap and readily available that they have just about replaced the old noisy teleprinters.

We use a **modem** for Amateur Radio digital communications. Modem is short for *mo*dulator-*dem*odulator. The modem accepts information from your computer and uses the information to modulate a transmitter. The modulated transmitter produces a signal that we send out over the air. When another station receives the signal, the other station uses a similar modem to demodulate the signal. The modem then passes the demodulated signal to a computer. The computer processes and displays the signal.

Sometimes hams use an older teleprinter instead of a computer. The teleprinter converts and displays information from the modem. A complete radioteletype station must have a computer or teleprinter, a modem and a transmitter. The modem connects between the computer and the transmitter, as shown in Figure 5-9.

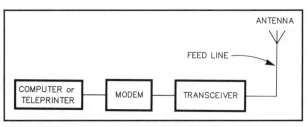

Figure 5-9—A typical radioteletype station. The modem connects between the transceiver and the computer or teleprinter.

PACKET RADIO

Packet radio uses a **terminal node controller (TNC)** as an interface between your computer and transceiver. We might call a TNC an "intelligent" modem. The TNC accepts information from your computer and breaks the data into small pieces called *packets*. Along with the information from your computer, each packet contains addressing, error-checking and control information.

The addressing information includes the call signs of the station sending the packet and the station the packet is being sent to. The address may also include call signs of stations that are being used to relay the packet. The receiving station uses the error-checking information to determine whether the received packets contain any errors. If the received packet contains errors, the receiving station asks for a retransmission. The retransmission and error checking continue until the receiving station gets the packet with no errors.

Breaking up the data into small parts allows several users to share a channel. Packets from one user are transmitted in the spaces between packets from other users. The addressing information allows each user's TNC to separate packets for that station from packets intended for other stations. The addresses also allow packets to be relayed through several stations before they reach their final destination. The error-checking information in each packet assures perfect copy.

A TNC connects to your station the same way a modem does. The TNC goes between the radio and the computer, as shown in Figure 5-10.

Figure 5-10—In a packet radio station, the terminal node controller (TNC) connects between the transceiver and a computer.

[Turn to Chapter 12 and study the questions numbered N7B06 through N7B11. If you have any problems, review this last section.]

If you're preparing for the Novice exam, that's all for this chapter. Don't worry if you're a bit confused by some of the terms. We covered a lot of information fairly quickly. We'll go over a lot of the operating information later in the book. For now, become familiar with the terms, and study the block diagrams carefully. If you are preparing only for the Novice exam, you don't have to continue with this chapter.

If you are preparing for the Technician exam, continue with this chapter *after* you review the questions in Chapter 12 listed above.

Additional Practical-Circuit Information For The Technician Exam

KEY WORDS

Amplifier—A device usually employing electron tubes or transistors to increase the voltage, current or power of a signal. The amplifying device may use a small signal to control voltage and/or current from an external supply. A larger replica of the small input signal appears at the device output.

Band-pass filter—A circuit that allows signals to go through it only if they are within a certain range of frequencies. It attenuates signals above and below this range.

Beat-frequency oscillator (BFO)—An oscillator that provides a signal to the product detector. In the product detector, the BFO signal and the IF signal are mixed to produce an audio signal.

Detector—The stage in a receiver in which the modulation (voice or other information) is recovered from the RF signal.

Filter—A circuit that will allow some signals to pass through it but will greatly reduce the strength of others.

Frequency modulation (FM)—The process of varying the frequency of an RF carrier in response to the instantaneous changes in the modulating signal.

High-pass filter—A filter that allows signals above the cutoff frequency to pass through. It attenuates signals below the cutoff frequency.

Intermediate frequency (IF)—The output frequency of a mixing stage in a superheterodyne receiver. The subsequent stages in the receiver are tuned for maximum efficiency at the IF.

Low-pass filter—A filter that allows signals below the cutoff frequency to pass through and attenuates signals above the cutoff frequency.

Modulate—To vary the amplitude, frequency, or phase of a radio-frequency signal.

Modulation—The process of varying a radio wave in some way to send information.

Phase modulation (PM)—Varying the phase of an RF carrier in response to the instantaneous changes in the modulating signal.

Radio-frequency (RF) carrier—A constant-amplitude, unmodulated, radio-frequency signal.

Reactance modulator—A device capable of modulating an ac signal by varying the reactance of a circuit in response to the modulating signal. (The modulating signal may be voice, data, video, or some other kind depending on what type of information is being transmitted.)

Variable-frequency oscillator (VFO)—An oscillator used in receivers and transmitters. The frequency is set by a tuned circuit using capacitors and inductors. The frequency can be changed by adjusting the components in the tuned circuit.

In this section we will discuss low-pass, high-pass and band-pass filter circuits. We'll show you block diagrams of complete transmitters and receivers, and investigate how the stages connect to make them work.

Keep in mind that entire books have been written on each topic covered in this section. You may not understand some of the circuits from our brief discussion. It would be a good idea to consult some other reference books. ARRL's *Understanding Basic Electronics* is a good starting point. *The ARRL Handbook for Radio Amateurs* includes plenty of detail about these topics. Even that book won't tell you everything about a topic, though. The discussion in this chapter will help you understand the circuits well enough to pass your Technician class license exam, however.

FILTERS

Harmonic interference to entertainment equipment is a problem some hams face. Harmonics are multiples of a given frequency. Your transmitter radiates undesired harmonics along with your signal. In the high-frequency amateur bands (3.5 to 29.7 MHz), the frequency you're transmitting on is much lower than the TV or FM channels. Some of your harmonics may fall within the home entertainment bands. The entertainment receiver cannot distinguish between the TV or FM signals that are supposed to be there and your harmonics, which are not. If your harmonics are strong enough, they can interfere with the broadcast signal. Modern Amateur Radio transceivers have filters built in to *attenuate* (reduce) harmonic radiation so they seldom create harmonic-related interference.

Harmonic interference must be cured at your transmitter. It is your responsibility as a licensed amateur. You must see that harmonics from your transmitter are not strong enough to interfere with other services. As mentioned before, all harmonics generated by your transmitter must be attenuated well below the strength of the fundamental frequency. If harmonics from your transmitting equipment exceed these limits, you are at fault.

You can usually tell harmonic interference when you see it on a TV set. This type of interference shows up as crosshatch or a herringbone pattern on the TV screen. See Figure 5-11. Unlike overload, interference from radiated harmonics seldom affects all channels. Rather, it may bother the one channel that is frequency-related to the band you're on. Generally, the low TV channels (channels 2-6) are most affected by harmonics from amateur transmitters operating below 30 MHz. Channels 2 and 6 are especially affected by 10-meter transmitters, and channels 3 and 6 experience trouble from 15-meter transmitters.

There are several possible cures for harmonic interference. We will discuss a few of them here. Try each step in the order we introduce them. Chances are good that your problem will be quickly solved. You should be familiar with three **filters**: the **low-pass filter**, the **high-pass filter** and the **band-pass filter**.

Figure 5-11—Harmonic interference may be visible as cross-hatching of the TV screen. More severe interference can completely destroy the picture. The width of the cross-hatch lines will vary, depending on the transmitter frequency.

Filters pass certain frequencies and block others. All modern radio communication devices use filter circuits. These filter circuits allow various kinds of equipment to operate on different frequencies without interfering with each other. Under certain conditions, some equipment may need a little extra help. A transmitter may need extra filtering of the output signal to reduce interference to television receivers. In other cases, the transmitter may already be well filtered (clean), but a TV receiver may need extra filtering of its input signals. A basic understanding of filters and their applications will often allow you to solve interference problems.

LOW-PASS FILTERS

The first step you should take is installation of a **low-pass filter** in the transmission line between your transmitter and antenna. A low-pass filter is one that passes all frequencies below a certain frequency, called the *cutoff frequency*. We measure the filter cutoff frequency by putting a variable-frequency signal into the filter. The input signal power is kept constant while the frequency is increased. At the same time we measure the filter output power. At some frequency the output begins to decrease. When the output power has decreased to 1/2 the input power, we have found the cutoff frequency. Frequencies above the cutoff frequency are *attenuated,* or significantly reduced in amplitude. See Figure 5-12.

The cutoff frequency depends on the design of the low-pass filter. The capacitors are chosen to provide a path to ground for the unwanted high frequencies. The inductor has a value that passes the low frequencies you want, but blocks the higher frequencies.

One other important trait that must be considered in the design of a low-pass filter is the *characteristic impedance.* This is the circuit impedance for which the filter is designed. Most external low-pass filters intended for amateur use will have a characteristic impedance of 50 ohms.

The low-pass filter is always connected between the transmitter and the antenna, as close to the transmitter as possible. The filter should have the same impedance as the feed line connecting the transmitter to the antenna. Filters should be used only in a feed line with a low standing-wave ratio. The cutoff frequency has to be higher than the highest frequency used for transmitting. A filter with a 45-MHz cutoff frequency would be fine for the high-frequency (HF) bands. This filter would significantly attenuate 6-meter (50-MHz) signals, so it should not be used for 50 MHz or higher-frequency operation. Most modern transmitters have a low-pass filter built into their output circuitry to prevent excess harmonic radiation.

HIGH-PASS FILTERS

Sometimes entertainment devices experience interference from amateur transmissions even though the transmitting device is operating properly. The harmonics can be well below levels necessary to prevent interference. This happens

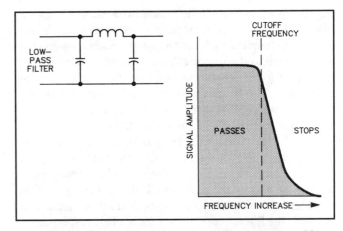

Figure 5-12—A low-pass filter schematic diagram and its output-versus-frequency curve. The low-pass filter passes signals below the cutoff frequency and attenuates signals above the cutoff frequency.

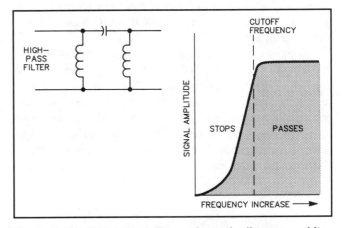

Figure 5-13—A high-pass filter schematic diagram and its output-versus-frequency curve are shown. The high-pass filter attenuates only those signals below the cutoff frequency.

because the design of the entertainment device is inadequate when it is operating in the presence of strong signals. In this situation, it may be necessary to reduce the level of the amateur signal reaching the entertainment device (TV or FM stereo receiver, or VCR). The desired higher-frequency signals must be allowed to pass unaffected. A **high-pass filter** can do this.

A high-pass filter passes all frequencies above the *cutoff frequency*, and *attenuates* those below it. See Figure 5-13. A high-pass filter should be connected to a television set, stereo receiver or other home-entertainment device that is being interfered with. It will attenuate the signal from an amateur station. The inductors have a value that allows them to conduct the lower frequencies to ground. The capacitor tends to block these low-frequency signals. At the same time it allows the higher-frequency television or FM-broadcast signals to pass through to the receiver. This is useful in reducing the amount of lower-frequency signal (at the ham's operating frequency). The lower-frequency signal might overload a television set, causing disruption of reception.

For best effect, a high-pass filter should be connected as close to the television-set tuner as possible. Put it inside the TV if it is your own set and you don't mind opening the cabinet. You can also attach it directly to the antenna terminals on the back of the television. If the set belongs to your neighbors, have them contact a qualified service technician to install the filter. That way you are not held responsible if something goes wrong with the set later on.

BAND-PASS FILTERS

A **band-pass filter** is a combination of a high-pass and low-pass filter. It passes a desired range of frequencies while rejecting signals above and below the *pass band*. This is shown in Figure 5-14.

Band-pass filters are commonly used in the **intermediate-frequency (IF)** stage of a receiver to provide different degrees of rejection. Very narrow filters are used for CW reception. Wider filters are switched in for SSB and AM double-sideband reception. (SSB and AM are explained later.) The band-pass filter allows the receiver IF stages to select signals within a certain frequency range and block signals outside that range.

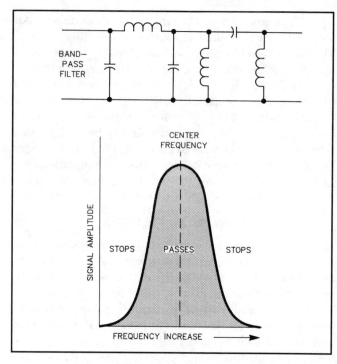

Figure 5-14—A band-pass filter schematic diagram and its output-versus-frequency curve are shown. The band-pass filter passes only those signals within the passband.

[Now turn to Chapter 13 and study exam questions T7A01, T7A02 and T7A03. Review this section if you have trouble with any of these questions.]

TRANSMITTERS

Amateur transmitters range from the simple to the elaborate. They generally have two basic stages, an *oscillator* and a *power amplifier (PA)*. This chapter introduces the CW transmitter and the FM transmitter.

Separate receivers and transmitters have been replaced by transceivers in most amateur stations. A transceiver combines circuits necessary for receiving and transmitting in one package. Some circuit sections may perform both transmitting and receiving functions. Other sections are dedicated to transmit-only or receive-only functions. To simplify some of the following information, we will speak of transmitter or receiver circuits. The principles of operation are the same in transceivers. The transmitter topics apply to transceivers in the transmit mode. Receiver topics apply to transceivers in the receive mode.

An oscillator produces an ac waveform with no input

except the dc operating voltages. You may be familiar with audio oscillators, like a code-practice oscillator. An RF oscillator can be used by itself as a simple low-power transmitter, but the power output is very low. See Figure 5-15. In a practical transmitter the signal from the oscillator is usually fed to one or more **amplifier** stages.

CW TRANSMITTERS

A block diagram of a simple amateur transmitter is shown in Figure 5-16A. It produces CW (continuous-wave) radio signals when the key is closed. The simple transmitter consists of a crystal oscillator followed by a driver stage and a power amplifier. A crystal oscillator uses a quartz crystal to keep the frequency of the radio signal constant.

To transmit information, we must **modulate** the radio

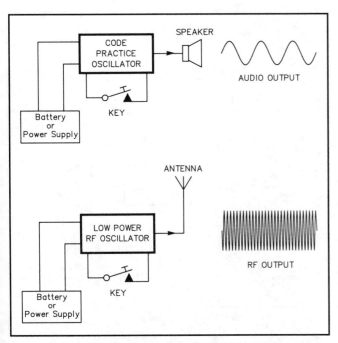

Figure 5-15—Audio- and radio-frequency oscillators are shown. Even a low-power RF oscillator connected to a good antenna can send signals hundreds of miles.

Figure 5-16—Part A shows a block diagram of a basic crystal-controlled CW transmitter. B shows a simple VFO-controlled CW transmitter.

signal. This means we are making some change to the radio signal, and adding information to it. Pressing the key of a CW transmitter and holding it down produces a **radio-frequency (RF) carrier**. This is also called an *unmodulated* carrier. An RF carrier is a constant-amplitude, unmodulated, radio-frequency signal. When you use the key to turn the signal on and off to produce Morse code dots and dashes, however, you modulate the signal. **Modulation** is the process of varying an RF carrier in some way to send information.

Crystal oscillators may not be practical in all cases. A different crystal is needed for each operating frequency. After a while, this becomes quite expensive, as well as impractical. If we use a **variable-frequency oscillator (VFO)** in place of the crystal oscillator, as shown in Figure 5-16B, we can change the transmitter frequency whenever we want.

The telegraph key may connect to one or more transmitter sections. Figure 5-16 shows the key connected to both the driver and power amplifier stages. You would not normally key the oscillator stage because this may cause the oscillator to change frequency slightly.

All transmitters also require some source of operating voltage. We have left the power supply block off the Figure 5-16 drawings for simplicity.

Great care in design and construction is required if the stability of a VFO is to compare with that of a crystal oscillator. Stability means the ability of a transmitter to remain on one frequency without drifting. The frequency-determining

components of a VFO are very susceptible to changes in temperature, supply voltage, vibration and changes in the amplifiers following the VFO. These factors have far less effect on the frequency of a vibrating quartz crystal.

FM TRANSMITTERS

When a radio signal or carrier is modulated, some characteristic of the radio signal is changed in order to convey information. We transmit information by modulating any property of a carrier. For example, we can modulate the frequency or phase of a carrier. Frequency modulation and phase modulation are closely related. The phase of a signal cannot be varied without also varying the frequency, and vice versa. **Phase modulation (PM)** and **frequency modulation (FM)** are especially suited for channelized local UHF and VHF communication. They feature good audio fidelity and high signal-to-noise ratio.

A simple FM transmitter is shown in Figure 5-17. Remember that a circuit containing capacitance and inductance is resonant at some frequency. A resonant circuit in the feedback path of an oscillator can control the oscillator frequency. By changing the resonant frequency of the tuned circuit, we can change the oscillator frequency.

A capacitor microphone is just a capacitor with one movable plate (the diaphragm). When you speak into the microphone, the diaphragm vibrates and the spacing between

Figure 5-17—
This diagram shows a simple FM transmitter. The frequency of the oscillator is changed by changing the capacitance in the resonant circuit. The vibrating plate in the microphone is part of the total capacitance in the resonant circuit.

the two plates of the capacitor changes. When the spacing changes, the capacitance changes. We can connect the microphone to the resonant circuit in our oscillator. Speaking into the microphone varies the oscillator frequency. Notice that if the microphone on your FM transmitter stops working, you will transmit an *unmodulated* **RF carrier**.

We can add frequency multipliers to bring the oscillator frequency up to our desired operating frequency. Adding an amplifier to increase the power provides a simple FM transmitter.

In practical FM systems, the carrier frequency is varied or modulated by changes in voltage that represent information to be transmitted. This information may originate from a microphone, a computer modem or even a video camera. The carrier frequency changes in proportion to the rise and fall of the modulating voltage.

In other words, a modulating voltage that is becoming more positive increases the carrier frequency. As the voltage becomes more negative, the carrier frequency decreases. For some purposes, reversing this relationship will work just as well (FM voice transmitters, for instance). The main point is, the carrier frequency is increased and decreased by the modulating signal voltage. The amount of increase or decrease depends on how much the voltage of the modulating signal changes. Speaking loudly into the microphone causes larger variations in carrier frequency than speaking in a normal voice. Speaking *too* loudly, though, may cause distortion.

One way to shift the frequency of the oscillator is to use a **reactance modulator**. A reactance modulator uses a vacuum tube or transistor. It is connected so that it changes either the capacitance or inductance of the oscillator resonant

Figure 5-18—A varactor diode (D1) can be used to frequency modulate an oscillator. The capacitance of the varactor diode changes when the bias voltage varies.

circuit. The changes occur in response to an input signal.

Modern FM transmitters may use a *varactor diode* to modulate the oscillator. A varactor diode changes capacitance when its bias voltage changes. You can connect a varactor diode to a crystal oscillator as shown in Figure 5-18. The varactor diode will "pull" the oscillator frequency slightly when the audio input changes.

Phase Modulation

One problem with direct-modulated FM transmitters is frequency stability. Designs that allow the oscillator frequency to be easily modulated may have an unfortunate side effect. It becomes more difficult to minimize unwanted frequency shifts arising from changes in temperature, supply voltage, vibration and so on. The frequency multiplying stages also multiply any drift or other instability problems in the oscillator. With phase modulation, the modulation takes place after the oscillator stage. **Phase modulation** produces what is called *indirect FM*.

The most common method of generating a phase-modulated telephony signal is to use a reactance modulator. A vacuum tube or a transistor is connected so that it changes either the capacitance or inductance of a resonant circuit in response to an input signal. The RF carrier is passed through this resonant circuit. Changes in the resonant circuit caused by the reactance modulator cause phase shifts in the RF carrier. Figure 5-19 shows

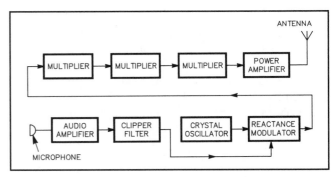

Figure 5-19—This is a block diagram of a phase-modulated transmitter, showing the reactance modulator.

the connection of a reactance modulator.

[Now turn to Chapter 13 and study questions T7A05, T7A08 and T7A11. Also study questions T8A09, T8A10 and T8B01 through T8B03. Review this section as needed.]

RECEIVERS

The main purpose of any radio receiver is to change radio-frequency signals (which we can't hear or see) to signals that we can hear or see. A good receiver can *detect* weak radio signals. It separates them from other signals and interference. Also, it stays tuned to one frequency without drifting. The ability of a receiver to detect weak signals is called *sensitivity*. *Selectivity* is the ability to separate (select) a desired signal from undesired signals. *Stability* is a measure of the ability of a receiver to stay tuned to a particular frequency. In general, then, a good receiver is very sensitive, selective and stable.

Amateur receivers can be simple or complex. You can build a simple solid-state receiver that will work surprisingly well. *The ARRL Handbook for Radio Amateurs* has receiver plans, including sources for parts and circuit boards.

DETECTION

The **detector** is the heart of a receiver. It is where we collect the information we want from the signal. "Crystal sets" using galena crystals and "cat's whiskers" were an early form of *amplitude modulation (AM)* detector. AM is generated by varying the amplitude of an RF signal in response to a microphone or other signal source. The cat's whisker is just a thin, stiff piece of wire. With the galena crystal it forms a point-contact diode. See Figure 5-20.

Today, however, more sensitive and selective receivers are required. In addition, crystal sets cannot receive single sideband (SSB) and CW signals properly. The crystal set shows how simple a receiver can be. It is an excellent example of how detection works. Every receiver has some type

Figure 5-20—Here is an example of a galena crystal and a cat's whisker, the detector in a crystal set.

of detector. Figure 5-21A shows another simple receiver. Diode D1 does the same job as the cat's-whisker detector.

Some later AM receivers had circuits similar to Figure 5-21B. This is a tuned-radio-frequency (TRF) receiver. An incoming signal causes current to flow from the antenna, through the resonant tuned circuit, to ground. The current induces a voltage with the same waveform in L2. The L1-C1 circuit resonates at the frequency of the incoming signal and tends to reject signals at other frequencies. The diode rectifies the RF signal, allowing only half the waveform to pass through. Capacitor C2 fills in and smooths out the gaps between the cycles of the RF signal. Only the audio signal passes on to the headphones. Amplification can improve the sensitivity of this receiver, but other receiver types have much better selectivity.

DIRECT-CONVERSION RECEIVERS

The next step up in receiver complexity is shown in Figure 5-22. Direct-conversion means the RF signal is converted directly to audio, in one step. The incoming signal is combined with one from a **variable-frequency oscillator (VFO)** in the *mixer* stage. We mix an incoming signal at 7040 kHz with a signal from the VFO at 7041 kHz. The output of the mixer contains signals at 7040 kHz, 7041 kHz, 14,081 kHz and 1 kHz (the original signals and their sum and difference frequencies). One of these signals (1 kHz) is within the range of human hearing. We use an audio amplifier and hear it in the headphones.

By changing the VFO frequency, other signals in the range of the receiver can be converted to audio signals. Direct-conversion receivers are capable of providing good usable reception with relatively simple, inexpensive circuits. Their main disadvantages are microphonics and ac hum.

Microphonics are caused by vibrating circuit components when the receiver is tuned or adjusted. A receiver must amplify microvolt signals to several volts, so they can be heard on a speaker or in headphones. Direct-conversion receivers do most of their amplification at AF (audio frequencies). A large amount of audio amplification is needed to build up the mixer (detector) output signal. Unfortunately, the movement of components generates signals (like a microphone) that are also amplified. Better receiver designs avoid microphonics by sharing the amplification among RF, IF (intermediate frequency) and AF stages.

Hum is induced from the 60-Hz ac power lines in buildings and houses. The 60-Hz ac can sometimes modulate the VFO signal. You will then hear a hum or buzz in the headphones.

SUPERHETERODYNE RECEIVERS

A block diagram of a simple superheterodyne receiver for CW and SSB is shown in Figure 5-23. The first mixer produces a signal at the **intermediate frequency** or **IF**, typically 9 MHz in a modern receiver. All the amplifiers after the first mixer are designed for peak efficiency at the IF. The superhet receiver solves the selectivity/bandwidth problem by converting all signals to the same IF before filtering and amplification. To receive SSB and CW signals, a second mixer, called a product detector, is used. The product detector mixes the IF signal with a signal from the **beat-frequency oscillator (BFO)**. The BFO converts the

Figure 5-21—Illustration A shows a simple AM receiver using a crystal detector. It consists only of a wire antenna, detector diode, capacitor, earphone and earth ground. The circuit at B has a tuned circuit that helps separate the broadcast-band signals, but otherwise operates the same as the circuit at A.

Figure 5-22—A block diagram of a direct-conversion receiver. This receiver converts RF signals directly to audio, using only one mixer.

Figure 5-23—A block diagram of a superheterodyne SSB/CW receiver. One mixer converts the incoming signal to the intermediate frequency (IF) and another mixer, called a product detector, recovers the audio or Morse code and converts it to an audio signal.

received-signal information from the intermediate frequency to the audio range. The product-detector output contains audio that can be amplified and sent to a speaker or headphones.

FM RECEIVERS

An FM superheterodyne receiver is similar, but with a few different stages. It has a wider bandwidth filter and a different type of detector. One common FM detector is the frequency discriminator. The discriminator output varies in amplitude as the frequency of the incoming signal changes. You can see that it performs the opposite job of the frequency or phase modulator we studied earlier in this chapter.

Most FM receivers also have a limiter stage between the IF amplifier and the detector. The limiter makes the receiver less sensitive to amplitude variations and pulse noise than AM or SSB/CW receivers. As its name suggests, the limiter output remains almost constant when the signal level fluctuates. Noise pulses are also amplitude-modulated signals. The limiter does not pass them on to the detector. This feature makes FM popular for mobile and portable communications. A comparison between an AM receiver and an FM receiver is shown in Figure 5-24.

[This completes your study of this chapter. Now turn to Chapter 13 and study questions T7A04, T7A06, T7A07, T7A09 and T7A10. Review any material you're unsure of before you go on to the next chapter.]

Figure 5-24—AM and FM receivers use similar stages, with a few variations.

KEY WORDS

Editor's note: No exam questions are covered in this chapter. The key words are often used when describing equipment. They are included to help you learn them.

Amplitude modulation (AM)—A method of combining an information signal and an RF (radio-frequency) carrier. In double-sideband voice AM transmission, we use the voice information to vary (modulate) the amplitude of an RF carrier. Shortwave broadcast stations use this type of AM, as do stations in the Standard Broadcast Band (540-1600 kHz). Few amateurs use double-sideband voice AM, but a variation, known as single sideband, is very popular.

Bandwidth—The range of frequencies that will pass through a given filter.

Beat-frequency oscillator (BFO)—A receiver circuit that provides a signal to the detector. The BFO signal mixes with the incoming signal to produce an audio tone for CW reception. A BFO is needed to copy CW and SSB signals.

Continuous wave (CW)—Morse code telegraphy.

Frequency modulation—A method of combining an information signal with a radio signal (see **Amplitude modulation**). As you might suspect, we use voice or data to vary the frequency of the transmitted signal. FM broadcast stations and most professional communications (police, fire, taxi) use FM. VHF/UHF FM voice is the most-popular amateur mode.

General-coverage receiver—A receiver used to listen to a wide range of frequencies. Most general-coverage receivers tune from frequencies below the standard-broadcast band to at least 30 MHz. These frequencies include the shortwave-broadcast bands and the amateur bands from 160 to 10 meters.

Ham-bands-only receiver—A receiver designed to cover only the bands used by amateurs. Usually refers to the bands from 80 to 10 meters, sometimes including 160 meters.

Lower sideband (LSB)—The common single-side-band operating mode on the 40, 80 and 160-meter amateur bands.

Multimode transceiver—Transceiver capable of SSB, CW and FM operation.

Offset—The 300- to 1000-Hz difference in CW transmitting and receiving frequencies in a transceiver. For a repeater, offset refers to the difference between its transmitting and receiving frequencies.

Receiver—A device that converts radio waves into signals we can hear or see.

Receiver incremental tuning (RIT)—A transceiver control that allows for a slight change in the receiver frequency without changing the transmitter frequency. Some manufacturers call this a clarifier (CLAR) control.

Resolution—The space between markings on a receiver dial. The greater the frequency resolution, the easier it is to separate signals that are close together.

Rig—The radio amateur's term for a transmitter, receiver or transceiver.

Selectivity—The ability of a receiver to separate two closely spaced signals.

Sensitivity—The ability of a receiver to detect weak signals.

Shack—The room where an Amateur Radio operator keeps his or her station equipment.

Single sideband (SSB)—A common mode of voice operation on the amateur bands. SSB is a form of amplitude modulation.

Ticket—A commonly used name for an Amateur Radio license.

Transceiver—A radio transmitter and receiver combined in one unit.

Transmitter—A device that produces radio-frequency signals.

Transmit-receive (TR) switch—A device used for switching between transmit and receive operation. The TR switch allows you to share one antenna between a receiver and a transmitter.

Upper sideband (USB)—The common single-sideband operating mode on the 20, 17, 15, 12 and 10-meter HF amateur bands, and all the VHF and UHF bands.

Variable-frequency oscillator (VFO)—A circuit used to control the frequency of an amateur transmitter.

Chapter 6

Selecting Your Equipment

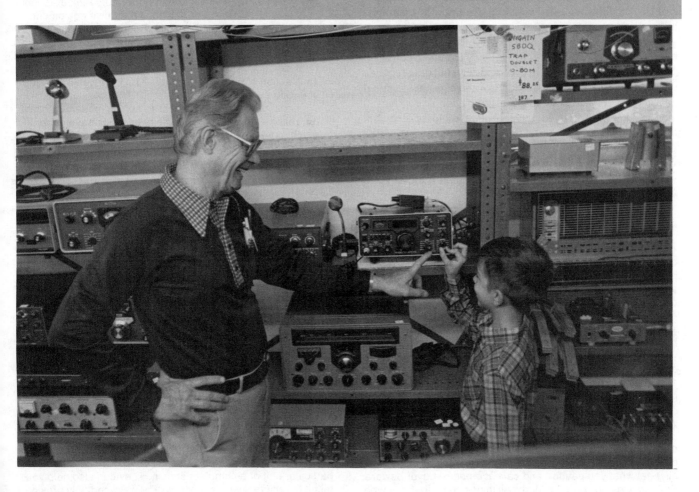

aving an amateur license and no station is a little like having a driver's license but no car. Without a station, your **ticket** (a ham's name for an Amateur Radio license) is just another piece of paper. You can probably arrange to use a friend's equipment or operate from a club station when your license first arrives. Very soon, you'll want to have your own station.

Most hams look with pride at their operating position. When hams meet, conversation always turns to the station. Whenever hams visit each other, the **shack** is usually the first stop on the tour. Many amateurs who develop friendships on the air eventually swap pictures of their stations. It's no wonder, really. Hams devote many hours to their hobby, and they spend most of these hours in their shacks. The shack is where a ham operates on the air, repairs equipment, makes improvements and experiments with new projects.

As a newcomer to Amateur Radio, you may find the wide assortment of equipment, antennas and available acces-

sories confusing at first. You'll wonder, "Why is this antenna better than that one?" or "What features does this **transceiver** offer? Do I really need them all?"

You have to decide what your goals are. Take a look at your available resources (how much money you have to spend). Do a little research and then choose the equipment that best suits your needs. Actually, selecting your station equipment can be very easy, if you know what you want it to do. The information in this chapter will help you select a radio that will provide you with many hours of enjoyable operating.

To help you select station equipment, ARRL publishes a collection of Product Reviews from *QST*, our monthly membership journal. *The ARRL Radio Buyer's Sourcebook* includes dozens of in-depth reviews and technical information about HF, VHF and UHF radios, and accessories. Sections of this book also explain equipment features and specifications that are important to enjoyable operating in more detail than we can cover them in this chapter.

LOOK BEFORE YOU LEAP

Before you go out and purchase a room full of equipment, find out how big the room is! Try to get some idea of how much space you will have for your shack. Chapter 8 has more detailed information about where to locate your station. Some hams use a corner of a bedroom or den, and some use a fold-out shelf in a closet. Available space may be an important consideration in what equipment you purchase.

When you know how much space you have, you're ready to select your equipment. There are many factors to consider. "Specs" (specifications) are very important, but don't choose your equipment on that basis alone. Some **rigs** are technical marvels but very difficult and frustrating for a new ham to operate. Other gear may look great but have a lot of technical problems. The most important consideration is what works for you. Choose equipment that will be enjoyable and comfortable to operate.

There are many different places to look for equipment.

SHOULD YOU BUILD YOUR OWN EQUIPMENT?

There was a time when almost all hams built their own rigs. The cost was small, and the operators took great pride in their work. Many hams still build their own equipment, either from used parts or as kits. Commercially made equipment is more affordable now than it used to be, but some hams still build their own equipment. We hope you will try building at least some part of your station as well. The satisfaction of being able to say, "I built it myself!" is a joy you will never forget.

Most beginners won't want to build a modern transmitter, receiver or transceiver. The electronics have gotten much more complicated over the years, and this complexity can frustrate an inexperienced builder. If you have a limited ham radio budget, buy used equipment.

USED EQUIPMENT

What used rig should you buy? The answer to that question depends largely on available resources and personal preferences. Older issues of amateur magazines can provide a lot of information on equipment you may see on the used market. The New Products and Product Review columns in QST, and The ARRL Radio Buyer's Sourcebook are a good start. Many local libraries carry back issues of QST. Local hams may also have some QSTs. If you can't find the issues you need, you can purchase many of them from ARRL HQ. In addition, several other product guides and product-review anthologies are available. They summarize features and performance of recently manufactured equipment.

Don't just rush out and buy the first piece of equipment you can find. Buying used radio equipment is much like buying a used car: You really should "kick the tires" a bit. Examine the equipment closely (as well as you can). Don't get snowed by salesmanship. A couple of old adages apply: "You (generally) get what you pay for" and caveat emptor—buyer beware. Perhaps we should add a third: "All things come to those who wait." Have patience!

Surplus Equipment

There's still some WW II government surplus equipment on the market, and you may also see some newer surplus gear. Generally speaking, a beginner who doesn't have an Elmer should stay away from this equipment.

Much of the military equipment is big and heavy. You might also have to find some special connectors or build a power supply to use it. Some units produce large amounts of television interference (TVI), and that's one thing we can all do without!

Many units that appear to be bargains will require quite a bit of work to get them on the air. Many hams enjoy using surplus gear, but it's not cost-effective for the beginner. You might try a surplus rig later, when you have gained some experience and want to do some tinkering.

Tube Availability

You might wonder about the practicality of buying tube-type equipment. After all, modern equipment uses transistors and integrated circuits, and many companies have stopped making tubes. Tubes are still available, but they are becoming expensive. A few electronic supply houses still stock tubes. Many amateurs have plenty of tubes (often free for the asking) in their "junk boxes." Some excellent transceivers made in the 1970s used tubes in the power-amplifier stages. These tubes usually are readily available.

Where Can I Find Used Gear?

There are many different sources of used ham gear. You can buy gear from equipment retailers and local hams. You may see what you want listed in the pages of ham magazines (see the QST Ham Ads) or used-equipment flyers. Other sources are auctions, flea markets, hamfests and even garage sales.

If you belong to an Amateur Radio club, ask your instructor and other club members. They can often provide you with some leads and other useful information about buying used gear. The club members can help you select a particular unit or a complete station. They can also help you test it to ensure it works properly. They will certainly provide some helpful hints on connecting all the pieces.

One of the safest ways to purchase used equipment is to buy it from a local retailer. Often the dealer sets aside a section of the ham shop for used equipment. When you buy from such a source, there's usually some sort of warranty (30 days is typical) on the equipment. If a problem comes up within that time, you can bring it back to the retailer for repairs. Ask about the warranty before you buy.

Many dealers route the used equipment they've received through their service shop. This ensures that the gear is working properly before they place it on the used-equipment shelf. Some dealers will even allow you to operate the equipment before you buy it. This is the best way to determine the rig's capabilities and condition.

When you buy used gear from an individual, you generally have no guarantees. Ask to see the unit in operation. You might even wish to take along a more knowledgeable ham who can help you make the decision. A local ham can be your best source of information on used equipment.

You may already know what you want or what you need. The hard part is deciding which of those used rigs provides the best value for your hard-earned dollars. An "old-timer" can be an invaluable source of information. Chances are, he or

Check with local hams to see what they use. Find out what they like about certain pieces of gear and what problems they have had. Learn from their experience. They are as proud of their shacks as you will be of yours. When you are deciding what you want, there is no substitute for sitting down and listening to a rig. (For the name of a local radio club, contact ARRL Headquarters.)

Another place to look is your local radio store. Check the telephone book for ham radio retailers in your area. Most large metropolitan areas have at least one. At the store, you can see and compare several of the newest pieces of equipment. In addition, most ham stores have antennas set up so you can listen to equipment you might buy, even if you don't have your license yet.

The popular ham magazines carry many ads in each issue. When you look through any issue of *QST*, you can find ads from many manufacturers. Local dealers and large mail-order companies also advertise in *QST*. Many of the ads list features and specifications, and will give you a

rough idea of what the equipment costs.

QST has an advertising acceptance policy that can help protect you. If a piece of equipment doesn't live up to its manufacturer's claims, you won't see it advertised in *QST*. It's a good idea to study the ads and have some equipment in mind when you visit the local ham store.

SIDEBANDS

There are two kinds of **single sideband: upper sideband (USB) and lower sideband (LSB)**. Hams usually use LSB on 160, 80 and 40 meters, and USB on 20, 17, 15, 12 and 10 meters. Upper sideband is also used on VHF and UHF. Some radios automatically switch to the correct sideband for the operating frequency. Others have separate USB and LSB positions on the mode switch. You will always use the USB position for 10-meter and VHF/UHF SSB operation.

she may have at one time owned a similar piece of equipment. With the seller's permission, your ham friend could operate the equipment and note any potential problems—your future headaches!

Local hamfests, flea markets and auctions are some of the better opportunities to see a quantity of used equipment. For the careful buyer, they may also be the source of some excellent equipment. But you have to know what you're looking for. What are the capabilities of the equipment, its current market value, cost of repairs and availability of repair parts?

At a hamfest or flea market, you may not be able to test equipment before you buy it. A thorough visual inspection by an experienced eye will generally suffice if the price is right, however. Again, help from an experienced amateur is a wise choice. You can usually tell a lot from the external appearance of equipment. If it looks physically abused, chances are it has been treated badly on the air as well.

There are some simple precautions you should take if you're buying gear from another ham through the mail. Try to be

sure you'll have the right to return the equipment if you don't like what you receive. Shipment by truck freight with the right of inspection permits you to examine the package contents before you accept delivery. If you don't like what you receive, simply refuse delivery. You may be asked to pay by bank check or money order, rather than personal check.

There is one other very important point to keep in mind when you buy used equipment from any source. Be sure you get an owner's manual with the radio! Some hams may even have the service or shop manual. A shop manual can be a valuable addition, because it generally has more complete service procedures and troubleshooting guidelines.

If you are not getting at least an owner's manual with the radio, be cautious about making the purchase. Manuals for some pieces of equipment are available from the manufacturer. Several companies sell manuals for used equipment; see Table 6-1.

Table 6-1

Sources of Equipment Manuals

Brock Publications (Swan, Cubic, Siliconix)
PO Box 5004
Oceanside, CA 92054
619-757-0372

Collins S-Line: See VISTA Technology, Inc

Cubic Communications: See Brock Publications

R. L. Drake Co
P.O. Box 3006
Miamisburg, OH 45342
513-866-3211

Hallicrafters Manuals
Ardco Electronics†
PO Box 95, Dept Q
Berwyn, IL 60402

Hammarlund Manuals
Pax Manufacturing Co
100 East Montauk Hwy
Lindenhurst, NY 11757
516-957-7200

HI Manuals, Inc††
PO Box 802Q
Council Bluffs, IA 51502

Howard W. Sams Co
Photofact Division
2647 Waterfront Parkway East Drive
Indianapolis, IN 46214-2041
800-428-7267

ICOM America, Inc
2380 116th Ave NE
Bellevue, WA 98004
206-454-7619

Kenwood USA Corporation
PO Box 22745
2201 E Dominguez St
Long Beach, CA 90801-5745

Siliconix Manuals: See Brock Publications

Swan Manuals: See Brock Publications

Ten-Tec, Inc
PO Box 8010
Sevierville, TN 37862
800-883-7373

VISTA Technology, Inc
(Collins Radio manuals)
3041 Rising Springs Ct
Bellbrook, OH 45305
513-426-6700

Yaesu USA
17210 Edwards Rd
Cerritos, CA 90701
310-404-2700

† Write for manual prices (specify model)

†† HI Manuals will not answer requests for information unless you include a $5 research fee.
　Order their catalog ($2) to see if they have the manual you need.

HF EQUIPMENT

TRANSCEIVER OR SEPARATES?

Although this section refers to HF equipment, much of the information applies to VHF/UHF transceivers, too. Even if you're planning to operate only on VHF/UHF, read this section before you go on to the section on VHF/UHF equipment later in this chapter.

Figure 6-1—The ICOM IC-728 is a good example of a trend toward less complicated transceivers. You can work the world with this rig and a good antenna.

(A)

(B)

Figure 6-2—(A) The Heath HR-1680 and HX-1681 "twins." The transmitter is a CW-only rig designed to cover portions of the 80, 40, 20, 15 and 10-meter bands. At B, a pair of Kenwood "twins" produced in the 1970s. The receiver is all solid-state. The transmitter uses solid-state devices except for the driver and final-amplifier stages, which use tubes.

One of the first decisions you'll face is whether to use a **transceiver** or a separate **receiver** and **transmitter**. When you make this decision, keep your space limitations in mind. A transceiver, as the name implies, combines both a transmitter and receiver in one package. That means you can fit a transceiver in a smaller space than that required by most separates. In fact, some newer transceivers require less than one foot of desk space. See Figure 6-1. Transceivers are easier to set up. Usually, you just connect a microphone and CW key, attach an antenna, plug in the ac-power cord and you're ready to go.

With separates, you may have to do a little wiring between the transmitter and receiver. You'll also need a **transmit-receive (TR) switch** to switch the antenna between the two units.

Transceivers are easier to operate than a separate receiver and transmitter. A single control sets both the transmit and receive frequency. With one frequency control to set, you're sure to be transmitting and receiving on the same frequency. Until recently, some manufacturers offered "twins"—separates interconnected so they can act as a transceiver. See Figure 6-2.

Separate transmitter-receiver combinations were popular because they offered better performance and flexibility than transceivers of the time. Modern transceivers have all the features once offered only in separates.

A primary advantage of a transceiver is that it automatically transmits and receives on the same frequency, which takes some care and practice with a separate receiver and transmitter. A simple transceiver suitable for Novice operation is shown in Figure 6-3. Although it operates only on 10 meters, it can be a good value for a new Novice.

Modern transceivers are at least equal to, and usually better than, a separate receiver and transmitter. Most offer split-frequency capability. Another feature offered in most new transceivers is a bank of frequency memories that allow you to switch quickly from one frequency to another. Figure 6-4 shows two modern, full-featured transceivers.

If you do decide to buy an older rig, try to avoid "boat anchors." This is what hams call radios that require two people simply to get them in the trunk of your car! While many of the older radios were excellent performers in their time, the saying "They don't build 'em like they used to" is fortunately true. Modern, solid-state radios offer far better reliability and performance. Right now, you want a rig you can use, not one you have to nurse back to health after every contact! After a little looking, you'll probably decide that either a transceiver or separate "twins" will work well for you.

PRICE

An important consideration for most of us is price. Don't be dismayed if you can't afford a new transceiver right away. Plenty of good, used transceivers are available at reasonable prices.

Figure 6-3—The Radio Shack HTX-100 is a small, inexpensive 10-meter transceiver.

(A)

(B)

Figure 6-4—Two modern transceivers with lots of "bells and whistles." (A) The ICOM IC-781 uses a computer-monitor type display in place of a digital readout. (B) The Yaesu FT-1000D allows you to listen on two frequencies at the same time. All the knobs and switches on these rigs can be intimidating. If you read the instruction manual carefully you should have no trouble understanding everything, however.

Figure 6-5—The Yaesu FT-890 is a marvel of modern miniaturization. The rig includes a general-coverage receiver and 100-watt transmitter with a built-in electronic keyer. All this fits in a box the size of a portable typewriter.

Figure 6-6—An example of a transceiver with an analog circular dial.

When you decide what kind of gear you want, you'll have to examine specific features. The next few sections describe some of the things to look for. Remember that many of the features explained here apply to separate transmitters and receivers as well as transceivers.

FREQUENCY DISPLAY

Of all the controls on a rig, you use the frequency control most often. The frequency display includes the knobs, dials, gears and readout. Make certain the mechanism works freely, the dial markings are understandable, and you feel comfortable operating it.

Two different types of tuning-dial mechanisms are available on recently manufactured transceivers and separates: Digital and analog.

Most manufacturers now use digital displays (Figure 6-5). They are generally quite accurate, and are easier to use than analog dials.

The analog circular dial (Figure 6-6) was very common before digital dials became popular. The dial is usually calibrated in kilohertz. A fixed pointer indicates where the dial is tuned.

Circular dials vary in **resolution**. The farther apart the markings on the dial, the better the resolution. The better the resolution, the easier it is to tell what frequency you're on. A dial with markings every 1 kHz or less is best, although some may be marked only every 5 kHz. The megahertz part of the display is marked on the panel at the band switch.

Choose the dial mechanism you are most comfortable reading and tuning. Make sure the knob operates smoothly. Avoid radios that feel sloppy and slip or skip—that's part of your comfort and accuracy.

RECEIVER SELECTIVITY AND SENSITIVITY

The receiver is the most critical part of the station. In this section, *receiver* applies to either a separate receiver or the receiver section of a transceiver.

Whether you buy separates or a transceiver, there are two important specifications you should know about: selectivity and sensitivity. **Selectivity** is the ability of a receiver to separate two closely spaced signals. This determines how well you can receive one signal that is very close to another. **Sensitivity** is the ability of a receiver to detect weak signals. Selectivity is more important than sensitivity. The ability to isolate the signal you are receiving from all the others nearby directly affects how much you will enjoy your time on the air.

We use **bandwidth** to measure selectivity. Bandwidth is nothing more than how wide a range of frequencies you hear with the receiver tuned to one frequency. With a 6-kHz bandwidth, you can hear signals as far as 3 kHz above and below where you are tuned.

If you can't hear signals more than 1200 Hz above or below where your receiver is tuned, your receiver has a bandwidth of 2400 Hz or 2.4 kHz. Narrower bandwidth means better selectivity. Narrow bandwidth makes it easier to copy one signal when another is close in frequency.

Special filters built into modern equipment determine the selectivity. Generally, these filters contain quartz crystals arranged to provide a specific selectivity. Some receivers come with several filters. You can choose the filter that gives the best reception.

Look for a selectivity of 500 Hz or less for **CW** operation. (CW is short for **continuous wave**, or Morse-code telegraphy.) Receivers designed for **single-sideband (SSB)** voice operation come with a standard filter selectivity of 2.4 to 2.8 kHz. (SSB is a common mode of voice operation on the amateur bands. SSB is a form of **amplitude modulation** or **AM**.) This filter is usable on CW, but a 500-Hz filter is much better. If the receiver has provision for adding narrow-bandwidth accessory filters, you can buy them later. While selectivity is very important, it comes with a price! The more filters a receiver has and the better their quality, the more the receiver or transceiver will cost.

If you're interested in operating RTTY, packet, AMTOR, or any other "digital" mode, you can also make use of a 500-Hz filter *if the receiver allows it to be used in FSK mode* (see Chapter 9 for more information on these modes). Some receivers, including those in modern transceivers, only

Figure 6-7—Audio filters. At left is an Autek QF1-A. The large filter at right is a JPS Communications NIR-10. The smaller filter above it is the j-Com Magic Notch, a switched-capacitor notch filter.

allow use of the SSB filter when operating these modes. A few minutes spent listening to the receiver and reading the manual will tell you what filters are available when the different modes are selected.

To receive HF single-sideband (SSB), CW, RTTY, AMTOR or packet signals, your receiver must have a **beat-frequency oscillator (BFO)**. Almost all HF transceivers do, but many VHF/UHF transceivers don't. On VHF and UHF, RTTY and packet signals are usually sent with FM rigs, so you won't need a BFO for those modes on those bands.

An *audio filter* (Figure 6-7) is an inexpensive way to add selectivity to a receiver that has only an SSB filter. An audio filter doesn't work against interference as well as a crystal filter, but it may mean the difference between maintaining a QSO or losing one.

Audio filters are simple to construct (see *The ARRL Handbook for Radio Amateurs*). If you don't want to build a filter, you can purchase one of the many available commercial units. Audio filters are easy to use. They plug into the receiver headphone or speaker jack, and you plug your headphones or a speaker into the filter.

Another selectivity feature available on some receivers is a notch filter. This filter can be used to cut out, or notch, a specific frequency from within the received bandwidth. A notch filter is handy when you're trying to receive a signal that is very close in frequency to another signal. By adjusting the notch control, you can effectively eliminate the unwanted signal.

OTHER RECEIVER FEATURES

You may find other features on receivers that improve their performance or make them easier to use. Most have a meter that shows the strength of the received signal. Most of the time, the first thing the operator you are working wants to know is his or her signal strength. These meters, called *S meters*, will tell you. They are also useful when you are comparing two antennas. Or, you can use the S meter when you are trying to rotate a directional antenna for maximum signal strength.

Some older receivers have a *crystal calibrator*. They produce a signal every 25 or 100 kHz. You can use the calibrator to make sure the dial or display is accurate. Most equipment with a digital display does not have or need a crystal calibrator.

Besides the filters already mentioned, some receivers have *noise blankers*. They are mostly useful for filtering ignition noise from cars and trucks. Some blankers can help reduce power-line noise as well. Some receivers have filters between the intermediate stages. These may be called variable IF filters, or the receiver may have a control labeled PASSBAND TUNING or VARIABLE BANDWIDTH TUNING. An RF gain control is sometimes helpful when interference is heavy.

Receiver incremental tuning (RIT) allows you to shift the receiver frequency over a limited range without affecting the transmitter frequency. Although transceivers theoretically transmit and receive on exactly the same frequency, there may

Figure 6-8—The Ten-Tec Argosy II is a transceiver available on the used market. This solid-state, CW/SSB transceiver makes operation as uncomplicated as possible.

actually be a slight frequency difference, or **offset**. This offset can vary from transceiver to transceiver. RIT lets you retune the receiver slightly. This feature also enables you to compensate if the station you are working has a different offset or is "drifting" (changing frequency slowly with time—a common occurrence with "boat anchors").

Most HF transceivers now feature **general-coverage receivers**. Unlike **ham-bands-only receivers**, a general-coverage receiver lets you listen to shortwave broadcast stations and utility stations transmitting RTTY, facsimile, Morse code and voice.

All these features can add to your operating enjoyment. They may not be necessities if you are on a tight budget, however. Frequency coverage, tuning mechanism, frequency resolution, selectivity and sensitivity are the most important things to look for.

THE TRANSMITTER

If you are following our advice and considering a transceiver, you'll find many common transmitter features among them. Almost all transceivers offer voice operation. Voice capability is important if you want to operate voice or data on the 10-meter Novice band (most data modes use the voice part of the radio).

Power output varies somewhat from rig to rig. Most transceivers produce about 100 watts output, which is less than the 200 watts output that Novices are allowed to use. This may sound bad, but it isn't. Most of the time, if 100 watts won't make the contact, neither will 200. One-hundred watts will even drive a high-power linear amplifier, if you decide to get one when you upgrade to General or a higher class license. Figure 6-8 shows a simple operating position, with an inexpensive transceiver as the station rig. This transceiver produces less than 100 watts, but it has been used to contact over 100 countries!

A shortcoming of most rigs made before 1980 is their inability to operate on the 17, 12, and 30-meter bands. When you upgrade to the General or higher class license, you may want to use those bands. You'll also be able to use much larger portions of the 80, 40, 15 and 10-meter bands, plus the 160- and 20-meter bands. The rig you buy may cover bands you can't use right now, but they'll be there, waiting for you when you upgrade. If you later decide to sell the rig, they're sure to make it more valuable than a rig that doesn't have them.

A typical separate ham HF transmitter runs between 75 and 200 watts output from the final amplifier. It operates **upper sideband (USB)**, **lower sideband (LSB)** and CW, and has **variable-frequency oscillator (VFO)** control. The transmitter will cover the 80, 40, 20, 15 and 10-meter amateur bands. Some also include the 160-meter band.

Figure 6-9—The ICOM IC-735 has many desirable features for a home-station rig, but is small enough for mobile and portable use.

Figure 6-10—The Kenwood TS-830S transceiver uses 6146 tubes in the final-amplifier section.

Newer transceivers also cover the 30, 17 and 12-meter bands. Either will be more than adequate for all the operating you're likely to do as a Novice. It will also serve you well after you earn a higher-class license. Figure 6-9 shows a compact 160-10 meter transceiver suitable for mobile or home station use.

As with the receiver, you should choose a transmitter that is easy to operate. Look for the same features in the transmitter tuning dial as you would in a receiver dial.

Modern transceivers or transmitters with vacuum-tube final amplifiers develop more than 100 watts PEP output. The workhorse of the final-amplifier-tube family is the 6146. It's tough and can take a beating—such as prolonged key-down periods or antenna mismatches. Given a little time to cool down, the 6146 will bounce right back as good as new. The Kenwood TS-830S is a popular transceiver that uses 6146 tubes in its final amplifier. It should be available at reasonable prices on the used market. This transceiver is shown in Figure 6-10.

Tube-type transmitters use two basic types of keying circuits: cathode keying and grid-block keying. If you use a hand key, the voltage polarity at the key jack is of little consequence. If you use an electronic keyer with a transistor switch in the output circuit, polarity must be considered.

Keyers that use relays with floating (ungrounded) contacts in the output circuit are more flexible since key-jack voltage polarity is unimportant. But relay contacts can stick and the trend is to use transistorized output stages in modern keyers.

Cathode keying presents a positive voltage to the transmitter key jack. In cathode keying, the key line opens and closes the cathode circuit of one or more tubes in the transmitter. When the key is up, there is an open circuit in the tube cathode. This effectively shuts off the tube so it draws no plate or screen current; thus, no RF output.

Grid-block keying uses a negative voltage applied to the grid(s) of the keyed tube(s). This voltage, usually at a reduced amplitude, is also present at the transmitter key jack. With the key open, the bias voltage cuts off the tube. Closing the key removes or reduces the bias and the tube conducts.

As a rule, older transmitters used cathode keying while more modern tube equipment uses grid-block keying. Fully transistorized rigs may use either positive or negative polarity at the key jack.

Before you connect a keyer to any rig, you should check the key-jack voltage and polarity. Make sure your keyer works with the rig. Check your equipment manual. It should tell you the type of keying circuit used in your transmitter, and the key-jack voltage. If you built the keyer, the construction article should tell you the appropriate keying voltage.

There are ways to change the keyer output circuit. You can use a transistor or optoisolator, but these techniques are beyond the scope of this book. A more-experienced amateur may be able to help you if you face this problem. Also, The *ARRL Handbook for Radio Amateurs* contains more information.

OTHER TRANSMITTER FEATURES
Microphone

To talk on SSB, you will need to add a microphone to your transceiver. The microphone connects to the microphone jack on your radio.

Individual voices and microphones have different characteristics. Your rig should have a microphone gain control you can adjust for a clean signal. Most SSB transmitters have an automatic level control (ALC) meter to help you determine the correct microphone-gain setting. See your equipment manual for instructions on adjusting everything to the right levels.

Speech Processor

Speech processing increases the average power of a single-sideband signal. Used properly, a speech processor can greatly improve the readability of a signal. Misused, however, it can severely degrade the audio quality and make the signal more difficult to understand.

Almost all SSB transmitters and transceivers built since 1980 have speech processors as standard equipment. There is also a variety of accessory speech processing equipment for transmitters without built-in processing.

On some rigs, there is no adjustment for the speech processor; there is just an on/off control. Other gear has one or more variable controls to set the speech processing level. Always check your equipment instruction manual for information on setup and operation. When you think you've got everything adjusted properly, ask other amateurs for on-the-air checks.

Voice-Operated Switch (VOX)

Most SSB transceivers and transmitters have a voice-operated switch (VOX). The VOX switches the rig into trans-mit automatically when you speak into the microphone, then back to receive when you stop talking. VOX is handy because it allows you to listen during pauses and lets you keep both hands free.

There are usually three VOX controls: gain, delay and anti-VOX. *VOX gain* sets the sensitivity. Adjust this control so the VOX keys the transmitter when you speak in a normal voice.

VOX delay sets the interval between when you stop talking and when the transceiver switches back to receive. Most hams set this control so the rig switches after a short pause in speech.

Anti-VOX works with VOX gain to keep receiver audio from keying the VOX. If improperly set, speaker audio can key the transmitter. You could set the VOX gain so speaker audio wouldn't key the transmitter, but if you did, you'd have to shout to activate the VOX. Anti-VOX circuitry allows the VOX to ignore audio from the speaker, yet respond when you speak into the microphone.

VHF AND UHF EQUIPMENT

Whether you are working toward the Novice or the Technician license, you have a wide variety of VHF and UHF equipment to choose from. Many transceivers for these bands operate only in the **frequency modulation (FM)** mode. Other gear can be used only for SSB and CW work. For both FM and SSB/CW operation, you need a third type: the multimode transceiver, like the one in Figure 6-11. You have to know what modes you'd like to try before you can choose a rig. Unlike five- or six-band HF transceivers, most VHF and UHF equipment operates on only one band. Two- and three-band rigs are becoming more popular, however.

Also consider where you'll use the equipment. Some VHF/UHF equipment is best for home station operation. Other, more compact, units work well for mobile operation. Portable, hand-held transceivers can be used anywhere.

Although manufacturers sometimes design equipment for a certain application, you can use it wherever you want. If you like, you can use a hand-held transceiver in a car, or a mobile rig at your home station. There are, however, some trade-offs to consider when you want to use a rig at home and in your car or portable.

BASE-STATION AND MOBILE EQUIPMENT

FM base stations often consist of a multimode transceiver or a mobile transceiver. Multimode base-station transceivers often have built-in ac power supplies and are larger than FM-only rigs.

Mobile rigs are often smaller than base-station trans-

Figure 6-11—The ICOM IC-375A 222-MHz transceiver. This rig can be used on FM, SSB and CW.

ceivers, and they operate from a 13.8-V automobile electrical system. See Figure 6-12. You'll need an external, accessory power supply to use one at home.

FM mobile transceivers usually have all the features found in base-station equipment. Power output for mobile and base-station equipment is typically in the 10- to 25-watt range. The rig's physical size and the power supply are the most important things to consider. Other than size and power requirements, there is no particular advantage to either a base-station unit or a mobile unit for FM work. Many amateurs buy

Figure 6-12—Modern VHF FM hand-held rigs are usually very small, like the Kenwood TM321A shown here.

Figure 6-13—The Kenwood TS-790A is a multiband, multimode transceiver. It operates on the 144- to 148-MHz and 420- to 450-MHz bands. Operation on 1240 to 1300 MHz is possible with an optional module.

a mobile transceiver and external power supply, and use the same radio at home and in the car.

TRANSVERTERS

There are fewer radios available for VHF/UHF SSB and CW work than for FM. Much SSB and CW operation is done with a linear transverter used with a regular 10-meter HF

transceiver. A transverter converts signals from one band to another. For example, a 222- to 28-MHz transverter receives a signal at 222.1 MHz and converts it to 28.1 MHz for reception on an HF receiver. Likewise, the transverter accepts a 28.1-MHz signal from your HF transmitter and converts it for transmission on 222.1 MHz.

The transverter has no external controls. The transceiver operates exactly like it does on HF; the only difference is that you're transceiving on 222 MHz instead of 28. Most transverters have 10 to 25 watts RF output. Transverters are available for all VHF/UHF bands, as well as several microwave (above 1000 MHz) bands.

MULTIBAND AND MULTIMODE TRANSCEIVERS

Multiband transceivers combine more than one VHF/UHF band in one box. Optional modules for other bands may be available. For example, the unit shown in Figure 6-13 comes supplied with modules for 144 to 148 and 420 to 450 MHz. A separate 1270- to 1300-MHz module is available at extra cost. This rig has the additional ability to transmit on one band and receive on another, at the same time. Amateur Radio repeater satellites require users to transmit and receive on separate bands, which is easy if you have two rigs. With transceivers like this one, you need only one rig.

The transceiver shown earlier in Figure 6-11 is also a **multimode transceiver**. It can be used on FM and SSB voice, and also on CW. Not all multiband rigs are multimode, and some multimode rigs operate on only one band.

HAND-HELD TRANSCEIVERS

Hand-held VHF or UHF FM transceivers (often called HTs, an abbreviation for "Handie-Talkie," a Motorola trademark) are very popular. Such equipment is the most compact of the lot. See Figure 6-14. HTs are self-contained stations. They use battery power and have a built-in antenna, speaker and microphone.

Most hand-held units can be used with an external 12-V supply and an external antenna. Some offer accessory external speakers and microphones, which are handy for mobile or home-station use.

To reduce size and conserve battery power, HTs usually have lower power output than mobile or base-station rigs. A typical HT has 1 or 2 watts output; a high-power HT may produce 5-7 watts.

FM TRANSCEIVER FEATURES

VHF and UHF FM transceivers have many features not always found on HF rigs. While many of these features are not essential, they are useful. All FM transceivers have a *squelch* control that quiets the receiver when no signal is present. Squelch lets you leave your transceiver on without having to listen to noise when the frequency is not in use.

Figure 6-14—The Kenwood TH31BT and the Yaesu FT-109H 222-MHz FM hand-held transceivers. This type of rig is a complete portable station, with an autopatch tone pad and a flexible vertical antenna.

Most crystal-controlled transceivers require two crystals for each repeater or simplex channel. One crystal is for the transmitter (repeater input frequency) and one is for the receiver (repeater output frequency). For simplex operation, you need two crystals for the same frequency. Some hams find an inexpensive, used, crystal-controlled rig for a packet-radio station.

Synthesized transceivers offer almost unlimited frequency flexibility. Going from simplex to repeater operation is as easy as flipping a switch. The transceiver comes factory programmed for the proper repeater frequency split for the band. You dial up the repeater output frequency and the transceiver automatically switches to the repeater input frequency when you transmit.

It may be inconvenient to tune a synthesized radio all over the band to change frequency. This is true if you normally operate on only a few repeater or simplex frequencies, perhaps separated by hundreds of kilohertz. For added convenience, synthesized transceivers often have memories to store your favorite frequencies. You store them in the memory and recall them as needed. For example, memory 1 may be for the local repeater input on 222.32 MHz, memory 2 for the 223.5 MHz simplex frequency, and so on.

Synthesized transceivers sometimes offer automatic *scanning* up and down the band, or among repeater channels. This feature is handy if you're traveling and want to find an active repeater.

There are other FM transceiver features, sometimes standard and sometimes available as options. *Tone pads* for autopatch use and *tone-burst* or *subaudible-tone generators* for repeaters that require such tones for access are standard features on most new FM transceivers. (See Chapter 9 for more information on using tone pads and subaudible tones.) An autopatch provides a connection between the amateur station and the telephone line. A tone pad is a 12- or 16-button keypad used to generate audio tones similar to those used on TouchTone telephone systems. Some transceivers feature *wideband receivers* that let you listen to signals outside the amateur bands. These transceivers are useful for public-service work during emergencies.

FOR MORE INFORMATION

We've tried to give you a few guidelines in this chapter. There are lots of rigs out there, both new and used. Take your time and try to get as much information as you can. Talk to other hams and read the Product Reviews in *QST*. Advertisements in *QST* will tell you what's available. Remember that everyone has an opinion, though. What they like might not be what you like.

Most of all, don't worry! If you buy a used rig and you don't like it, you can probably sell it for almost as much as you paid for it. Start small and simple. Ham radio is fun; part of the fun is dreaming about new equipment and upgrading your station as you upgrade your license.

All newer transceivers use a synthesizer to control the operating frequency, while many older ones use VFOs or crystals. Synthesized and most VFO-type transceivers can operate on any frequency in the band. Crystal-controlled radios have a front-panel control to allow you to switch between two or more repeater or simplex channels.

If you live in an area with many repeaters, you'll want a synthesized transceiver. You should also consider this type if you plan to use repeaters in other areas, perhaps on business or vacation trips. If you add the cost of crystals (about $15 per channel) to the cost of a second-hand crystal-controlled transceiver, the total cost may be more than that of a synthesized transceiver.

KEY WORDS

Antenna—A device that picks up or sends out radio waves.

Balun—Contraction for balanced to unbalanced. A device to couple a balanced load to an unbalanced source, or vice versa.

Beam antenna—A directional antenna. A beam antenna must be rotated to provide coverage in different directions.

Characteristic impedance—The opposition to electric current that an antenna feed line presents. Impedance includes factors other than resistance, and applies to alternating currents. Ideally, the characteristic impedance of a feed line is the same as the transmitter output impedance and the antenna input impedance.

Coaxial cable—Coax (pronounced kó-aks). A type of feed line with one conductor inside the other.

Dipole antenna—See **Half-wave dipole**. A dipole need not be ½ wavelength long.

Directivity—The ability of an antenna to focus transmitter power into certain directions. Also its ability to enhance received signals from specific directions.

Director—An element in front of the driven element in a Yagi and some other directional antennas.

Driven element—The part of an antenna that connects directly to the feed line.

Feed line (feeder)—See **Transmission line**.

Gain—A measure of the **directivity** of an antenna, as compared with another antenna such as a dipole.

Half-wave dipole—A basic antenna used by radio amateurs. It consists of a length of wire or tubing, opened and fed at the center. The entire antenna is ½ wavelength long at the desired operating frequency.

Matching network—A device that matches one impedance level to another. For example, it may match the impedance of an antenna system to the impedance of a transmitter or receiver. Amateurs also call such devices a Transmatch, impedance-matching network or match box.

Omnidirectional—Antenna characteristic meaning it radiates equal power in all compass directions.

Open-wire feed line—Parallel-conductor feeder with air as its primary insulation material.

Parallel-conductor feed line—Feed line with two conductors held a constant distance apart.

Polarization—Describes the electrical-field characteristic of a radio wave. An antenna that is parallel to the surface of the earth, such as a dipole, produces horizontally polarized waves. One that is perpendicular to the earth's surface, such as a quarter-wave vertical, produces vertically polarized waves.

Quarter-wavelength vertical antenna—An antenna constructed of a quarter-wavelength long radiating element placed perpendicular to the earth.

Reflector—An element behind the driven element in a Yagi and some other directional antennas.

Resonant frequency—The desired operating frequency of a tuned circuit. In an antenna, the resonant frequency is one where the feed-point impedance contains only resistance.

RF burn—A flesh burn caused by exposure to a strong field of radio-frequency (RF) energy.

Standing-wave ratio (SWR)—Sometimes called VSWR. A measure of the impedance match between the feed line and the antenna. Also, with a Transmatch in use, a measure of the match between the feed line from the transmitter and the antenna system. The system includes the Transmatch and the line to the antenna. VSWR is the ratio of maximum voltage to minimum voltage along the feed line. Also the ratio of antenna impedance to feed-line impedance when the antenna is a purely resistive load.

SWR meter—A device used to measure SWR.

Transmission line—The wires or cable used to connect a transmitter or receiver to an antenna.

Vertical antenna—A common amateur antenna, usually made of metal tubing. The radiating element is vertical. There are usually four or more radial elements parallel to or on the ground.

Wavelength—Often abbreviated λ. The distance a radio wave travels in one RF cycle. The wavelength relates to frequency. Higher frequencies have shorter wavelengths.

Yagi antenna—The most popular type of amateur directional (beam) antenna. It has one driven element and one or more additional elements.

Chapter 7

Choosing an Antenna

The "antenna farm" at W1AW (ARRL Headquarters) is a bit bigger than you'll need as a Novice or Technician. Large beam antennas on tall towers almost always ensure the best communications, however.

We know that a transmitter generates radio-frequency energy. We convert this electrical energy into radio waves with an **antenna**. An antenna may be just a piece of wire or other conductor designed to *radiate* the energy. An antenna converts current into an electromagnetic field (radio waves). The radio waves spread out or *propagate* from the antenna. You might relate their travel to the ever-expanding waves you get when you drop a pebble in water. Waves from an antenna radiate in all directions, though, not just in a flat plane.

It also works the other way. When a radio wave crosses an antenna, it generates a voltage in the antenna. That voltage isn't very strong, but it's enough to create a small current. That current travels through the **transmission line** to the receiver. The receiver detects the radio signal. In short, the antenna converts electrical energy to radio waves and radio waves to electrical energy. This process makes two-way radio communication possible with just one antenna.

Your success in making contacts depends heavily on your antenna. A good antenna can make a fair receiver seem like a champ. It can also make a few watts sound like a whole lot more. Remember, you'll normally use the same antenna to transmit *and* receive. Any improvements to your antenna make your transmitted signal stronger, and increase the strength of the signals you receive.

Assembling an antenna system gives you a chance to be creative. You may discover, for example, that property size or landlord restrictions rule out a traditional antenna. If so, you can innovate. *Low Profile Amateur Radio* is an ARRL publication full of ideas for just these conditions. *The ARRL*

Antenna Book, *The ARRL Antenna Compendiums*, and similar publications also offer lots of antenna suggestions. This chapter contains a few suggestions, too.

Some antennas work better than others. Antenna design and construction have kept radio amateurs busy since the days of Marconi. You'll probably experiment with different antenna types over the years. Putting up a better antenna is an inexpensive but rewarding way to improve your station.

WAVELENGTHS

We sometimes talk about antenna lengths in wavelengths. A **wavelength** relates to the operating frequency. When you construct an antenna for one particular amateur band, you cut it to the proper wavelength. Often you'll see the Greek letter lambda (λ) used as an abbreviation for wavelength. For example, ½ λ means "one-half wavelength."

Most popular ham antennas are less than 1 λ long. (A very popular antenna is a ½-λ **dipole antenna**. You'll learn how to build one later in this chapter.) There is a simple relationship between operating frequency and wavelength. The wavelength is shorter at the higher frequencies. Wavelength is longer at the lower frequencies.

If numbers interest you, use this equation to find the wavelength for a specific frequency.

$$\lambda \text{ (in feet)} = \frac{984}{f \text{(in MHz)}} \qquad \text{(Eq 7-1)}$$

This equation gives the wavelength in feet. The frequency is given in megahertz (MHz). Let's say we wanted to know the wavelength for 7.15 MHz. We divide 984 by 7.15, and the answer is about 137.6 feet.

$$\lambda \text{ (in feet)} = \frac{984}{f \text{(in MHz)}} = \frac{984}{7.15} = 137.6$$

Whenever we talk about an antenna, we specify its design frequency or the amateur band it covers. We could talk about an "80-meter dipole," for example, as an antenna intended for operation in the 80-meter band. Antennas are tuned circuits. A simple antenna such as a dipole or a ¼-λ vertical has a resonant **frequency**. Such antennas do best at their resonant frequency, as do most other tuned circuits.

To change the resonant frequency of a tuned circuit, you vary the capacitor value or the inductor value. You can change the resonant frequency of an antenna by changing its length, which affects its capacitance *and* inductance.

FEED LINES

To get RF energy from your transmitter to an antenna you use **transmission line**. A transmission line is a special cable or arrangement of wires. Such lines commonly go by the name **feed line**, or **feeders** for short. They feed power to the antenna, or feed a received signal from the antenna to the receiver.

CHARACTERISTIC IMPEDANCE

One electrical property of a feed line is its **characteristic impedance**. In Chapter 3 you learned that resistance is an opposition to electric current. Impedance is another form of resistance to electric current. Impedance includes factors other than resistance, however.

The spacing between line conductors and the type of insulating material determines the characteristic impedance. Characteristic impedance is important because we want the feed line to take all the transmitter power and feed it to the antenna. For this to occur, the transmitter (source) must have the same impedance as its load (the feed line). In turn, the feed line must have the same impedance as its load (the antenna).

We can use special circuits called **matching networks** if any of these impedances are different. *Network* just refers to a combination of inductors and capacitors to form a special

circuit. Still, careful selection of a feed line can minimize such matching problems.

COAXIAL CABLE

Several different feeder types are available for amateur use. The most common is **coaxial cable**. Called "coax" (pronounced kó-aks) for short, this feed line has one conductor inside the other. It's like a wire inside a flexible tube. The center conductor is surrounded by insulation, and the insulation is surrounded by a wire braid called the shield. The whole cable is then encased in a tough vinyl outer coating, which makes the cable weatherproof. See Figure 7-1. Coax comes in different sizes, with different electrical properties. Figure 7-1 also shows other types of coaxial cables used by amateurs.

The most common types of coax have either a 50-ohm or 72-ohm **characteristic impedance**. Coax designated RG-58, RG-8 and RG-213 are 50-ohm cables. Some coax designations may also include a suffix such as /U, A/U or B/U, or bear the label "polyfoam." Feed line of this type may be used with most antennas. Cables labeled RG-59 or RG-11 are 72-ohm lines. Many hams use these types to feed **dipole antennas**.

Figure 7-1—Types of coaxial cable used by amateurs. Abbreviated coax, it has a center conductor surrounded by insulation. The second conductor, called the shield, goes around that. Plastic insulation covers the entire cable.

The impedance of a ½-λ dipole far from other objects is about 73 ohms. Practical dipoles placed close to the earth, trees, buildings, etc, have an input impedance closer to 50 ohms. In any case, the small impedance mismatch caused by using 50- or 72-ohm cable as an antenna feeder is unimportant.

In choosing the feed line for your installation, you'll have a trade-off between electrical characteristics and physical properties. The RG-58 and RG-59 types of cable are about ¼ inch in diameter, comparatively lightweight, and reasonably flexible. RG-8, RG-213 and RG-11 are about ½ inch in diameter, nearly three times heavier, and considerably less

flexible. As far as operation goes, RG-8, RG-213 and RG-11 will handle much more power than RG-58 and RG-59.

Any line that feeds an antenna absorbs a small amount of transmitter power. That power is lost, because it serves no useful purpose. (The lost power warms the feed line slightly.) The loss occurs because the wires are not perfect conductors, and the insulating material is not a perfect insulator. Signal loss also increases slightly when the SWR is greater than 1:1, so we try to keep the SWR below 2:1 if possible.

Better-quality coaxial cables have lower loss than poor-quality cables. More of the transmitter power is lost as heat in a poor-quality coaxial cable. You usually get better quality coax if you stick with name brands. You can also look at the shield braid on the coax. Better-quality cables have more complete coverage of the center insulator. If you can easily see through the holes in the braid, you should probably select another cable.

The larger coax types, RG-8, RG-213 and RG-11, have less signal loss than the smaller types. If your feed line is less than 100 feet long, you probably won't notice the small additional signal loss, at least on the HF bands. This, combined with light weight and flexibility, is why many HF operators find the smaller coax better suited to their needs. Also, the smaller feed line costs about half as much per foot as the larger types.

On the VHF/UHF bands, however, you will find the losses in RG-58 and RG-59 more noticeable, especially if your feed line is longer than about 50 feet. On these bands, most amateurs use RG-213 coax or even lower-loss special coaxial cables.

Coaxial cable has several advantages as a feed line. It is readily available, and is resistant to weather. Most common amateur antennas have characteristic impedances of about 50 ohms. Coax can be buried in the ground if necessary. It can be bent, coiled and run next to metal with little effect. Its major drawback is its cost.

PARALLEL-CONDUCTOR FEED LINE

Parallel-conductor feed line is another popular line type for use below 30 MHz. The most familiar example of this feeder is the 300-ohm ribbon used for TV antennas. It has two parallel conductors encased along the edges of a strip of plastic insulation. We often call this kind of line *twin lead*. See Figure 7-2A.

Open-wire feed line is a third type of parallel-conductor line. It contains two wires separated by plastic spacer rods. There is a rod every few inches along the feeder to maintain a uniform wire separation. The primary insulation is air. See Figure 7-2B. Often called **ladder line**, this type usually has a characteristic impedance between 450 and 600 ohms. The conductors can be bare wire, or they might be insulated with plastic. Ladder line can handle much higher power than twin lead.

You can make ladder line yourself. It is also available commercially, but may be difficult to find. A near equivalent is a cross between twin lead and ladder line. The wires are

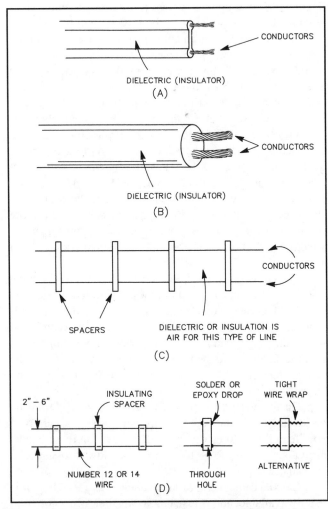

Figure 7-2—Parallel-conductor feed lines. At A, common 300-ohm TV twin lead. B shows 72-ohm transmitting twin lead, designed to carry higher power levels. C shows twin lead that uses air as the dielectric, with insulating spacers every few inches. Ladder line is its common name. D shows the construction of home-built ladder line.

encased along the edges of a plastic ribbon, but the ribbon is not solid. Instead, it has punched rectangular holes along its length, leaving air as the primary insulation material. The common variety of this line has a 450-ohm impedance. For the same conductor spacing, this line has slightly more loss than ladder line. For most amateur work, the difference is negligible, however.

Parallel-conductor lines have some disadvantages. For example, they cannot be coiled, or run next to metal drain pipes and gutters without adverse effects. Another drawback to these lines is their characteristic impedance of 300 ohms or higher. It cannot be connected directly to most transmitters. Thus, you need an impedance-**matching network** if you use any type of parallel-conductor feed line. Connect such a network between your transmitter and your feed line. If you use a matching network, even inexpensive TV twin lead can be used as your feed line. The need for a matching network is the main reason parallel-conductor lines are not used on VHF/UHF. Matching networks are more difficult to build for these frequencies, so most amateurs use coax instead.

The major advantage of ladder line is its very low loss. This means that for the same feeder length, more of your transmitter power will get to your antenna.

It is even possible to use a single wire from your rig to the antenna. We call this a *single-wire feed line*. With this feed arrangement, the feeder often becomes a part of the antenna. It can radiate outdoors, of course, but it can also radiate right in the shack. The impedance of this type of antenna system is seldom within the tuning range of most transmitters. You will need an impedance-matching network to connect between the rig and the feed line.

Many new amateurs worry needlessly about feed-line length. If you use coaxial cable between a rig and a dipole or vertical antenna, the cable can be any reasonable length. A good length to start with is simply the distance between the rig and the antenna! The feed line cannot be any shorter, and any extra length is probably unnecessary.

[Now study the questions in Chapter 12 numbered N9C01 through N9C09. Review this section if you have difficulty with any of the questions.]

[**If you are studying for the Technician exam**, you should also study question T9B09 in Chapter 13. Review this section if you have difficulty with that question.]

IMPEDANCE-MATCHING NETWORKS

Your transmitter won't operate very well if connected to a mismatched feed line. Let's say you want to use parallel-conductor feed lines on your dipoles. Your transmitter probably has an output circuit designed for a 50-ohm load. But that's not what the load will be with this antenna system. An impedance-**matching network** will provide the proper impedance correction. For some mismatches, a suitable net-

work might contain only an inductor and a capacitor.

A *Transmatch* is a special type of matching network. Transmatches contain variable matching components (inductors and capacitors), and often a band switch. They offer the flexibility of matching a wide range of impedances over a wide frequency range. With a Transmatch, it is possible to use one antenna on several bands. For example, you might

Figure 7-3—With a Transmatch you have flexibility in designing your antenna system. You can use the antenna on several bands, and the length isn't critical. You can use a dipole fed by coax (A) or twin lead (B). With a Transmatch, the dipole legs can be any length, although they should be as long as possible.

Figure 7-4—At A, the schematic diagram for a versatile Transmatch circuit. B shows a homemade Transmatch constructed from the circuit of A.

use a center-fed wire antenna with inexpensive 300-ohm twin lead. Each band will use its own Transmatch settings: one combination for 80, one for 40, one for 15, and one for 10 meters.

Connect the Transmatch between the antenna and the **SWR meter**, as Figure 7-3 shows. (An SWR meter measures the impedance mismatch on the feed line between the two

pieces of equipment it is connected to.) Adjust the controls on your Transmatch for minimum SWR. Don't worry if you can't achieve a perfect match (1:1). Anything lower than 2:1 will work just fine.

Figure 7-4A shows the schematic diagram for a versatile Transmatch circuit. Part B shows a homemade Transmatch built from this circuit.

BALUNS

A center-fed wire with open ends, such as a dipole, is a *balanced* antenna. In a balanced center-fed antenna, the current flowing into one half of the antenna is equal to the current in the other half. The two currents are also opposite in *phase*. (*Phase* refers to the relative positions of two points on a wave, or on two different waves at a particular instant of time. If the two currents are opposite in phase, one is in the positive half of the cycle when the other is in the negative half cycle.) You can think of a balanced antenna as one where neither side connects to ground. A balanced antenna is balanced with respect to ground.

If we feed the antenna with a parallel-conductor line, we keep balance throughout the system. Parallel-conductor feed line is balanced: Neither side connects to ground. You can connect this feed line directly to a dipole antenna.

If we feed a dipole at the center with coax, we upset the system balance. One side of the antenna connects to the coax inner, or center, conductor. The other side connects to the coax shield. The shield connects to ground, so coax is an *unbalanced line*. This unbalanced condition may allow some antenna current to flow down the outside of the coax braid from the antenna. This can lead to several antenna problems, and should be avoided.

You don't have to go searching for 72-ohm twin lead to have a balanced dipole, however. You can feed it with ordinary 300-ohm twin lead. (To do this, you'll also need a Transmatch at the transmitter end of the line.) Another way is to feed your antenna with coaxial line and use a device called a **balun**. Balun is a contraction for *bal*anced to *un*balanced. You install the balun at the antenna feed point.

Different types of baluns are commercially available, or you can make your own. A common type is a balun transformer, which uses wires wound on a toroidal core. Besides providing a balance, these baluns can transform an impedance, such as from 50 to 75 ohms. Another type is a bead balun. Several ferrite beads go over the outside of the coax, one after the other. The beads tend to choke off any RF current that might otherwise flow on the outside of the shield.

You can easily make another type of choke balun from the coax itself. At the antenna feed point, coil up 10 turns of coax into a roll about 6 inches in diameter. Tape the coax turns together. The inductance of the coiled turns tends to choke off RF currents on the shield.

Feeding a dipole with parallel-conductor line doesn't assure a balanced antenna. You must also consider how the line connects to the transmitter. Most transmitters have a coaxial-style connector. Suppose you are using 450-ohm parallel-conductor feed line. To feed such a balanced antenna system, you would need to install a balun at the transmitter end of the line (or use a special balanced matching network). Unless you use a balanced matching network, you need a balun at the transmitter end of the line if you are using any type of parallel-conductor feed line.

[Study questions in Chapter 12 numbered N9C10 and N9C11. Review this section if you have difficulty answering these questions.]

[**If you are studying for the Technician exam**, also turn to Chapter 13 and study questions T9B10 and T9B11. Review this section if you have any difficulty.]

STANDING-WAVE RATIO (SWR)

If an antenna system does not match the characteristic impedance of the transmitter, some of the transmitter energy is reflected from the antenna. The power traveling from the transmitter to the antenna is called *forward power*. When that power reaches the antenna in an unmatched system, some of the power is reflected back down the feed line toward the transmitter. Some of the power is also radiated from the antenna, which is what you want to happen. The power that returns to the transmitter from the antenna is called *reflected power*.

The forward power and the reflected power passing each other on the feed line cause voltage standing waves on the line. When this happens, the RF voltage and current are not uniform along the line. An **SWR meter** measures the voltage **standing-wave ratio**. This is the ratio of the maximum voltage on the line to the minimum voltage. (These two points

will always be ¼ λ apart.) Lower SWR values mean a better impedance match exists between the transmitter and the antenna system. If a perfect match exists, the SWR is 1:1. Your SWR meter thus gives a relative measure of how well the antenna system impedance matches that of the transmitter.

Modern transmitters are designed to match 52-ohm coaxial lines and antennas. Most commercial antennas are designed to have nearly the same characteristic impedance *when properly adjusted*. So, if your SWR is not close to 1:1, it means your antenna is not properly adjusted for the frequency you are using. We adjust the antenna for minimum SWR (not always 1:1) somewhere in the middle of the band of interest. On other frequencies the SWR may be higher. If the antenna is properly assembled, an SWR of 2:1 or less is probably all right.

If you are using a matching network, you can probably

NOT FOR TECHNICIANS ONLY

SWR: What Does It Mean?

You already know that SWR is defined as the ratio of the maximum voltage to the minimum voltage in the standing wave:

$$SWR = \frac{E_{max}}{E_{min}}$$

An SWR of 1:1 means you have no reflected power. The transmission line is said to be "flat." If the load is completely resistive (no reactance), then the SWR can be calculated. Divide the line characteristic impedance by the load resistance, or vice versa. Use whichever gives a value greater than one:

$$SWR = \frac{Z_0}{R} \text{ or } SWR = \frac{R}{Z_0}$$

where
Z_0 = characteristic impedance of the transmission line
R = load resistance (not reactance)

For example, if you feed a 100-ohm antenna with 50-ohm transmission line, the SWR is 100/50 or 2:1. Similarly, if the impedance of the antenna is 25 ohms the SWR is 50/25 or 2:1.

When a high SWR exists, losses in the feed line are increased. This is because of the multiple reflections from the antenna and transmitter. Each time the transmitted power has to travel up and down the feed line, a little more energy is lost as heat.

This effect is not so great as some people believe, however. Some line losses are less than 2 dB (such as for 100 feet of RG-213 or RG-58 cables up to about 30 MHz).

The standing-wave ratio is the ratio of the voltage amplitude at point A to the voltage amplitude at point B, or the ratio of the current amplitude at point B to the current amplitude of point A.

The SWR would have to be greater than 3:1 to add an extra decibel of loss because of the SWR. (A high SWR will cause the output power to drop drastically with many solid-state transceivers. The drop in power is caused by internal circuits that sense the high SWR and automatically reduce output power to protect the transceiver.)

A transmission line should be terminated in a resistance equal to its characteristic impedance. Then maximum power is delivered to the antenna, and transmission line losses are minimized. This would be an ideal condition. Such a perfect match is seldom realized in a practical antenna system.

adjust it so the SWR meter reads 1:1. The SWR on the transmission line between the tuner and the antenna will, however, be different. A matching network can cover up the mismatch, but it does not eliminate it.

SWR METERS

The most common **SWR meter** application is tuning an antenna to resonate on the frequency you want to use. (This discussion applies if you connect the feeder directly to the transmitter output, with no Transmatch.)

An SWR reading of 2:1 or less is quite acceptable. A reading of 4:1 or more is unacceptable. This means there is a serious mismatch between your antenna and your feeder.

To use the SWR meter, you transmit through it. (You must have a license to operate a transmitter! What if your license has not arrived by the time you are ready to test your antenna? Just invite a licensed ham over to operate the transmitter).

How you measure the SWR depends on your type of meter. Some SWR meters have a SENSITIVITY control and a FORWARD-REFLECTED switch. If so, the meter scale usually

gives you a direct SWR reading. To use the meter, first put the switch in the FORWARD position. Then adjust the SENSITIVITY control and the transmitter power output until the meter reads full scale. Some meters have a mark on the meter face labeled SET or CAL. The meter pointer should rest on this mark. Next, set the selector switch to the REFLECTED position. Do this without readjusting the transmitter power or the meter SENSITIVITY control. Now the meter pointer shows you the SWR value.

Most SWR meters are designed for operation on the high-frequency (HF) bands. You may be able to use your SWR meter on the VHF bands if you are able to adjust the meter for a full-scale reading in the SET or CAL position. The readings may not be as accurate as the readings you obtain on the HF bands, but it may give an indication of the impedance match to your VHF antenna.

You can also measure SWR with a wattmeter, which measures RF power in watts. Your wattmeter may have meters to read both forward and reflected power. If not, it should have a switch or another way to change from forward to reflected power readings. To compute the SWR with a wattmeter, first note both the forward and reflected power for a given transmitter setting. Then consult a graph provided with the meter to find the corresponding SWR.

Most RF wattmeters are accurate over a limited frequency range. A wattmeter designed for HF operation (3 to 30 MHz) probably will not be accurate at VHF or UHF.

Find the **resonant frequency** of an antenna by connecting the meter between the feed line and your antenna. This technique will measure the relative impedance match between your antenna and its feed line. Measure the SWR at different frequencies across the band. Ideally, you will measure the lowest SWR at the center of the band, with higher readings at each end.

Sometimes it isn't practical to put your SWR meter at the antenna feed point, between the feed line and the antenna. Most hams just put the meter in their shack, at the transmitter, and make measurements there. You should realize that this is a compromise, however. You are measuring the relative impedance match between your transmitter and the *antenna system*, which includes the feed line.

If you are using a Transmatch with your antenna system, you should place the SWR meter between the transmitter and the Transmatch. The meter then indicates when you have adjusted the Transmatch to provide the best impedance match to your transmitter.

Sometimes your antenna may resonate far off frequency. See Figure 7-5. In this situation you will not get a "dip" in SWR readings with frequency. Instead, your readings will increase as you change frequency from one end of the band to the other. For example, you might read 2.5:1 at the low-frequency band edge, and 5:1 at the high-frequency end. This means antenna resonance is closer to the low-frequency end of the band than the high. It also means that resonance is below the low-frequency band edge. For a dipole or vertical antenna, this condition exists when the antenna is too long. Trimming the length will correct the problem.

Suppose the readings were 5:1 at the low-frequency end of the band and 2.5:1 at the high end. Here, the antenna is too short. Adding to the antenna length will correct this problem. Adjusting the antenna length for resonance in this way is what we call *tuning the antenna*.

This method works for dipoles or vertical antennas. It does not show antenna resonance if you have a matching network between the SWR meter and the antenna. Nor does it show resonance for antenna systems that include a matching network at the antenna. An SWR meter *will* show when you have adjusted the matching network properly, however. Use the settings that give you the lowest SWR at your preferred operating frequency.

FINDING ANTENNA PROBLEMS WITH AN SWR METER

Sometimes problems occur when you first install an antenna, or only after the weather batters your antenna for weeks, months or years.

It's handy to have an SWR meter or power meter to help diagnose antenna problems. This section tells how to inter-

Figure 7-5—To determine if your antenna is cut to the right length, measure the SWR at different points in the band. Plot these values and draw a curve. Use the curve to see if the antenna is too long or too short. Adjusting the length will bring the lowest SWR to the desired frequency. Here the SWR is higher at the low end of the band. This antenna needs to be longer.

pret SWR meter readings to solve specific problems. This information applies to any type of antenna. We assume the antenna you're using normally provides a good match to your feed line at the measurement frequency.

One common problem is a loose connection where the feed line from your station attaches to the antenna wire. Splices or joints are another possible failure point. Your SWR meter will tell if the problem you're experiencing is a poor connection somewhere in the antenna. Observe the SWR reading. It should remain constant. If it is erratic, fluctuating markedly, chances are you have a loose connection. This problem is very easy to see on windy days.

If your SWR reading is unusually high, greater than 10:1 or so, you probably have a worse problem. *Caution:* Do not operate your transmitter with a very high SWR any longer than it takes to read the SWR! The problem could be an open connection or a short circuit. The most likely failure point is at the antenna feed point. The problem might be at the connector attaching your feed line to your transmitter. Carefully check your connections and your feed line for damage. You can also get unusually high SWR readings if the antenna is far from the correct length. This would happen if you try to operate your antenna on the wrong amateur band!

Most hams leave an SWR meter in the line all the time. Any sudden changes in the SWR mean you have a problem, such as a broken wire or bad connection.

[Now turn to Chapter 12 and study questions N4C01 through N4C11. Review this section if you have difficulty with any of these questions.]

[**If you are preparing for the Technician exam,** also study questions T9B06, T9B07 and T9B08 in Chapter 13. Review this section if these questions give you any difficulty.]

PRACTICAL ANTENNAS

Hams use many different kinds of antennas. There is no one best kind. Beginners usually prefer simpler, less expensive types. Some hams with more experience have antenna systems that cost thousands of dollars. Others have antennas that use several acres of property!

THE HALF-WAVE DIPOLE ANTENNA

Probably the most common amateur antenna is a wire cut to ½ λ at the operating frequency. The feed line attaches across an insulator at the center of the wire. This is the **half-wave dipole**. We often refer to an antenna like this as a **dipole antenna**. (*Di* means two, so a dipole has two equal parts. A dipole could be a length other than ½ λ.) The total length of a half-wavelength dipole is ½ λ. The feed line connects to the center. This means that each side of this dipole is ¼-λ long.

Use Equation 7-2 to find the total length of a ½-λ dipole for a specific frequency. Notice that the frequency is given in megahertz and the antenna length is in feet for this equation.

$$\text{Length (in feet)} = \frac{468}{f(\text{MHz})} \qquad \text{(Eq 7-2)}$$

Equation 7-2 gives us the following approximate lengths for ½-λ dipoles.

Wavelength	Frequency	Length
80 meters	3.725 MHz	125.6 feet
40 meters	7.125 MHz	66 feet
15 meters	21.125 MHz	22 feet
10 meters	28.150 MHz	16.6 feet
10 meters	28.475 MHz	16.4 feet
2 meters	146.0 MHz	3.25 feet
1.25 meters	223 MHz	2.1 feet = 25 inches

Equation 7-2 is only accurate for frequencies up to around 30 MHz. One reason for this frequency limitation is that the element diameter is a larger percentage of the wavelength at VHF and higher frequencies. Another reason

is because of effects commonly known as *end effects*. So the values given here for 2 and 1.25 meters are only for comparison purposes. These lengths might serve as a starting point for building antennas for 2 meters or 1.25 meters, but they will probably be too long.

Figure 7-6A shows the construction of a basic ½-λ dipole antenna. Parts B through D show enlarged views of how to attach the insulators. You can use just about any kind of copper or copper-clad steel wire for your dipole. Most hardware or electrical supply stores carry suitable wire. Ordinary house wire or two-conductor small-appliance wire (called "zip cord"), can work well. Avoid using zip cord on 40 and 80 meters, however.

House wire and stranded wire will stretch with time, so a heavy gauge copper-clad steel wire is best. This wire consists of a copper jacket over a steel core. Such construction provides the strength of steel combined with the excellent conducting properties of copper. You can sometimes find copper-clad steel wire at a radio store. This wire is used for electric fences to keep farm animals in their place, so another place to try is a farm supply store.

Remember, you want a good conductor for the antenna, but the wire must also be strong. The wire must support itself *and* the weight of the feed line connected at the center.

We use wire gauge to rate wire size. Larger gauge numbers represent smaller wire diameters. Conversely, smaller gauge numbers represent larger wire diameters. Although you can make a dipole antenna from almost any size wire, 12 or 14 gauge is usually best.

Cut your dipole according to the dimension found by Equation 7-2, but leave a little extra length to wrap the ends around the insulators. You'll need a feed line to connect it to your transmitter. For the reasons mentioned earlier, the most popular feed line for use with dipole antennas is coaxial cable. When you shop for coax, look for some with a heavy braided shield. If possible, get good quality cable that has at least 95 percent shielding. If you stick with name brand cable, you'll

Figure 7-6—Simple half-wave dipole antenna construction. B and C show how to connect the wire ends to various insulator types. D shows the feed-line connection at the center.

get a good quality feed line. Figure 7-7 shows the steps required to prepare the cable end for attachment to the antenna wires at the center insulator.

The final items you'll need for your dipole are three insulators. You can purchase them from your local radio or hardware store (Figure 7-8). You can also make your own insulators from plastic or Teflon blocks. See Figure 7-9. One insulator goes on each end and another holds the two wires together in the center. Figure 7-10 shows some examples of how the feed line can attach to the antenna wires at a center insulator.

Dipole antennas send radio energy best in a direction that is 90° to the antenna wire. For example, suppose you install a dipole antenna so the ends of the wire run in an east/west direction. This antenna would send stronger signals in north and south directions. A dipole also sends radio energy straight up and straight down. Of course the dipole also sends some energy in directions off the ends of the wire, but these signals won't be as strong. So you will be able to contact stations to the east and west with this antenna, but you may find that the signals are stronger with stations to the north and south.

[Now turn to Chapter 12 and study questions N9A01, N9A03, N9A04 and N9A07. Also study question N9B09. Review this section if you need to.]

ANTENNA LOCATION

Once you have assembled your dipole, find a good place

to put it. *Never* put your antenna under, or over the top of electrical power lines. If they ever come into contact with your antenna, you could be electrocuted. Avoid running your antenna parallel to power lines that come close to your station. Otherwise you may receive unwanted electrical noise. Sometimes power-line noise can cover up all but the strongest signals your receiver hears. You'll also want to avoid running your antenna too close to metal objects. These could be rain gutters, metal beams, metal siding, or even electrical wiring in the attic of your house. Metal objects tend to shield your antenna, reducing its capability.

The key to good dipole operation is height. How high? One wavelength above ground is good, and this ranges from about 35 feet on 10 meters to about 240 feet on 80 meters. On the 2-meter band, one wavelength is only about 7 feet. You should still try to install the antenna higher, to get it clear of buildings and trees. Of course very few people can get their antennas 240 feet in the air, so 40 to 60 feet is a good average height for an 80-meter dipole. Don't despair if you can get your antenna up only 20 feet or so, though. Low antennas can work well. Generally, the higher above ground and surrounding objects you can get your antenna, the greater the success you'll have. You'll find this to be true even if you can get only part of your antenna up high.

Normally you will support the dipole at both ends. The supports can be trees, buildings, poles or anything else high enough. Sometimes, however, there is just no way you can

Figure 7-7—Preparing coaxial cable for connection to antenna wire. A—Remove the outer insulation with a sharp knife or wire stripper. If you nick the braid, start over. B—Push the braid in accordion fashion against the outer jacket. C—Spread the shield strands at the point where the outer insulation ends. D—Fish the center conductor through the opening in the braid. Now strip the center conductor insulation back far enough to make the connection and tin (flow solder onto) both center conductor and shield. Be careful not to use too much solder, which will make the conductors inflexible. Also be careful not to apply too much heat, or you will melt the insulation. A pair of pliers used as a heat sink will help. The outer jacket removed in step A can be slipped over the braid as an insulator, if necessary. Be sure to slide it onto the braid before soldering the leads to the antenna wires.

Figure 7-8—Various commercially made antenna insulators.

Figure 7-9—Some ideas for homemade antenna insulators.

put your dipole high in the air at both ends. If you're faced with this problem, you have two reasonably good alternatives. You can support your dipole in the middle or at one end.

If you choose to support the antenna in the middle, both ends will droop toward the ground. This antenna, known as an *inverted-V dipole*, works best when the angle between the wires is no less than 90°. See Figure 7-11. If you use an inverted-V dipole, make sure the ends are high enough that no one can touch them. When you transmit, the high voltages present at the ends of a dipole can cause an **RF burn**. Yes, radio energy can burn your skin.

If you support your antenna at only one end, you'll have what is known as a *sloper*. This antenna also works well. As with the inverted-V dipole, be sure the low end is high enough to prevent anyone from touching it.

If you don't have the room to install a dipole in the standard form, don't be afraid to experiment a little. You can get away with bending the ends to fit your property, or even

Figure 7-10—Some dipole center insulators have connectors for easy feeder removal. Others have a direct solder connection to the feed line.

Figure 7-11—These are simple but effective wire antennas. A shows a horizontal dipole. The legs can be drooped to form an inverted-V dipole, as at B. C illustrates a sloping dipole (sloper). The feed line should come away from the sloper at a 90° angle for best results. If the supporting mast is metal, the antenna will have some directivity in the direction of the slope.

making a horizontal V-shaped antenna. Many hams have enjoyed countless hours of successful operating with antennas bent in a variety of shapes and angles.

On the 6- and 2-meter bands, dipoles for FM or packet operation work much better if they are installed vertically. Now you need only one support. The coax should come away from the antenna at a right angle for as far as possible, so it doesn't interfere with the radiation from the antenna. Dipoles are not often used on frequencies above the 2-meter band, as other, simpler, antennas work better. We describe one such antenna later in this chapter.

ANTENNA INSTALLATION

After you've built your antenna and chosen its location, how do you get it up? There are many schools of thought on putting up antennas. Can you support at least one end of your antenna on a mast, tower, building or in an easily climbed tree? If so, you have solved some of your problems. Unfortunately, this is not always the case. Hams use several methods to get antenna support ropes into trees. Most methods involve a weight attached to a rope or line. You might be able to tie a rope around a rock and throw it over the intended support. This method works for low antennas. Even a major league pitcher, however, would have trouble getting an antenna much higher than 40 feet with this method.

A better method is to use a bow and arrow, a fishing rod or even a slingshot to launch the weight and rope. See Figures 7-12 and 7-13. You'll find that strong, lightweight fishing line is the best line to attach to the weight. (Lead fishing weights are a good choice.) Regular rope is too heavy to shoot any great distance. When you have successfully cleared the supporting tree, remove the weight. Then tie the support rope to the fishing line and reel it in.

If your first attempt doesn't go over the limb you were hoping for, try again. Don't just reel in the line, however. Let the weight down to the ground first and take it off the line. Then you can reel the line in without getting the weight tangled in the branches.

Figure 7-12—There are many ways to get an antenna support rope into a tree. These hams use a bow and arrow to shoot a lightweight fishing line over the desired branch. Then they attach the support rope to the fishing line and pull it up into the tree.

Figure 7-13—Another method for getting an antenna support into a tree. Small hose clamps attach a casting reel to the wrist bracket of a slingshot. Monofilament fishing line attached to a 1-ounce sinker is easily shot over almost any tree. Remove the sinker and rewind the line for repeated shots. When you find a suitable path through the tree, use the fishing line to pull a heavier line over the tree.

You can put antenna supports in trees 120 feet and higher with this method. As with any type of marksmanship, make sure all is clear downrange before shooting. Your neighbors will not appreciate stray arrows, sinkers or rocks falling in their yards!

When your support ropes are in place, attach them to the ends of the dipole and haul it up. Pull the dipole reasonably tight, but not so tight that it is under a lot of strain. Tie the ends off so they are out of reach of passersby. Be sure to leave enough rope so you can let the dipole down temporarily if necessary.

Just one more step and your antenna installation is complete. After routing the coaxial cable to your station, cut it to length and install the proper connector for your rig. Usually this connector will be a PL-259, sometimes called a UHF connector. Figure 7-14 shows how to attach one of these fittings to RG-8 or RG-11 cable. Follow the step-by-step instructions exactly as illustrated and you should have no trouble. Be sure to place the coupling ring on the cable *before* you install the connector body! If you are using RG-58 or RG-59 cable, use an adapter to fit the cable to the connector. Figure 7-15 illustrates the steps for installing the connector

Figure 7-14—The PL-259 or UHF connector is almost universal for amateur HF work. It is also popular for equipment operating in the VHF range. Steps A through E illustrate how to install the connector properly. Despite its name, the UHF connector is rarely used on frequencies above 225 MHz.

1) Cut end of cable even. Remove vinyl jacket 3/4"—don't nick braid. Slide coupling ring and adapter on cable.

2) Fan braid slightly and fold back over cable.

3) Position adapter to dimension shown. Press braid down over body of adapter and trim to 3/8". Bare 5/8" of center conductor. Tin exposed center conductor.

4) Screw the plug assembly on adapter. Solder braid to shell through solder holes. Solder conductor to contact sleeve.

5) Screw coupling ring on assembly.

Figure 7-15—If you use RG-58 or RG-59 with a PL-259 connector, you should use an adapter, as shown here. Thanks to Amphenol Electronic Components, RF Division, Bunker Ramo Corp, for this information.

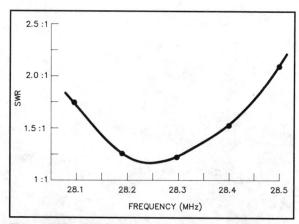

Figure 7-16—This graph shows how the SWR might vary across an amateur frequency band. The point of lowest SWR here is near the center of the band, so no further antenna-length adjustments are necessary.

with adapter. The PL-259 is standard on most rigs. If you require another kind of connector, consult your radio instruction manual or *The ARRL Handbook for Radio Amateurs* for installation information.

Tuning the Antenna

When you build an antenna, you cut it to the length given by an equation. This length is just a first approximation. Nearby trees, buildings or large metal objects and height above ground all affect the antenna resonant frequency. An SWR meter can help you determine if you should shorten or lengthen the antenna. The correct length provides the best impedance match for your transmitter.

The first step is to measure the SWR at the bottom, middle and top of the band. On 80 meters, for example, you would check the SWR at 3.626, 3.700 and 3.724 MHz. (A friend with a higher license class can help you check the SWR

over a wider frequency range.) Graph the readings, as shown in Figure 7-16. You could be lucky—no further antenna adjustments may be necessary, depending on your transmitter.

Many tube-type transmitters include an output tuning network. They will usually operate fine with an SWR of 3:1 or less. Most solid-state transmitters (using all transistors and integrated circuits) do not include such an output tuning network. These no-tune radios begin to shut down—the power output drops off—with an SWR much higher than 1.5:1. In any event, most hams like to prune their antennas for the lowest SWR they can get at the center of the band. With a full-size dipole 30 or 40 feet high, your SWR should be less than 2:1. If you can get the SWR down to 1.5:1, great! It's not worth the time and effort to do any better than that.

If the SWR is lower at the low-frequency end of the band, your antenna is probably too long. Making the antenna *shorter* will *increase* the resonant frequency. Disconnect the transmitter and try shortening your antenna at each end. The amount to trim off depends on two things. First is which band the antenna is operating on, and, second, how much you want to change the resonant frequency. Let's say the antenna is cut for the 80-meter band. You'll probably need to cut 8 or 10 inches off each end to move the resonant frequency 50 kHz. You may have to trim only an inch or less for small frequency changes on the 10-meter band. Measure the SWR again (remember to recheck the calibration). If the SWR went down, keep shortening the antenna until the SWR at the center of the band is less than 2:1.

If the SWR is lower at the high-frequency end of the band, your antenna was probably too short to begin with. If so, you must add more wire until the SWR is acceptable. Making the antenna *longer* will *decrease* the resonant frequency. Before you solder more wire on the antenna ends, try attaching a 12-inch wire on each end. Use alligator clips, as Figure 7-17 shows. You don't need to move the insulators

Figure 7-17—If your antenna is too short, attach an extra length of wire to each end with an alligator clip. Then shorten the extra length a little at a time until you get the correct length for the antenna. Finally, extend the length inside the insulators with a soldered connection. (See Figure 7-18.)

Figure 7-18—When splicing antenna wire, remember that the connection must be strong mechanically and good electrically.

yet. Clip a wire on each end and again measure the SWR. Chances are the antenna will now be too long. You will need to shorten it a little at a time until the SWR is below 2:1.

Once you know how much wire you need to add, cut two pieces and solder them to the ends of the antenna. When you add wire, be sure to make a sound mechanical connection before soldering. Figure 7-18 shows how. Remember that these joints must bear the weight of the antenna and the feed line. After you solder the wire, reinstall the insulators at the antenna ends, past the solder connections.

If the SWR is very high, you may have a problem that can't be cured by simple tuning. A very high SWR may mean that your feed line is open or shorted. Perhaps a connection isn't making good electrical contact. It could also be that your antenna is touching metal. A metal mast, the rain gutter on

your house or some other conductor would add considerable length to the antenna. If the SWR is very high, check all connections and feed lines, and be sure the antenna clears surrounding objects.

Now you have enough information to construct, install and adjust your dipole antenna. You'll need a separate dipole antenna for each band you expect to operate on. Sometimes a 40-meter dipole will also work on 15 meters, though. Check the SWR on 15 meters—you may have a two-for-one dipole!

[Before you go on to the next section turn to Chapter 12 and study questions N9A08 through N9A11. Review this section if you have difficulty with any of these questions.]

MULTIBAND DIPOLE OPERATION

The single-band dipole is fine if you operate on only one band. If you want to operate on more than one band, however, you could build and install a dipole for each band. What if you don't have supports for all these dipoles? Or what if you don't want to spend the money for extra coaxial cable? The popular and inexpensive multiband dipole described here will solve your problems nicely.

A single-band dipole can be converted into a *multiband antenna* without too much difficulty. All you need to do is connect two additional ¼ -λ wires for each additional band you want to use. Each added wire connects to the same feed line as the original dipole. The result is a single antenna system, fed with a single coaxial cable, that works on several different bands without adjustment. There is one potential problem with this antenna, though. The antenna will radiate signals on two or more bands simultaneously, so make sure your transmitter is adjusted properly. A poorly adjusted transmitter may produce harmonics of the desired output. If so, energy from your transmitter may show up on more than one band. The FCC takes a dim view of such operation!

Three-Band Dipole

You can build a three-band dipole for 80, 40 and 15 meters from ladder line. To build this antenna, you'll need a 100-foot roll of this line, three insulators and a coax feed line.

This antenna construction is similar to that for a regular dipole. Carefully remove the line from the spool and lay it on the ground. Take care to avoid twists and kinks in the wire. See Figure 7-19. At 33 feet, 6 inches from one end (X), cut *one* of the two wires. At 63 feet, 8 inches from the same end, cut the *other* wire. Remove the plastic spacers between these cuts, separating the open-wire line into two pieces. Measuring from the other end (Y), cut the shorter wire at 33 feet, 6 inches and the longer one at 63 feet, 9 inches. Figure 7-19A shows how the two antenna halves should look at this point.

Reverse the position of the two halves, as Figure 7-19B shows. Now prepare the wire ends for connection to the feed line. Sandpaper the protective coating off both wires at the ends. (X and Y identify these ends in the drawing.) Connect the wire and a piece of coaxial cable in the same manner as described for a single-band dipole. Waterproof the connec-

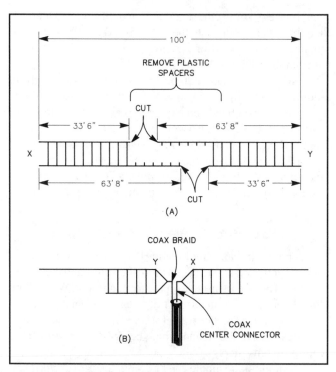

Figure 7-19—At A, cut the two lengths of ladder line as shown. Reverse the two sections to make the three-band dipole antenna, as shown at B.

tion with tape. Spray the taped connection with clear lacquer or coat it with liquid rubber or silicone sealant for added protection against weather. Attach insulators to the ends, and your antenna is complete.

All information on antenna location and installation for a single-band dipole also applies to this multiband antenna. Use the same procedure described earlier to tune the antenna for the lowest SWR. Just remember that you must adjust the SWR on two bands. Adjust both the 80-meter part (the longer wire) and the 40/15-meter part (the shorter wire).

Another multiband antenna is shown in Figure 7-20. The legs for this antenna should each be ¼ λ at the lowest frequency you want to use. In other words, if you want to operate on all bands, 80 through 10 meters, each leg should be about 63 feet long. Feed this antenna with open-wire feed line (either 300- or 450-ohm), or TV twin lead. It requires a Transmatch at the station end.

THE QUARTER-WAVE VERTICAL ANTENNA

The **quarter-wavelength vertical antenna** is simple and popular. It requires only one support and can be very effective. On the HF bands (80-10 meters) it is often used for DX work. Vertical antennas send radio energy equally well in all compass directions. This is why we sometimes call them *nondirectional antennas* or **omnidirectional** *antennas*.

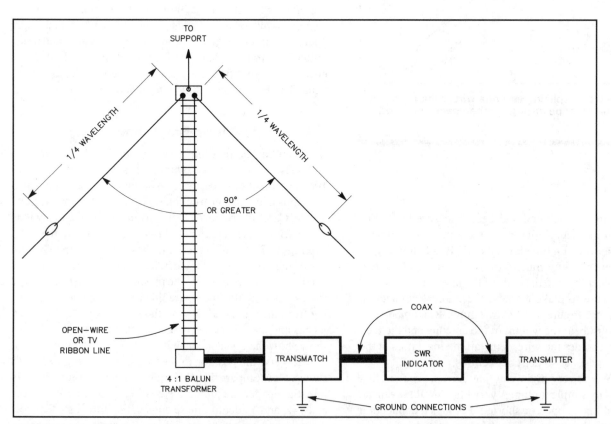

Figure 7-20—A half-wave dipole antenna can be made to work on several bands. Use twin-lead feed line and a Transmatch.

Figure 7-21—The quarter-wave vertical antenna has a center radiator and four or more radials spread out from the base. These radials form a ground plane.

They also tend to concentrate the signals towards the horizon. Vertical antennas do not generally radiate strong signals straight up, like horizontal dipoles do. Because they concentrate the signals towards the horizon, we say they have **gain** as compared to a dipole. Gain always refers to a comparison with another antenna. A dipole is one common comparison antenna for stating antenna gain.

Figure 7-21 shows a simple vertical antenna you can make. This **vertical antenna** has a radiator that is ¼-λ long. Use Equation 7-3 to find ¼ λ for the radiator. The frequency is given in megahertz and the length is in feet in this equation.

$$\text{Length (in feet)} = \frac{234}{\text{f (in MHz)}} \qquad \text{(Eq 7-3)}$$

Equation 7-3 gives us the following approximate lengths for the radiator and each ground radial of a ¼-λ vertical.

Wavelength	Frequency	Length
80 meters	3.700 MHz	63.24 feet
40 meters	7.125 MHz	32.8 feet
15 meters	21.125 MHz	11.1 feet
10 meters	28.150 MHz	8.3 feet
10 meters	28.4 MHz	8.2 feet
6 meters	52.5 MHz	4.5 feet
2 meters	146.0 MHz	1.6 feet (19.25 inches)
1.25 meters	223.0 MHz	1.05 feet (12.6 inches)
70 cm	440.0 MHz	0.53 feet (6.4 inches)
23 cm	1282.5 MHz	0.18 feet (2.2 inches)

Equation 7-3 is only accurate for frequencies up to around 30 MHz. One reason for this frequency limitation is because the ratio of wavelength to element diameter is larger at VHF and higher frequencies. Another reason is because of effects commonly known as *end effects*. So the values given here for the VHF and UHF bands are only for comparison purposes. These lengths might serve as a starting point for building antennas for 2 meters, 1.25 meters, 70 cm or 23 cm, but they will probably be too long.

As with ½-λ dipoles, the resonant frequency of a ¼-λ vertical decreases as the length increases. Shorter antennas have higher resonant frequencies.

The ¼-λ vertical also has radials. For operation on 80 through 10 meters, the vertical may be at ground level, and the radials placed on the ground. The key to successful operation with a ground-mounted vertical antenna is a good radial system. The best radial system uses *many* ground radials. Lay them out like the spokes of a wheel, with the vertical at the center. Some hams have buried ground radial systems containing over 100 individual wires.

Ideally, these wires would be ¼ λ long or more at the lowest operating frequency. With such a system, earth or ground losses will be negligible. When the antenna is mounted at ground level, radial length is not very critical, however. Studies show that with fewer radials you can use shorter lengths, but with a corresponding loss in antenna efficiency.[1, 2] Some of your transmitter power does no more than warm the earth beneath your antenna. With 24 radials, there is no point in making them longer than about ⅛ λ. With 16 radials, a length greater than 0.1 λ is unwarranted. Four radials should be considered an absolute minimum. Don't put the radials more than about an inch below the ground surface.

Compared with 120 radials of 0.4 λ, antenna efficiency with 24 radials is roughly 63%. For 16 radials, the efficiency is roughly 50%. So it pays to put in as many radials as you can.

If you place the vertical above ground, you reduce earth losses drastically. Here, the wires should be cut to ¼ λ for the band you plan to use. Above ground, you need only a few radials—two to five. If you install a multiband vertical antenna above ground, use separate ground radials for each band you plan to use. These lengths are more critical than for a ground-mounted vertical. For elevated verticals, you should have a minimum of two radials for each band. You can mount a vertical on a pipe driven into the ground, on the chimney or on a tower.

The radials at the bottom of a vertical antenna mounted above ground form a surface that acts like the ground under the antenna. These antennas are sometimes called *ground-plane antennas*.

[1]J. O. Stanley, "Optimum Ground Systems for Vertical Antennas," *QST*, December 1976, pp 14-15.
[2]B. Edward, "Radial Systems for Ground-Mounted Vertical Antennas," *QST*, June 1985, pp 28-30.

Ground-plane antennas are popular mobile antennas because the car body can serve as the ground plane. You can place a magnetic-base whip antenna on the roof of your car, for example. This type of antenna is often used on the VHF and UHF bands for FM voice communication. Because it is a vertical antenna, it sends radio energy out equally well in all compass directions.

Once a vertical antenna is several feet above ground, there is little advantage in more height. (This assumes your antenna is above nearby obstructions.) For sky-wave signals, a height of 15 feet for the base is almost as good as 50. This isn't the case for a horizontal antenna, where height is important for working DX. Only if you can get the vertical up 2 or 3 wavelengths does the low-angle radiation begin to improve. Even then the improvement is only slight. At VHF and UHF, however, it pays to get the antenna up high. At these frequencies you want the antenna higher than even distant obstructions.

Most vertical antennas used at lower frequencies are ¼ λ long. For VHF and UHF, antennas are physically short enough that longer verticals may be used. A popular mobile antenna is a ⅝-λ vertical, often called a "five-eighths whip." This antenna is popular because it may concentrate more of the radio energy towards the horizon than a ¼-λ vertical. Mounted on the roof of a car, a ⅝-λ vertical may provide more **gain** than a ¼-λ vertical. Simply stated, "gain" means a concentration of transmitter power in some direction. A ⅝-λ vertical concentrates the power toward the horizon. Naturally, this is the most useful direction, unless you want to talk to airplanes or satellites. At 222 MHz, a ⅝-λ whip is only 28½ inches long.

Don't use any of the equations in this chapter to calculate the length of a ⅝-λ whip. The equations won't give the correct answer because of a variety of antenna factors. In addition, there is an impedance-matching network at the antenna feed point. See *The ARRL Antenna Book* for complete construction details.

A ⅝-λ vertical is great for mobile operation because it is omnidirectional. That means it radiates a signal equally well in all compass directions. This is especially useful for mobile operation because you change direction often. One minute you may be driving toward the repeater, and the next minute you may be driving away from it.

Vertical antennas that are ½-λ long can be used without ground radials. This may sometimes be a definite advantage because it takes less wire and occupies less horizontal space. To find the size of a ½-λ vertical antenna, double the lengths given in the table of ¼-λ verticals, above.

Commercially available vertical antennas need a coax feed line, usually with a PL-259 connector. Just as with the dipole antenna, you can use RG-8, RG-11 or RG-58 coax. The instructions that accompany the antenna should provide details for attaching both the feed line and the ground radials.

Some manufacturers offer *trap verticals*. Traps are tuned circuits that change the antenna electrical length. They allow the antenna to work on several bands, making it a *multiband antenna*. Some manufac-

Figure 7-22—Commercial "trap vertical" antennas generally look something like this. These antennas operate on several bands.

turers even offer 20- to 30-foot-high vertical antennas that cover all HF bands. Figure 7-22 shows one such antenna.

[Now turn to Chapter 12 and study questions numbered N9A02, N9A05 and N9A06. Also study questions N9B07 and N9B08. Review this section if you have difficulty with any of these questions.]

[**If you are studying for the Technician exam**, turn to Chapter 13 now. Study questions T9A09, T9A10 and T9A11. Review this section if you have trouble answering any of these questions.]

VERTICAL ANTENNAS FOR 146, 222 AND 440 MHz

For FM and packet radio operation with nearby stations, the ease of construction and low cost of a ¼-λ vertical make it an ideal choice. Three different types of construction are shown in Figures 7-23 through 7-26; the choice of construc-

tion method depends on the materials available and the desired style of antenna mounting.

The 146-MHz model shown in Figure 7-23 uses a flat piece of sheet aluminum, to which radials are connected with machine screws. A 45° bend is made in each of the radials. This bend can be made with an ordinary bench vise. An SO-239 chassis connector is mounted at the center of the aluminum plate with the threaded part of the connector facing down. The vertical portion of the antenna is made of no. 12 copper wire soldered directly to the center pin of the SO-239 connector.

The 222-MHz version, Figure 7-24, uses a slightly different technique for mounting and sloping the radials. In this case the corners of the aluminum plate are bent down at a 45° angle with respect to the remainder of the plate. The four radials are held to the plate with machine screws, lock washers and nuts. A mounting tab is included in the design of this antenna as part of the aluminum base. A compression type of

Figure 7-23—These drawings show the dimensions for the 146-MHz groundplane antenna. The radials are bent down at a 45° angle.

Figure 7-24—Dimensional information for the 222-MHz groundplane antenna. Lengths for A, B, C and D are the total distances measured from the center of the SO-239 connector. The corners of the aluminum plate are bent down at a 45° angle rather than bending the aluminum rod as in the 146-MHz model. Either method is suitable for these antennas.

FREQUENCY (MHz)	A (INCHES)	B (INCHES)
146	19 - 5/16	18 - 11/16
225	12 - 5/8	12
445	6 - 3/8	5 - 3/4

Figure 7-25—Simple groundplane antenna for the 146, 222 and 440-MHz bands. The vertical element and radials are ³⁄₃₂- or ¹⁄₁₆-in. brass welding rod. Although ³⁄₃₂-in. rod is preferred for the 146-MHz antenna, no. 10 or 12 copper wire can also be used.

Figure 7-26—A 440-MHz groundplane antenna constructed using only an SO-239 connector, no. 4-40 hardware and ¹⁄₁₆-in. brass welding rod.

Figure 7-27—A simple Novice vertical antenna for use on 10 and 15 meters.

hose clamp could be used to secure the antenna to a mast. As with the 146-MHz version, the vertical portion of the antenna is soldered directly to the SO-239 connector.

A very simple method of construction, shown in Figures 7-25 and 7-26, requires nothing more than an SO-239 connector and some 4-40 hardware. A small loop formed at the inside end of each radial is used to attach the radial directly to the mounting holes of the coaxial connector. After the radial is fastened to the SO-239 with no. 4-40 hardware, a large soldering iron or propane torch is used to solder the radial and the mounting hardware to the coaxial connector. The radials are bent to a 45° angle and the vertical portion is soldered to the center pin to complete the antenna. The antenna can be mounted by passing the feed line through a mast of ¾-inch ID plastic or aluminum tubing. A compression hose clamp can be used to secure the PL-259 connector, attached to the feed line, in the end of the mast. Dimensions for the 146, 222, and 440-MHz bands are given in Figure 7-25.

If these antennas are to be mounted outside it is wise to apply a small amount of RTV sealant or similar material around the areas of the center pin of the connector to prevent the entry of water into the connector and coax line.

A SIMPLE NOVICE VERTICAL

Figure 7-27 shows a simple, inexpensive vertical antenna for 10 and 15 meters. The antenna requires very little space and it's great for working DX. A few materials make up the entire antenna.

12-foot piece of clean 2 × 2 pine from the local lumberyard

20 feet of flat four-wire rotator control cable

20 feet of regular TV twin lead

Several TV standoff insulators

Cut the twin lead into lengths of 11 feet 3 inches, and 8 feet 6 inches. Remove 1 inch of insulation from both wires

Figure 7-28— For each length of twin lead, strip back the insulation from both ends, connect the two wires, and solder them.

Figure 7-29—Cut some flat TV rotator-control cable to make the radials.

at both ends on each twin-lead length. Wrap the two wires together securely at each end, as Figure 7-28 shows, and solder. Cover one end of each length of twin lead with electrical tape.

Mount the TV standoff insulators at regular intervals on opposite sides of the 2 × 2. These will support the two pieces of twin lead. Separate the control cable into two pieces of two-conductor cable. Do this by slitting the cable a small amount at one end, then pulling the two pieces apart like a zipper.

Carefully cut the rotator cable as Figure 7-29 shows. Separate the pieces between the center cuts to make four identical sets of two-conductor cable. These make up the radials for the antenna system. Strip about 1 inch of insulation

from each wire on the evenly cut end. The unevenly cut ends will be away from the antenna.

Attach a suitable length of RG-58 coaxial cable to the antenna. Connect the coax center conductor to the two wires on the 2 × 2. Then attach the cable braid to all the radial ends soldered together. Figure 7-30 shows the antenna construction. Make all the connections waterproof.

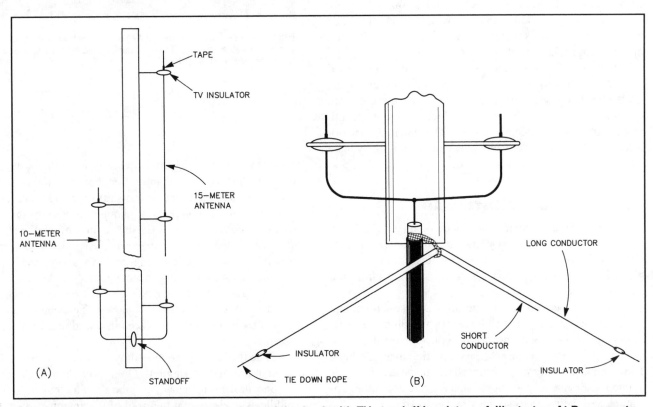

Figure 7-30—Attach the twin lead to the sides of the 2 × 2 with TV standoff insulators. A illustrates. At B, connect the feed line to both twin-lead lengths at the base.

**Figure 7-31—
With a
Transmatch,
you can use a
wire cut to a
convenient
random length
as an antenna.**

Now install the 2 × 2. You can do this in different ways. You could clamp it with U bolts to a TV mast, or hang it with a hook over a high branch. Or you might mount it on your house, at the side, near the roof peak. Mount the antenna as high as possible. Hang the radials at about a 45° angle away from the antenna base. For example, you could clamp the wood to your chimney with TV chimney-mount hardware. Let the radials follow the roof slope. Tie them off at the four corners. Unless you are surrounded by buildings or high hills, this antenna should perform well on 10 and 15 meters.

RANDOM-LENGTH WIRE ANTENNAS

If you can't install either a dipole or a vertical, you can still get on the air. Try a *random-length wire antenna*. See Figure 7-31. As the name implies, the antenna requires no specific length. As a rule of thumb, you should make your random-length antenna as long as possible. If you live in an apartment, you might use an antenna running along the ceiling in a few rooms. On the other hand, you may be able to string up a long length of wire outdoors. Use small wire (no. 22 to no. 28) if you want an antenna with low visibility. The only disadvantage of small wire is that it breaks easily. You may have to replace it often.

Random-length antennas are versatile: They can be used almost anywhere. But they do have one major disadvantage. Unlike the dipole and vertical, which can be fed directly from the transmitter through coaxial cable, a random-length antenna requires a matching network. The matching network is required because the antenna impedance is not likely to be 50 ohms.

BEAM ANTENNAS

Although generally impractical on 80 meters, and very large and expensive even on 40 meters, *directional antennas* often see use on 15 and 10 meters. The most common directional antenna that amateurs use is the **Yagi antenna**, but there are other types, too.

Generally called **beam antennas**, these directional antennas have two important advantages over dipole and vertical systems. First, a beam antenna concentrates the transmitted signal more in one compass direction than in others. The antenna provides **gain** or **directivity** in the direction it is pointed. Gain makes your signal sound stronger to other operators, and their signals sound stronger to you.

The second important advantage is that the antenna sup-

Figure 7-32—Yagi beams at W1AW. A single beam sits atop the tower at left. Three stacked beams adorn the taller tower at the right.

presses signals coming from directions other than where you point it. This reduces the interference from stations in other directions, and increases your operating enjoyment.

A graph of an antenna's gain and directivity shows its *radiation pattern*. Figure 7-32 shows some of the Yagi beams at W1AW, the station located at ARRL Headquarters in Newington, Connecticut. Figure 7-33 shows the typical radiation pattern of a Yagi beam.

A Yagi beam antenna has several elements attached to a central *boom*, as Figure 7-34 shows. The elements are parallel to each other and are placed in a straight line along the boom.

The feed line connects to only one element. We call this element the **driven element**. On a three-element Yagi like the one shown in Figure 7-34, the driven element is in the middle. The element at the front of the antenna (toward the favored direction) is a **director**. Behind the driven element is

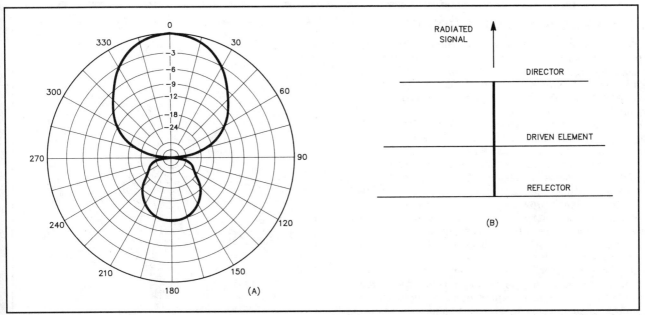

Figure 7-33—Typical radiation pattern for a Yagi beam antenna. The inset shows the direction of beam pointing. The transmitted signal is stronger in the forward direction than in others.

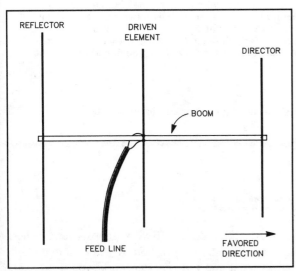

Figure 7-34—A three-element Yagi antenna has a director, a driven element and a reflector. A boom supports the elements.

the **reflector** element. The driven element is about ½ λ long at the antenna design frequency. The director is a bit shorter than ½ λ, and the reflector a bit longer.

Yagi beams can have more than three elements. Seldom is there more than one reflector. Instead, the added elements are directors. A standard 4-element Yagi would have a reflector, a driven element and two directors.

Because beams are directional, you'll need something to turn them. A single-band beam for 10 or 15 meters can be mounted in the same manner as a TV antenna. You can use a TV mast, hardware and rotator. You could plan to buy a large triband beam for use on 10 and 15 meters as a Novice. It will cover 20 meters when you upgrade to a General or higher license class. For such a big antenna you'll need a heavy-duty mount and rotator. You can get good advice about the equipment you'll need from your instructor or from local hams.

[Now turn to Chapter 12 and study questions N9B01 through N9B06. Review this section if you have any difficulty answering these questions.]

CONVERTING CB ANTENNAS

It is usually a simple matter to convert a Citizen's Band (CB) antenna to work on the Novice 10-meter band. This is because the 11-meter Citizens Band at 27 MHz is close in frequency to the Novice 10-meter band. Since most CB antennas are ¼-λ verticals, the formula for the correct length is quite simple—234 divided by the frequency in megahertz (Equation 7-3). The antenna length is given in feet. For example, 234 divided by 27 MHz equals 8.67 feet or 8 feet, 8 inches.

To resonate (or tune) such an antenna on 10 meters, you shorten it to the required length. Using Equation 7-2, ½ λ for 28.1 MHz works out to be 16 feet, 8 inches. The length for a ¼-λ vertical, using Equation 7-3, is 8 feet 4 inches. This means removing 8 inches from a dipole, or 4 inches from a full-sized vertical (not loaded with a coil). Before shortening any CB antenna, try the antenna on 10 meters. If your transmitter loads and tunes up, it may be a waste of time to prune the antenna. It may work well just as it is!

ANTENNA RADIATION PATTERNS

Often one desirable antenna feature is **directivity**. Directivity means the ability to pick up signals from one direction, while suppressing signals from other unwanted directions. Going hand in hand with directivity is **gain**. Gain tells how much signal a given antenna will pick up as compared with that from another antenna, usually a dipole.

An antenna that has directivity should also have gain. These two antenna properties are useful not only for picking up or receiving radio signals, but also for transmitting them. An antenna that has gain will boost your transmitted energy in the favored direction while suppressing it in other directions.

When you mention gain and directivity, most amateurs envision large antenna arrays, made from aluminum tubing, with many elements. Simple wire antennas can also be very effective, however, as illustrated in this antenna radiation pattern. Such patterns reveal both the gain and the directivity of a specific antenna.

Let's say we connect the antenna to a transmitter and send. The pattern shows the relative power received at a fixed distance from the antenna, in various compass directions. If you connect the antenna to a receiver, the pattern shows how the antenna responds to signals from various directions. In the direction where the antenna has gain, the incoming signals will be enhanced. The incoming signals will be suppressed in other directions.

Here is an important point to remember. You can never have antenna gain in one direction without a loss (signal suppression) in one or more other directions. Never! Another way to think of this is that an antenna cannot create power. It can only focus or beam the power supplied by the transmitter.

We call the long, thin lobes in a pattern the *major lobes*. The smaller lobes in a pattern are *minor lobes*. One or more major lobes mean directivity. An antenna with less directivity than this one would have fatter lobes. An antenna with no directivity at all would have a pattern that is a perfect circle. A theoretical antenna called an isotropic radiator has such a pattern. Radiation patterns are a very useful tool in measuring antenna capability.

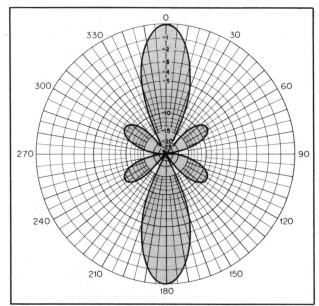

The calculated or theoretical radiation pattern of an extended double Zepp antenna. In its favored directions, this antenna exhibits roughly 2 decibels of gain over a half-wavelength dipole. This would make a 200-watt signal as strong as 317 watts into a dipole. An extended double Zepp antenna may be made with a horizontal wire hanging between two supports. (The wire is 1.28-λ long at the operating frequency and should be fed at the center with open-wire line.) The wire axis is along the 90/270-degree line shown in the chart.

ANTENNA POLARIZATION

For VHF and UHF base-station operation, most amateurs use a vertical antenna or a beam antenna. For VHF and UHF FM and repeater operation, almost everyone uses vertical **polarization**.

Polarization refers to the electrical-field characteristic of a radio wave. You can think of it as how the antenna is positioned. An antenna that is parallel to the earth's surface (like a dipole antenna) produces horizontally polarized radio waves. One that is perpendicular (at a 90° angle) to the earth's surface, such as a ¼-λ vertical, produces vertically polarized waves.

For SSB and CW VHF/UHF work, almost everyone uses a beam with horizontal polarization. If you plan to do both VHF weak-signal and FM repeater work, you'll probably need separate antennas for each. This is because at VHF and UHF, signal strength suffers greatly if your antenna has different polarization than the station you're trying to work.

A beam antenna is impractical for VHF and UHF FM mobile operation. Most hams use a vertical of some kind. Mobile antennas are available in several varieties. Most mount on the automobile roof or trunk lid, but some even mount on a glass window.

INSTALLATION SAFETY

No matter what antenna type you choose to build, you should remember a few key points about safety. If possible, ask someone with experience to help you plan and carry out your installation, especially if climbing is involved. If you use a slingshot or bow and arrow to get a line over a tree, use caution. Make sure you keep everyone away from the "downrange" area. Hitting a helper with a rock or a fishing sinker is not a good idea!

If you build a wire antenna, be sure the antenna is out of reach of passersby. The ends should be high enough that a person on the ground cannot touch them. Even at Novice power levels, enough voltage exists at the antenna ends when you transmit to cause **RF burns**. If you have a vertical antenna with its base at ground level, you might consider building a wooden safety fence around it. The fence should be at least 4 feet away from the antenna. Do not use a metal fence, as this will interfere with the proper antenna operation. If you have a tower, you should either fence it or put shields directly on the tower to prevent unauthorized climbing.

If you use a parallel-conductor feed line, such as open-wire line, you should use care to install the feed line so no one can touch the wires when you are transmitting. The voltage on this line could cause RF burns.

Antenna work sometimes requires that someone climb up on a tower, into a tree, or onto the roof. Never work alone! Work slowly, thinking out each move before you make it. The person on the ladder, tower, tree or roof top should always wear a safety belt, and keep it securely anchored. Before each use, inspect the belt carefully for damage such as cuts or worn areas. The belt will make it much easier to work on the antenna and will also prevent an accidental fall. A hard hat and safety glasses are also important safety equipment.

Never try to climb a tower carrying tools or antenna components in your hands. Carry what you can with a tool belt, including a long rope leading back to the ground. Then use the rope to pull other needed objects up to your workplace after securing your safety belt! It is helpful (and safe) to tie strings or lightweight ropes to all tools. You can save much time in retrieving dropped tools if you tie them to the tower. This also reduces the chances of injuring a helper on the ground.

Helpers on the ground should never stand directly under the work being done. All ground helpers should wear hard hats and safety glasses for protection. Even a small tool can make quite a dent if it falls from 50 or 60 feet. A ground helper should always observe the tower work carefully. Have you ever wondered why electric utility crews seem to have someone on the ground "doing nothing"? Now you know that for safety's sake, a ground observer with no other duties is free to notice potential hazards. That person could save a life by shouting a warning.

As mentioned earlier, be especially certain that your antenna is not close to any power lines. That is the only way you can be sure it won't come in contact with them!

When using a hand-held transceiver that runs more than a few watts, be careful! Always keep the antenna away from your head and away from others standing nearby. Some antennas are safer than others. A short, helically wound flexible antenna (called a "rubber duck") is not the safest kind. It concentrates its radiation in a small area near your head. A longer antenna, such as a ½-λ whip, would be safer for a hand-held. A ½-λ antenna concentrates its radiation at its center, which will be farther from your head.

[Now study the questions in Chapter 12 with numbers N9B10 and N9B11. Also study questions N4B08 through N4B11. Review this section if you have any problems.]

[**If you are preparing for the Technician exam**, turn to Chapter 13 and study questions T9C09 and T9C10. Review this section if these questions give you difficulty.]

HOW TO LIVE LONG ENOUGH TO UPGRADE

Keep antenna safety in mind when you're setting up your station. Antennas for the ham bands are often large, requiring care and attention to detail when installed. Here are two points to keep in mind when putting up your antenna.

1) Be sure your antenna materials and supports are strong—strong enough to withstand heavy winds without breaking.

2) Keep away from power lines!

If your antenna falls, it could damage your house, garage or property. If it falls into a power line, your house might end up like the unlucky CBer's shown in the photo. A windstorm knocked over his groundplane antenna, sending it into a 34,500-V power line. The resulting fire damaged his house extensively.

Safety pays! Accidents are avoidable, if you use good sense.

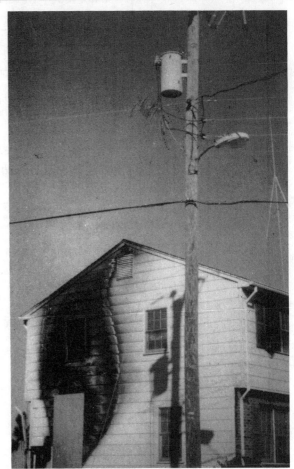

Photo by Charles Stokes, WB4PVT

Additional Antenna Information for the Technician Exam

KEY WORDS

Balanced line—Feed line with neither conductor at ground potential, such as open-wire line or ladder line.

Cubical quad antenna—An antenna built with its elements in the shape of four-sided loops.

Delta loop antenna—A variation of the cubical quad with triangular elements.

Horizontally polarized wave—An electromagnetic wave with its electric lines of force parallel to the ground.

Major lobe—The shape or pattern of field strength that points in the direction of maximum radiated power from an antenna.

Parasitic element—Part of a directive antenna that derives energy from mutual coupling with the driven element. Parasitic elements are not connected directly to the feed line.

Polarization—Describes the electrical-field characteristic of a radio wave. An antenna that is parallel to the surface of the earth, such as a dipole, produces horizontally polarized waves. One that is perpendicular to the earth's surface, such as a quarter-wave vertical, produces vertically polarized waves.

Unbalanced line—Feed line with one conductor at ground potential, such as coaxial cable.

Vertically polarized wave—A radio wave that has its electric lines of force perpendicular to the surface of the earth.

FEED-LINE ATTENUATION

As we have seen, transmission lines can be constructed in a variety of forms. Both parallel-conductor feed line and coaxial cable can be divided into two classes: those in which the majority of the space between the conductors is air, and those in which the conductors are embedded in and separated by a solid plastic or foam insulation (dielectric).

AIR-INSULATED FEED LINES

Air-insulated feed lines have the lowest loss per unit length (usually expressed in dB/100 ft). Adding a solid dielectric between the conductors increases the losses in the feed line. The power loss causes heating of the dielectric. (*Dielectric* is the technical term for an insulating material.) As frequency increases, conductor and dielectric losses become greater for coaxial and parallel-conductor feed lines. An advantage of this type of transmission line is that it can be operated at a high SWR and still retain its low-loss properties.

The characteristic impedance of an air-insulated parallel-conductor line depends on the diameter of the wires used in the feed line and the spacing between them. The greater the spacing between the conductors, the higher the characteristic impedance of the feed line. The impedance decreases, however, as the size of the conductors increases. The characteristic impedance of a feed line is not affected by the length of the line.

SOLID-DIELECTRIC FEED LINES

Transmission lines in which the conductors are separated by a flexible dielectric have several advantages over air-dielectric line: They are less bulky, maintain more uniform spacing between conductors, are generally easier to install and are neater in appearance. Both parallel-conductor and coaxial lines are available with this type of insulation.

One disadvantage of these types of lines is that the power loss per unit length is greater than air-insulated lines because of the dielectric. As the frequency increases, the dielectric losses become greater. The power loss causes heating of the dielectric. Under conditions of high power or high SWR, the dielectric may actually melt and cause short circuits or arcing inside the line.

We mentioned TV-type parallel-conductor line, commonly called twin lead. Twin lead consists of two no. 20 wires that are molded into the edges of a polyethylene ribbon about a half-inch wide. The presence of the solid dielectric lowers the characteristic impedance of the line as compared to the same conductors in air.

The fact that part of the field between the conductors exists outside the solid dielectric leads to operating disadvantages. Dirt or moisture on the surface of the ribbon tends to change the characteristic impedance. Weather effects can be minimized, however, by coating the feed line with silicone grease or car wax. In any case, the changes in the impedance will not be very serious if the line is terminated in its charac-

teristic impedance (Z_0). If there is a considerable standing-wave ratio, however, then small changes in Z_0 may cause wide fluctuations of the input impedance.

Parallel-conductor feed lines are **balanced lines** because neither conductor connects to ground. They can be connected directly to *balanced antennas*.

COAXIAL CABLE

The characteristic impedance of a coaxial line depends on the diameter of the center conductor, the dielectric constant of the insulation between conductors, the inside diameter of the shield braid and the distance between conductors. The characteristic impedance of coaxial cables increases for larger-diameter shield braids, but it decreases for larger-diameter center conductors.

The larger the diameter, the higher the power capability of the line because of the increased dielectric thickness and conductor size. In general, losses decrease as the cable diameter increases, because there is less power lost in the conductor.

Amateurs commonly use RG-8, RG-58, RG-174 and RG-213 coaxial cable. RG-8 and RG-213 are similar cables, and they have the least loss of the types listed here. RG-174 is only about ⅛ inch in diameter, and it has the highest loss of the cables listed here.

As the frequency increases, the conductor and dielectric losses become greater, causing more attenuation of the signal in the cable. Although the cost, size and weight are larger for RG-213, it is usually the best choice (of the cables mentioned) for a run of over 150 feet for frequencies up to 54 MHz (the amateur 6-meter band). Open-wire transmission line has the least loss of any feed line type commonly used by amateurs.

Of the common coaxial cables discussed here, RG-213 has the least loss. RG-58 has a bit more loss than RG-213, and RG-174 has the most loss of these cables. RG-174 is normally used for cables that connect sections of a transmitter or receiver, or for short interconnecting cables in a low-power system. Some amateurs use RG-174 cable as the feed line for a low-power portable station.

Extra cable length increases attenuation. When using coaxial cable, you should try to use a matched antenna and feed line. You should then be able to change feed line lengths without significantly affecting the antenna system. Then your feed line has only to be long enough to reach your antenna. A low SWR on the line means that the impedance "seen" by the transmitter will be about the same regardless of line length. You can cut off or shorten excess cable length to reduce attenuation of the signal caused by feed-line loss. (This does not apply to multiple antennas in phased arrays or line sections used for impedance-matching purposes.)

Attenuation is not affected by the characteristic impedance of a matched line if the spacing between the conductors in the coaxial cable is a small fraction of a wavelength at the operating frequency.

The outside shield of coaxial cable is normally connected to ground. Coax, then, is an **unbalanced line**. When connecting coaxial *(unbalanced)* feed lines to *balanced antennas*, many hams prefer to use a *balun*.

Coaxial-cable connectors are an important part of a coaxial feed line. Your choice of connectors normally depends on the matching connectors on your radios. Most HF radios and many VHF radios use SO-239 connectors. The mating connector is called a PL-259. (These are military-type designations.) The PL-259 is sometimes called a *UHF connector,* although they are not the best choice for the UHF bands. Figure 7-35C shows the SO-239 connector and its mating PL-259.

Many VHF and UHF hand-held radios use BNC connectors. These connectors are designed for use with RG-58 coax. They produce a low-loss connection that is also weatherproof. Figure 7-35A shows a pair of BNC connectors. BNC connectors are well-suited for use with hand-held-radio antennas because they require only a quarter turn to install or remove, yet they lock securely in place.

Many amateurs prefer N-type connectors for UHF equipment. N connectors are low-loss, high-power connectors. They are designed for use with RG-213 or RG-8 coaxial cable. N connectors have the lowest loss of the common coaxial connectors, which is why they are often used at UHF. Figure 7-35B shows a set of N connectors.

(A)　　　　**(B)**　　　　**(C)**

Figure 7-35—Some common coaxial-cable connectors. A shows a BNC connector pair. Many hand-held radios use BNC connectors. They are also popular when a weatherproof connector is needed for RG-58 sized cable. B shows a pair of N connectors. These are often used for UHF equipment because of their low loss. Part C shows an SO-239 chassis connector and its mating PL-259. Most HF equipment uses these connectors.

It is a good idea to check your coaxial connectors on a regular basis. Be sure they are clean and tight. If you suspect a bad solder connection, you should resolder the joints.

[Turn to Chapter 13 and study questions T9C01 through T9C08, and T9C11 before continuing with this chapter. Review this section if needed.]

ANTENNA POLARIZATION

The signal sent from an amateur station depends on the antenna type and how it is oriented. A horizontal antenna, parallel to the earth's surface (like a dipole), will produce a *horizontally polarized signal.* A Yagi antenna with horizontal elements will also produce a horizontally polarized signal. See Part A in the drawing.

A vertical antenna (perpendicular to the earth's surface) will produce a *vertically polarized signal.* A Yagi with vertical elements also produces vertical polarization. See Part B in the drawing.

Most communications on the HF bands (80 through 10 meters) use horizontal polarization. Polarization on the HF bands is not critical, however. As a signal travels through the ionosphere, its polarization can change.

On the VHF/UHF bands, most FM communications use vertical polarization. Here, the polarization is important. The signals retain their polarization from transmitter to receiver. If you use a horizontal antenna, you will have difficulty working through a repeater with a vertical antenna.

(A)　　　　　　　　(B)

The plane of the elements in a Yagi antenna determines the transmitted-signal polarization. If the elements are horizontal, as shown at A, the signal will have horizontal polarization. Vertical mounting, shown at B, produces a vertically polarized wave.

POLARIZATION OF ANTENNAS AND RADIO WAVES

An electromagnetic wave consists of moving electric and magnetic fields. Remember that a field is an invisible force of nature. We can't see radio waves, but we can show a representation of where the energy is in the electric and magnetic fields. We did this in Chapter 3 to show the magnetic flux around a coil, and the electric field in a capacitor. We can visualize a traveling radio wave as looking something like Figure 7-36. The lines of electric and magnetic force are at right angles (90°) to each other. They are also perpendicular (90°) to the direction of travel. These fields can have any position with respect to the earth.

Figure 7-36—A horizontally polarized electromagnetic wave. The electrical lines of force are parallel to the ground. The fields are stronger where the arrows are closer together. The points of maximum field strength correspond to the points of maximum amplitude for the voltage producing the RF signal.

HORIZONTAL POLARIZATION

Polarization is defined by the direction of the electric lines of force in a radio wave. A wave with electric lines of force that are parallel to the surface of the earth is a **horizontally polarized wave**.

The polarization of a radio wave as it leaves the antenna is determined by the orientation of the antenna. For example, a half-wavelength dipole parallel to the surface of the earth transmits a horizontally polarized wave.

Figure 7-37—The orientation of the elements in a Yagi antenna determines the polarization of the transmitted wave. If the elements are horizontal as shown at A, the wave has horizontal polarization. Vertical mounting, shown at B, produces a vertically polarized wave.

VERTICAL POLARIZATION

The electric lines of force can also be perpendicular to the earth. A wave with vertical electric field lines is a **vertically polarized wave**. A half-wavelength dipole perpendicular to the surface of the earth transmits a vertically polarized wave. A mobile whip antenna is mounted vertically on a car. It transmits a wave that is vertically polarized.

The quarter-wavelength vertical antenna is a popular HF antenna because is provides low-angle radiation when a beam or dipole cannot be placed far enough above ground. Low-angle radiation refers to signals that travel closer to the horizon, rather than signals that are high above the horizon. Low-angle radiation is usually better when you are trying to contact distant stations. Vertical antennas of any length radiate vertically polarized waves.

Polarization is most important when installing antennas for VHF or UHF. Propagation at these frequencies is mostly line of sight. The polarization of a VHF or UHF signal does not change from transmitting antenna to receiving antenna. Best signal reception occurs when both transmitting and receiving stations use the same polarization. The polarization of an HF signal may change many times as it passes through the ionosphere. Antenna polarization at HF is not as important.

Most VHF/UHF FM and data communications is done with vertically polarized antennas. Vertically polarized antennas are more popular for VHF/UHF FM operation, because the antennas used on cars are almost always verticals. Vertical antennas are more useful for repeaters and home-station use on these bands as well, because they are not directional. For long-distance FM work, a vertically polarized beam is the best antenna. Data communication on the VHF/UHF bands is also mostly done with vertically polarized antennas. VHF/UHF CW and SSB operation, however, is done mostly with horizontally polarized antennas.

Yagi beam antennas can have either horizontal or vertical polarization. If the antenna elements are parallel to the earth, or horizontal, the antenna will produce horizontally polarized waves.

If the elements are turned to be perpendicular to the

earth, the antenna will produce vertically polarized waves. Figure 7-37 shows two Yagi antennas. The antenna at A produces horizontally polarized waves and the one at B produces vertically polarized waves. Notice that the boom is horizontal in both cases.

Most man-made noise tends to be vertically polarized. Thus, a horizontally polarized antenna will receive less noise of this type than a vertical antenna will.

[Turn to Chapter 13 and study questions T9B01 through T9B05. Review the material in this section as needed.]

MORE ABOUT BEAM ANTENNAS

In most multiple-element antennas, the additional elements are not directly connected to the feed line. They receive power by mutual coupling from the *driven element*. The driven element is the element connected to the feed line. The additional elements then reradiate the power in the proper phase relationship. Proper phasing achieves *gain* or *directivity* over a simple half-wavelength dipole. These elements are called **parasitic elements**. Beam antennas that use parasitic elements are sometimes called *parasitic beam antennas*.

There are two types of parasitic elements. A *director* is generally shorter than the driven element. A director is located at the front of the antenna. A *reflector* is generally longer than the driven element. A reflector is located at the back of the antenna. See Figure 7-38. The direction of maximum radiation from a parasitic beam antenna is from the reflector through the driven element to the director. The term *major lobe* refers to the region of maximum radiation from a directional antenna. The major lobe is also sometimes referred to as the *main lobe*. Communication in different directions may be achieved by rotating the array in the *azimuthal*, or *horizontal, plane*, to point it in different compass directions.

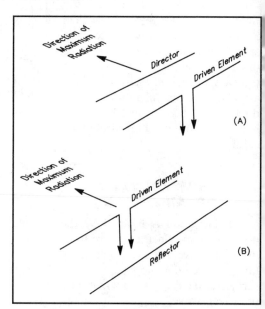

Figure 7-38—In a directional antenna, the reflector element is placed behind the driven element. The director goes in front of the driven element.

Yagi Arrays

The Yagi arrays in Figure 7-39 are examples of antennas that make use of parasitic elements to produce a *unidirectional radiation pattern*. (This means most of the radio signal goes in one direction.) A Yagi antenna has at least two elements. One element is a driven element. The other elements are directors and/or reflectors. These elements are usually parallel to each other and made of straight metal tubing. They line up with each other along the antenna *boom*.

Though typical HF antennas of this type have three elements, some may have six or more. Multiband Yagi antennas have many elements. Some elements work on some frequencies and others are used for different frequencies. The radiation pattern of the Yagi antenna is shown in Figure 7-33. This pattern indicates that the antenna will reject signals coming from the sides and back. It selects mainly those signals from a desired direction. There are several types of Yagi antennas.

The length of the driven element in the most common type of Yagi antenna is approximately an electrical half-wavelength. This means that Yagi antennas are most often used for the 20-meter band (14.0 to 14.35 MHz) and higher frequencies. Yagis for frequencies below 14 MHz are very large. They require special construction techniques, heavy-duty supporting towers and large rotators. Yagis for 40 meters are fairly common. Rotatable Yagis for 80 meters are few and far between. You can overcome the mechanical difficulties of large Yagis. Some amateurs build nonrotatable Yagi antennas for 40 and 80 meters, with wire elements supported on both ends.

In a three-element beam, the director is approximately 5% shorter than the driven element. The reflector is approximately 5% longer than the driven element. These lengths can vary considerably, however. The

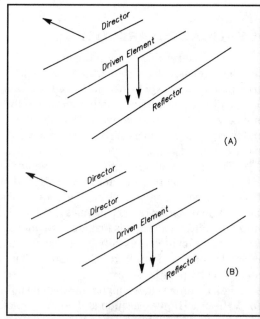

Figure 7-39—At A, a three-element Yagi. At B, a four-element Yagi with two directors.

Figure 7-40—The cubical quad antenna. The total length of the driven element is about one wavelength. This antenna can be fed directly with coaxial cable.

$$\text{Driven Element (Overall FT)} = \frac{1005}{f(MHz)}$$

$$\text{Reflector (Overall FT)} = \frac{1030}{f(MHz)}$$

Figure 7-41—The feed point of a quad antenna determines the polarization. Fed in the middle of the bottom side (as shown at A) or at the bottom corner (as shown at B), the antenna produces a horizontally polarized wave. Vertical polarization is produced by feeding the antenna at the side in either case.

Figure 7-42—A delta loop antenna. The total length of each element can be found using the same equations as used for a quad antenna. The antenna is fed with a gamma match.

$$\text{DRIVEN ELEMENT (OVERALL FT)} = \frac{1005}{f(MHz)}$$

$$\text{REFLECTOR (OVERALL FT)} = \frac{1030}{f(MHz)}$$

DELTA LOOP

actual lengths depend on the spacing between elements and the diameter of the elements. Whether the elements are made from tapered or cylindrical tubing also makes a difference.

CUBICAL QUAD ANTENNAS

The **cubical quad antenna** also uses parasitic elements. This antenna is sometimes simply called a *quad*. The two elements of the quad antenna are usually wire loops. The total length of the wire in the driven element is approximately one electrical wavelength. A typical quad, shown in Figure 7-40, has two elements—a *driven element* and a *reflector*. A two-element quad could also use a driven element with a *director*. You can add more elements, such as a reflector and one or more directors. The radiation pattern of a typical quad is similar to that of the Yagi shown in Figure 7-33.

The elements of the quad are usually square. Each loop is about an electrical wavelength long. Each side of the square is about ¼-λ long.

The polarization of the signal from a quad antenna can be changed. Polarization is determined by where the feed point is located on the driven element. See Figure 7-41. If the feed point is located in the center of a horizontal side, parallel to the earth's surface, the transmitted wave will be horizontally polarized. When the antenna is fed in the center of a vertical side, the transmitted wave is vertically polarized. We can turn the antenna 45 degrees, so it looks like a diamond. When the antenna is fed at the bottom corner, the transmitted wave is horizontally polarized. If the antenna is fed at a side corner, the transmitted wave is vertically polarized.

DELTA LOOP ANTENNAS

The **delta loop antenna**, shown in Figure 7-42, is a variation on the quad. A delta loop antenna has triangular elements, rather than square. The total loop length is still one wavelength. Divide the total length by 3 to find the length of each side of the elements. The radiation pattern of a delta loop is similar to that of the quad and Yagi, shown in Figure 7-33.

[Congratulations. This completes your study of antennas for your Technician exam. You will probably want to learn more about antennas later. After you pass the exam and have your license you'll be able to experiment with lots of antenna types. Now turn to Chapter 13 and study questions T9A01 through T9A08. Review this section if any of these questions give you difficulty.]

KEY WORDS

Ground connection—A connection made to the earth for electrical safety. This connection can be made inside (to a cold-water pipe) or outside (to a **ground rod**).

Ground rod—A copper or copper-clad steel rod that is driven into the earth. A heavy copper wire from the ham shack connects all station equipment to the ground rod.

Key-operated on-off switch—A good way to prevent unauthorized persons from using your station is to install a key-operated switch that controls station power.

Lightning protection—There are several ways to help prevent lightning damage to your equipment (and your house), among them unplugging equipment, disconnecting antenna feed lines and using a lightning arrestor.

RF safety—Preventing injury or illness to humans from the effects of radio-frequency energy.

Safety interlock—A switch that automatically turns off ac power to a piece of equipment when the top cover is removed.

Station grounding—Connecting all station equipment to a good earth ground improves both safety and station performance.

Chapter 8

Putting It All Together

Y ou have selected your equipment and put up an antenna. Now it's time to turn this collection into a radio station. Once you've connected the various pieces of gear, you'll be ready to begin making on-the-air contacts with your new license.

This chapter discusses how to set up a station, and why it's important to do so with safety in mind. In addition, we'll look at some common and useful station accessories that will make operating your station more enjoyable.

First, we'll discuss setting up a home station. Later, we'll take a look at installing a mobile station.

HOME STATIONS

LOCATION

First, give some thought to station location. Hams put their equipment in many places. Some use the basement or attic, while others choose the den, kitchen, closet or a spare bedroom. Some hams with limited space build their station

into a small closet. A fold-out shelf and folding chair form the operating position.

Where you put your station depends on the room you have available and on your personal tastes. There are, however, several things to keep in mind while searching for the best place. The photos show several ways amateur stations can be arranged.

One often-overlooked requirement for a good station location is adequate electrical service. Eventually you will have several pieces of station equipment and accessories, many of which will require power to operate. Be sure that at least one, and preferably several, electrical outlets are located near your future operating position. Be sure the outlets provide the proper voltage and current for your rig. Most modern radios require only a few amperes. You may run into problems, however, if your shack is on the same circuit as the air conditioner or washing machine. The total current drawn at any one time must not exceed rated limits. Someday you'll probably upgrade and may want to purchase a linear amplifier. If so, you should have a 240-volt line in the shack, or the capability to get one.

Another must for your station is a good **ground connection**. A good ground not only reduces the possibility of electrical shock but also improves the performance of your station. By connecting all of your equipment to ground, you will help to avoid stray radio-frequency (RF) current in the shack. Stray RF can cause equipment to malfunction. A good ground can also help reduce the possibility of interference. The wire connecting your station to an earth ground should be as short as possible.

How you arrange your station equipment is a matter of personal taste and available space. The most important considerations are safe operation and a pleasing and convenient layout.

Basement and first-floor locations generally make it easier to provide a good ground connection for your station. You can also find a way to put your ham shack on the second or third floor or even on the top floor of a high-rise apartment, for that matter.

Your radio station will require a feed line of some sort to connect the antenna to the radio. So you will need a convenient means of getting the feed line into the shack. There

are many ways of doing this, but one of the most effective requires only a window. Many hams simply replace the glass pane in a nearby window with a clear acrylic panel. The panel can be drilled to accept as many feed lines as needed. Special threaded "feedthrough" connectors make the job easy. If you decide to relocate your station, the glass pane can be replaced. This will restore the window to its original condition.

A simpler, more temporary approach is to pass the feed

lines through the open window and close it gently. Do this carefully, because crushed coaxial cable will give you nothing but trouble. Be sure to secure the window with a block of wood or some other "locking" device. Your new equipment, seen through an open window, may tempt an unwanted visitor!

Another important requirement for your station is com-

fort. The space should be large enough so you can spread out as needed. Operating from a telephone booth isn't much fun! Because you will probably be spending some time in your shack (an understatement!), be sure it will be warm in the winter and cool in the summer. It should also be as dry as possible. High humidity can cause such equipment problems as high-voltage arcing and switch-contact failure.

If your station is in an area often used by other family members, be sure they know what they shouldn't touch. You should have some means of ensuring that no "unauthorized" person can use your equipment. One way to do this is to install a **key-operated on-off switch** in the equipment power line. When the switch is turned off and you have the key in your pocket, you will be sure no one can misuse your station.

You will eventually want to operate late at night to snag the "rare ones" on 80 and 40 meters. Although a Morse code or voice contact is music to your ears, it may not endear you to a sleeping family. Putting your station in a bedroom shared with others may not be the best idea. You can keep the "music" to yourself, however, by using a good pair of headphones when operating your station.

WHAT IS A GOOD GROUND?

Since **station grounding** is a good idea for several reasons, it's worth discussing in detail. All station equipment should be tied to a good ground. Most amateurs connect their

Figure 8-1—An effective station ground looks like this: all equipment is bonded together with a strong conductor such as copper flashing or coaxial cable braid. It then ties into a good earth ground—an 8-foot ground rod located as close to the station as possible.

equipment to a **ground rod** driven into the ground as close to the shack as possible. For a few dollars you can purchase a ½- or ⅝-inch-diameter, 8-foot-long ground rod at any electrical supply store. (Eight feet is the shortest practical length for a station ground rod.) Drive this copper-clad steel rod into the ground outside of your house, as close to your station location as possible. (Don't settle for the short, thin "ground rods" sold by some discount electronics stores. The copper cladding on the outside of most of these steel rods is very thin. They begin to rust almost immediately when they are put in the ground.)

Run a heavy copper wire (number 8 or larger) from your shack and attach it to the rod with a clamp. You can purchase the clamp when you buy the ground rod. Heavy copper strap or flashing (sold at hardware or roofing-supply stores) is even better. The braid from a piece of RG-8 coaxial cable also makes a good ground cable. Figure 8-1 shows one method of grounding each piece of equipment in your station. It's important to connect the chassis of each piece of station equipment to an effective ground connection. Keep the cable between your station and the earth ground as short as possible.

The National Electrical Code requires all ground rods to be connected to form a single grounding system. These connections must use number 8 or larger copper wire. Have a qualified electrician perform this work if you are not familiar with the National Electrical Code and common wiring practices.

Some hams ground their station equipment by connecting the ground wire to a cold-water pipe. Caution is in order here. If you live in an apartment or have your shack in an attic, be careful. The cold-water pipe near your transmitter may follow such a long and winding path to the earth that it may not act as a ground at all! It may, in fact, act as an antenna, radiating RF energy—exactly what you don't want it to do.

Beware, too, of nonmetallic cold-water pipes. PVC and other plastic pipes are effective insulators. There may be a piece of copper water pipe running close to your station. If there is a piece of PVC pipe connected between that spot and where the water line enters your house, however, you will not have a ground connection!

When your station is not being used, it is also important to ground all antennas, feed lines and rotator cables for effective **lightning protection**. Lightning hazards are discussed later in the chapter.

[It's time to study the following questions in Chapter 12: N4A01, N4A03, N4A04 and N4A05, and N4A08 through N4A11. If you have questions after reviewing these, reread the material in this section.]

ARRANGING YOUR EQUIPMENT

Before you set everything up and hook up the cables, think about where you want each piece of equipment. While there is no one best layout for a ham station, some general rules do apply. Of course, the location you've chosen for your station may limit your choices a bit. For example, if you're going to put the station in the basement, you may have a lot

Figure 8-2—The "business side" of a typical station. Some radios have built-in antenna tuners and keyers.

of space. If you'll be using a corner of the bedroom or den, however, you may have to keep the equipment in a small area.

Generally, the piece of gear that requires the most adjustment is the transceiver (or receiver if you have one). Make sure you can conveniently reach its controls, keeping in mind which hand you're going to use to make the adjustments. It doesn't make much sense to put the transceiver on the left side of the desk or table if you're going to adjust it with your right hand. Once you've found the best location for the transceiver, you can position the rest of your equipment around it.

If you'll be using Morse code, placement of the sending device (key, keyer or keyboard) is also very important. It must be easy to reach with the hand you send with, and placed so your arm will be supported when you're using it. Try to position it away from anything dangerous, such as sharp corners and edges, rough surfaces and electrical wires.

If you have room for a desk or table long enough to hold all your gear, you might prefer to keep everything on one surface. If you're pressed for space, however, a shelf built above the desk top is a good solution. To ensure that there is adequate air circulation above your transceiver, leave at least 3 inches of space between the top of the tallest unit on the desktop and the bottom of your shelf. Try to restrict the shelf space to lightweight items that you don't have to adjust very often, such as the station clock, SWR meter and antenna-rotator control box.

After you have a good idea of where you want everything, you can start connecting the cables. Figure 8-2 shows how an amateur station, including some common accessories, can be connected. The "Not for Technicians Only" section of this chapter discusses these accessories in more detail.

MOBILE STATIONS

Many VHF/UHF-oriented hams start with a hand-held transceiver. Others buy a radio designed for mobile operation. Either type can be used in a car or truck (or boat or airplane). The hand-held radio gives maximum convenience, as it can be removed easily and used anyplace. The mobile radio is ideal for those who want the advantages of a permanent installation.

Hand-held radios can be powered from your vehicle's cigarette lighter. Most hand-helds have an optional cable available for this purpose. Operating from your vehicle's electrical system saves the radio's battery for "foot mobile" operation.

Installing a mobile transceiver requires some thought. One type of mobile radio has a detachable front panel. You mount the light, thin panel on or under the instrument panel, and hide the rest of it in the trunk. The more traditional (one-piece) type of transceiver needs extra support for mounting beneath the instrument panel. Use machine screws and nuts, not sheet-metal screws. Figure 8-3 shows a neat in-dash installation.

A clean (but often challenging) way of installing a mobile radio is to build it into the dashboard or console. Keep resale value in mind as you plan your installation: The person who buys your car may not appreciate the "benefit" of your mobile installation.

It's best to connect a permanently installed radio's power cable directly to the vehicle battery. This minimizes voltage drops and electrical noise. Although one side of the battery is connected to the car frame, you should run positive *and* negative wires, and install fuses in series with both the red and black leads. The fuses should be located *at the battery*. See Figures 8-4 and 8-5.

Figure 8-3—In-dash installation of a mobile transceiver is both convenient and attractive.

You'll want to prevent unauthorized persons from using your mobile radio once it's installed. You can do that easily by disconnecting and removing the microphone when you leave the vehicle.

[Now turn to Chapter 12 and study question N4A02. Review this section if you have any difficulty with this question.]

[**If you're preparing for the Technician exam**, also see question T4A06 in Chapter 13.]

Figure 8-4—To protect your equipment against being zapped, install fuses on both the positive and negative leads, close to the battery terminals.

Figure 8-5—A wiring diagram for a typical mobile transceiver installation.

SAFETY

It is important to arrange your home, portable or mobile station so it is safe for you and for any visitors. There are two types of potential hazards to keep in mind: electrical and RF. One noteworthy type of electrical hazard is lightning.

ELECTRICAL SAFETY

Wiring should be neat and out of the way. Make sure there is no way you can tangle your feet in loose wires. Don't leave any voltage (regardless of how low) exposed.

A main power switch allows you to turn all station equipment on or off at once. This saves needless wear on the equipment power switches and, more importantly, is a highly recommended safety item. Make sure every member of your family knows how to turn off the power to your workbench and operating position. If you ever receive an electrical shock and cannot free yourself, the main disconnect switch will help your rescuer come to your aid quickly.

You must take special safety precautions if young children can come into your shack. Locking your station into its own room is an ideal solution. Few of us have this luxury, however. There are other ways to secure your equipment so unauthorized persons won't be able to use it. For example, you can build your station into a closet or cabinet that can be locked. If the equipment is in a nonsecure area, install a key-operated power switch and keep track of the keys. Even a simple toggle switch, if well-hidden, can help keep your station secure.

The same principles apply if you set up your station for public display. FCC rules require that only a licensed control operator may put the station on the air. To ensure that unauthorized persons will not be able to transmit, you may have to remove a microphone or control cable temporarily while the equipment is left unattended.

Whether you use commercially built equipment or homemade gear, you should never operate the equipment without proper shielding over all circuit components. Dangerous voltages may be exposed on the chassis-mounted components. Therefore, all equipment should have a protective shield on the top, the bottom and all sides. An enclosure also prevents unwanted signals from entering a receiver, or from being radiated by a transmitter.

There should be a device that turns power off automatically if you remove the shielding. Such a **safety interlock** reduces the danger of electrical shock from high voltages when you open the cabinet.

In arranging your station, always think of safety as well as comfort and convenience. After you've arranged things where you think they should be, try to find fault with your layout. Don't feel satisfied with your station until you can't find anything to improve.

All the equipment in your station should be connected to earth ground to prevent electrical shocks. Earlier in this chapter, we talked about what makes a "good ground." Briefly, you should connect your equipment to the earth with as short a cable as possible. If you live in an apartment, you may have to use an indoor cold-water pipe rather than an outside ground rod.

Lightning Protection

The lightning hazard from an antenna is often exaggerated. Ordinary amateur antennas are no more likely to be hit by a direct strike than any other object of the same height in the neighborhood. Just the same, lightning does strike thousands of homes each year, so it doesn't hurt to be careful. When your station is not in use, you should ground the antenna and rotator cables and unplug your equipment. An ungrounded antenna can pick up large electrical charges from storms in the area. These charges can damage your equipment (particularly receivers) if you don't take precautions.

Most commercial beam and vertical antennas are grounded for lightning protection through the tower itself. Of course, the tower must be grounded, too. If you use a roof mount, run a heavy ground wire from the mount to a ground rod. Dipoles and end-fed wires are not grounded. Disconnect the antenna feed line from your equipment and use an alligator-clip lead to connect both sides of it to your station ground.

Storm clouds often carry dangerous electrical charges that are coupled into high objects (like amateur antennas) without visible lightning. You may be operating and suddenly hear a "SNAP!" in your antenna tuner, or your receiver may go dead. You can usually hear an increase in static crashes in your receiver well in advance of a thunderstorm. Be safe. When it sounds like a thunderstorm is headed your way, get off the air. If the weather forecast is for thunderstorms, don't operate! Snow and rain also generate static charges on antennas, but usually not enough to damage equipment.

The best protection against lightning is to disconnect all antennas and power cords when you aren't on the air. It takes time to hook up everything when you want to operate again, but you will protect your station and your home if you follow this simple precaution. By the way, power companies recommend that you unplug *all* electronic appliances, including TVs, VCRs and computers, when a storm threatens.

Why unplug your equipment if the antennas are disconnected? Lightning can still find its way into your equipment through the power cord. Power lines can act as long antennas, picking up sizable charges during a storm. Simply turning off the main circuit breaker is not enough—lightning can easily jump over the circuit breaker contacts and find its way into your equipment.

You may decide to leave your antennas connected and your equipment plugged in except during peak thunderstorm months. If so, you can still protect the equipment from unexpected storms. One simple step you can take is to install a grounding switch, as shown in Figure 8-6. A small knife switch will allow you to ground your feed line when you are not on the air. It will not disturb the normal operation of your

Figure 8-6—A heavy-duty knife switch can be used to connect the wires in your antenna feed line to ground. A clip lead to a ground wire can be used instead of the switch. This will prevent a large static build-up on your antenna. It will also prevent equipment damage caused by voltage on your antenna produced by a nearby lightning strike. There is no sure way to prevent a direct hit by lightning, however.

station (with the switch open, of course!) if the lead from the feed line to the switch is no more than a couple of inches long. An alligator clip can be used instead of the switch. Whatever you use, don't forget to disconnect the ground when you transmit. This precaution is useful only on the HF bands. The switch will cause high SWR if used at VHF and UHF.

Another device that can help protect your equipment in an electrical storm is a *lightning arrestor*. This device connects permanently between your feed line and the ground. When the charge on your antenna builds up to a large enough potential, the lightning arrestor will "fire." This shorts the charge to ground—not through your station. A lightning arrestor can help prevent serious damage to your equipment. Most, however, don't work fast enough to protect your station completely. Lightning arrestors are useful for commercial stations and public service (fire and police) stations that must remain on the air regardless of the weather. Unless you are actually handling emergency communications, you shouldn't rely on them alone to protect your equipment, home or life.

[Now turn to Chapter 12 and study questions N4A06, N4A07 and N4B07. Review this section if you have any difficulty with these questions.]

RF SAFETY

We all know basic safety precautions to follow in the ham shack. We know to pull the plug before taking the covers off a transceiver or power supply, for example. Another important concern, one often overlooked, is **RF safety**. This involves minimizing human exposure to strong radio-

frequency fields. These potentially dangerous fields occur near or around antennas.

Biological effects of RF exposure have been studied for several decades. We know that body tissues subjected to large amounts of RF energy may suffer heat damage. It is possible to receive an RF burn from touching an antenna that is being used for transmitting. You don't have to come into direct contact with an antenna to damage body tissues, however. Just being present in a strong RF field can cause problems.

Taken to extremes, we could compare the effects of RF exposure to the way a microwave oven cooks food. A typical microwave oven uses a 500-W RF source operating at 2450 MHz. The Novice 1270-MHz band is about half this frequency, and so can have similar effects. Of course, the microwave oven is designed to concentrate its RF power for heating food and is not directly comparable to Amateur Radio operations. Some studies have shown that persons who are exposed to strong RF fields over a period of time may be at increased risk to develop certain kinds of cancer.

There is no cause for alarm in most amateur installations. But be aware that exposure to strong RF fields—which we cannot see, smell, hear or touch—can cause health risks.

The amount of RF energy that the body absorbs depends on the radio frequency. The body absorbs RF energy most efficiently in the VHF range (30 to 300 MHz). Absorption is greatest if the antenna orientation is parallel to the body (vertically polarized). Sensitive parts of the body, such as the eyes, are particularly prone to damage from RF energy.

Most amateur operation is with relatively low RF power and is intermittent—the transmitter is not operating continuously. Hams spend more time listening than transmitting. If you use modes such as RTTY and FM, in which the RF carrier is present continuously at full power, you'll need to pay more attention to RF safety.

RF Safety Guidelines

Take the time to study and follow these general guidelines to minimize your exposure to RF fields. Most of these guidelines are just common sense and good amateur practice.

• Confine RF radiation to the antenna, where it belongs. Provide a good earth ground for your equipment. Particularly at VHF and UHF, poor-quality feed line and improperly installed connectors can be a source of unwanted RF radiation. Use only good-quality coaxial cable. Be sure the connectors are of good quality and are properly installed. Good-quality coaxial cable and connectors will also help reduce RF losses in your system.

• Don't operate RF power amplifiers or transmitters with the covers or shielding removed. This practice helps you avoid both electric shock hazards and RF safety hazards. A safety interlock prevents the gear from being turned on accidentally while the shielding is off. This is especially important for VHF and UHF equipment. When reassembling transmitting equipment, replace all the screws that hold the RF compartment shielding in place. Tighten all the screws securely before applying power to the equipment.

• In high-power operation in the HF and VHF region, keep the antenna away from people. Humans should not be allowed within 10 to 15 feet of vertical antennas. This is especially important with higher power, high-duty-cycle operation (such as FM or RTTY). Amateur antennas that are mounted on towers and masts, away from people, pose no exposure problem.

• Always install your antennas where people and animals cannot touch them.

• When using mobile equipment with 10-W RF power output or more, do not transmit if anyone is standing within 2 feet of the antenna.

• The best location for a VHF/UHF mobile antenna—from an RF safety standpoint and for the best radiated-signal pattern—is in the middle of the automobile roof. This position best protects the car's occupants.

• When using a hand-held transceiver with RF power output of several watts or more, maintain at least 1 to 2 inches separation between the antenna and your forehead to protect your eyes. It is recommended that hand-held radios have a power of no more than 7 watts.

• Never touch an antenna that has RF power applied. Be sure RF power is off and stays off before working on or adjusting an antenna. Also, make sure any nearby antennas are deactivated. Never have someone else transmit into the antenna and monitor the SWR while you are making adjustments. When matching an antenna, you should turn the transmitter off and make the adjustment. Then, back away to a safe distance before turning the transmitter on again to check your work.

• During transmissions, never point a high-gain UHF or microwave antenna (such as a parabolic dish) toward people or animals.

• Never look into the open end of a UHF or microwave waveguide feed line that is carrying RF power. Never point the open end of a UHF waveguide that is carrying RF power toward people or animals. Make sure that all waveguide connections are tightly secured.

The following additional safety guidelines were developed in response to scientific studies done in the mid to late 1980s. (From the October 1989 *QST* article, "Is Amateur Radio Hazardous to Our Health?" by Ivan A. Shulman, MD, WC2S.)

• Do not stand or sit close to your power supplies or linear amplifiers while operating, even when they are in standby mode.

• Stay at least 24 inches away from any power transformer, electrical fans or other source of high level 60-Hz magnetic fields while in operation.

• Do not tune up or operate a high-powered linear amplifier while the shields or covers are off.

• Run your transmission lines away from where you or other people sit in or near your shack.

• Properly terminated coaxial transmission feed lines should be used in preference to open-wire or end-fed antenna installations that come directly into the transmitter, as the RF radiated from a coaxial feed line is much lower.

• Use common sense about placing all antennas well away from yourself and others, especially for VHF, UHF and particularly microwave application. No one should be in the near field of an antenna.

• No person should be near any transmitting antenna while it is operating. This is especially true for mobile or ground-mounted vertical antennas. The use of indoor transmitting antennas that are close to people in a house or apartment should be reconsidered.

• Use the minimum power needed to make a QSO, especially if the antenna is less than 35 feet above the ground.

• Hand-held radios should be used on the lowest power setting needed to carry out communications.

• Hand-helds should be kept as far from the head as possible when operating. The use of a separate microphone or similar device is recommended.

• Transmissions using a hand-held radio should be kept as short as possible.

• Power density measurements should be made before running more than 25 watts in a VHF mobile installation, particularly if the antenna is rear-deck mounted and passengers may ride in the back seat. The safest mobile antenna location is in the center of the metal roof.

• The development of an accurate inexpensive power-density meter would be of major benefit to the Amateur Radio community so that RF power-density measurements could be taken in all radio installations. Because of the current high cost of such devices, groups of amateurs or clubs may wish to purchase one and share in its use.

• Soldering should only be done in a well ventilated area. A small fan should be used to blow away toxic fumes.

• When using toxic chemicals, such as when etching PC boards or repairing fiberglass, wear gloves and goggles, use proper tools, and avoid contact with any of the chemicals. If accidentally contaminated, wash off the compounds immediately with lots of water. Again, the importance of always working in a well ventilated area with personal protective covering cannot be overemphasized.

• Hazardous chemicals, such as those in the PCB class, are used in some capacitors and dummy loads. Use extreme care in handling these materials, and consult with the appropriate local authorities to determine the proper means of disposing of these chemicals in an environmentally responsible way.

[It's time to study questions N4B01 through N4B06 in Chapter 12. Review this section if you have difficulty answering any of these questions.]

Additional Station Set-up Information for the Technician Exam

KEY WORDS

Ammeter—A test instrument that measures current.

ANSI RF protection guide—Standards formulated by the American National Standards Institute that provide RF exposure guidelines.

Crystal calibrator—(See **Marker generator**.)

Directional wattmeter—(See **Wattmeter**.)

Dummy antenna—A station accessory that allows you to test or adjust transmitting equipment without sending a signal out over the air. Also called *dummy load*.

Marker generator—A high-stability oscillator that produces reference signals at known frequency intervals. It can be used to calibrate receiver and transmitter tuning dials. Also called a **crystal calibrator**.

Multimeter—An electronic test instrument used to measure current, voltage and resistance in a circuit. Describes all meters capable of making these measurements, such as the volt-ohm-milliammeter (VOM), vacuum-tube voltmeter (VTVM) and field-effect transistor VOM (FET VOM).

National Electrical Code—A set of guidelines governing electrical safety, including antennas.

Radio Frequency Protection Guide of the American National Standards Institute—Standards that recommend RF exposure limits for human beings.

Reflectometer—A test instrument used to indicate standing wave ratio (SWR) by measuring the forward power (power from the transmitter) and reflected power (power returned from the antenna system).

S meter—A meter in a receiver that shows the relative strength of a received signal.

Signal generator—A test instrument that produces a stable low-level radio-frequency signal. The signal can be set to a specific frequency and used to troubleshoot RF equipment.

SWR meter—(See **Reflectometer**.)

Voltmeter—A test instrument used to measure voltage.

Wattmeter—Also called a *power meter*, a test instrument used to measure the power output (in watts) of a transmitter. A **directional wattmeter** measures both forward and reflected power.

WWV and **WWVH**—Radio stations that broadcast the precise time on precise frequencies. Hams tune in to these stations to calibrate transceiver dials (and to set their clocks).

The remainder of this chapter covers the items related to setting up a station that you'll need to know for your Technician exam. These include electrical wiring safety, power-supply safety, test equipment, frequency accuracy, antenna measurements, station accessories and RF safety.

ELECTRICAL WIRING SAFETY

Your station equipment makes use of ac line voltage. This voltage can be dangerous. The equipment also generates additional potentially lethal voltages of its own. You should be familiar with some basic precautions. Your own safety and that of others depends on it.

POWER-LINE CONNECTIONS

In most residential systems, three wires come in from outside to the distribution board. Older systems may use only two wires. In the three-wire system, the voltage between the two "hot" wires is normally 240. The third wire is neutral and is grounded. Half of the total voltage appears between each of the hot wires and neutral, as shown in Figure 8-7. Lights, appliances and 120-V outlets are divided as evenly as possible between the two sides of this circuit. Half of the load connects between one hot wire and the neutral. The other half of the load connects between the other hot wire and the neutral.

Heavy appliances, such as electric stoves and most high-power amateur amplifiers, are designed for 240-V operation. Connect these across the two hot wires. Both ungrounded wires should be fused. The neutral wire should never have a fuse or switch in it.

High-power amateur amplifiers use the full 240 V. It only takes half as much current to supply the same power as with a 120-V line. The power supply circuit efficiency improves with the higher voltage and lower current. A separate 240-V circuit for the amplifier ensures that you stay within the capacity of the 120-V circuits in your shack. An amplifier drawing more current than the house wiring can handle could cause the house lights to dim. This is because the heavy current causes the power-line voltage to drop.

Three-Wire 120-V Power Cords

State and national electrical-safety codes require three-wire power cords on many 120-V tools and appliances. Two of the conductors (the "hot" and "neutral" wires) power the device. The third conductor (the safety ground wire) connects to the metal frame of the device. See Figure 8-8. The "hot" wire is usually black or red. The "neutral" wire is white. The frame/ground wire is green or sometimes bare.

Let's look at the power cord connections for a transmitter power supply. The power supply will have a transformer in it. When you attach the power cord, the black wire will attach to one end of the fuse. The white wire will go to the side

Figure 8-7—In home electrical systems, two wires carrying 240 volts are split between the house circuits carrying 120 volts each. Some heavy appliances use the full 240 V. As shown, a fuse should be placed in each hot wire, but no fuse or switch should be placed in the neutral wire.

Figure 8-8—Correct wiring technique for 120-V power cords and receptacles. The white wire is neutral, the green wire is ground and the black or red wire is the hot lead. Notice that the receptacles are shown as viewed from the back, or wiring side.

of the transformer primary winding without a fuse. The green (or bare) wire will attach to the chassis.

When plugged into a properly wired mating receptacle, a three-contact plug connects the third conductor to an earth ground. This grounds the appliance chassis or frame and prevents the possibility of electric shock. A defective power cord that shorts to the case of the appliance will simply blow a fuse. Without the ground connection, the case could carry the full line voltage, presenting a severe shock hazard. All commercially manufactured electronic test equipment and most ac-operated amateur equipment uses these three-wire cords. Adapters are available for use where older electrical installations do not have mating receptacles. The lug from the adapter must be attached under the cover-plate screw. The outlet (and outlet box) must be grounded for this to be effective. Power wires coming into the electric box inside a flexible metal covering provide grounding through the metal covering. The common name for this type of wire is armored cable.

A "polarized" two-wire plug has one blade that is wider than the other. The mating receptacle will accept the plug only one way. This ensures that the hot and neutral wires in the appliance connect to the appropriate wires in the house electrical system. Consider what happens without this polarized plug and receptacle. The power switch in the equipment will be in the hot wire when the plug is inserted one way. It will be in the neutral wire when inserted the other way. This can present a dangerous condition. It is possible for the equipment to be "hot" even with the switch off. With the switch in the neutral line, the hot line may be connected to the equipment chassis. An unsuspecting operator could form a path to ground by touching the case, and might be electrocuted or receive a nasty shock!

Wiring an outlet or lamp socket properly is important. The black (hot) wire should be connected to the brass terminal on the lamp socket or outlet. The white (neutral) wire should be connected to the white or silver-colored terminal. This will ensure that the proper blade of the plug connects to the hot wire (that the polarity is correct). This practice is especially important when wiring lamp sockets. The brass screw of the socket connects to the center pin in the socket. With this pin as the "hot" connection, there is much less shock hazard. Anyone unscrewing a bulb from a correctly wired socket has to reach inside the socket to get a shock. In an incorrectly wired socket, the screw threads of the bulb are "hot." A dangerous shock may result when replacing a bulb.

Current Capacity

There is another factor to take into account when you are wiring an electrical circuit. It is the current-handling capability (*ampacity*) of the wire. Table 8-1 shows the current-handling capability of some common in-wall residential wire sizes. The table shows that number 14 wire could be used for a circuit carrying 15 A. You must use number 12 (or larger) for a circuit carrying 20 A. Wires smaller than number 14

Table 8-1

Current-Carrying Capability of Some Common Wire Sizes

Copper Wire Size (AWG)	Allowable Ampacity (A)	Max Fuse or Circuit Breaker Size (A)	Permissible Load (A)[2]
6	55	50	40
8	40	40	32
10	30	30	24
12	25 (20)[1]	20	16
14	20 (15)[1]	15	12

[1]The National Electrical Code limits that fuse or circuit breaker size (and, as such, the maximum allowable circuit load) to 15 A for #14 AWG and to 20 A for #12 AWG conductors.

[2]Adjustments from these values may be possible. Consult your local electrical inspector, a licensed electrician or a registered electrical engineer for assistance.

---

may not be used for in-wall residential wiring.

To remain safe, don't overload the ac circuits in your home. The circuit breaker or fuse rating is the maximum load for the line at any one time. Simple addition can be used to calculate the current drawn on the circuit serving your ham radio equipment. Most equipment has the power requirements printed on the back. If not, the owner's manual should contain such information. Calculations must include any other household equipment or plugged-in appliances on the same line (lighting, air-conditioning, fans, and so on). The sum of all the power requirements connected to the same circuit must not exceed the permissible loads shown in Table 8-1.

Do not put a larger fuse in an existing circuit—too much current could be drawn (see Table 8-1 for the maximum fuse size for the existing wire size). The wires would become hot and a fire could result. Installing too large of a fuse or circuit breaker violates the **National Electrical Code** and most local ordinances. Such violations may also invalidate your fire insurance! The National Electrical Code is used across the US as the basis for electrical safety relating to power wiring and antennas.

One final precaution: It's important that all family members know the location of the main electrical box in your home, and how to turn it off. In case of electrical emergency, this precaution could save a life.

[Now turn to Chapter 13 and study questions T4A01 through T4A05 and question T4A07. Review this section if you have any difficulty.]

POWER-SUPPLY SAFETY

Safety must always receive careful consideration during the design and construction of any power supply. Power supplies can produce potentially lethal currents and voltages. Be careful to guard against accidental exposure to these currents and voltages. Use electrical tape, insulated tubing (spaghetti) or heat-shrink tubing to cover exposed wires. This includes component leads, component solder terminals and tie-down points. Whenever possible, connectors used to mate the power supply to the outside world should be of an insulated type. They should be designed to prevent accidental contact with the voltages present. AC power to the supply should be controlled by a clearly labeled front-panel switch. That way it can be seen and reached easily in an emergency.

All dangerous voltages in equipment should be made inaccessible. A good way to ensure this is to enclose all equipment in metal cabinets. That way no "hot" spots can be reached. Don't forget any component shafts that might protrude through the front panel. If a control shaft is hot, protect yourself from accidental contact by using an insulated shaft extension or insulated knob. Another precaution to take is to avoid wearing loose jewelry when working around potentially dangerous pieces of equipment, such as an amplifier.

Each metal enclosure must be effectively grounded. This is generally accomplished by connections to the equipment grounding conductor (green or bare wire) contained in the power cord or premises wiring system. If a connection is made to a separate ground rod, the ground rod must be bonded to the building grounding electrode system as described earlier in this chapter. Then if a failure occurs inside a piece of equipment, the metal case will never present a shock hazard. The fuse will blow or the circuit breaker will trip instead.

You should never underestimate the potential hazard when working with electricity. Table 8-2 shows some of the effects of electric current—as little as 100 mA, or 1/10 A, can be fatal! As the saying goes, "It's volts that jolts, but it's mills that kills." Low-voltage power supplies may seem safe, but even battery-powered equipment should be treated with respect. The minimum voltage considered dangerous to humans is 30 volts. These voltage and current ratings are only general guidelines. Automobile batteries are designed to provide very high current (as much as 200 A) for short periods when starting a car. This much current can kill you, even at 12 volts. You will feel pain if the shock current is in the range of 30 to 50 mA. (Question T4A10 indicates a current as small as 2 mA (1/500 A) can be painful.)

A few factors affect just how little voltage and current can be considered dangerous. One factor is skin resistance. The lower the resistance of the path, the more current that will pass through it. If you perspire heavily, you may get quite a bit more severe shock than if your skin was dry. Another factor is the path through the body to ground. As Figure 8-9 shows, the most dangerous path is from one hand to the other. This path passes directly through a person's heart. Even a very minimal current can cause heart failure and death. Current passing from one finger to another on the same hand will not have quite such a serious effect. For this reason, if you must troubleshoot a live circuit, keep one hand behind your back or in your pocket. If you do slip, the shock may not

Figure 8-9—The path from the electrical source to ground affects how severe an electrical shock will be. The most dangerous path (from hand to hand directly through the heart) is shown at A. The path from one finger to the other shown at B is not quite so dangerous.

Table 8-2

Effects of Electric Current Through the Body of an Average Person

Current (1 Second Contact)	Effect
1 mA	Just perceptible.
5 mA	Maximum harmless current.
10-20 mA	Lower limit for sustained muscular contractions.
30-50 mA	Pain
50 mA	Pain, possible fainting. "Can't let go" current.
100-300 mA	Normal heart rhythm disrupted. Electrocution if sustained current.
6 A	Sustained heart contraction. Burns if current density is high.

be as severe as if you were using both hands.

If you should discover someone who is being burned by high voltage, immediately turn off the power, call for help and give cardio-pulmonary resuscitation (CPR). Be sure all family members know how to turn off power at the main power switch. These measures could save the life of a friend or family member.

[Now turn to Chapter 13 and study exam questions T4A08 through T4A15. Review any topics necessary before proceeding.]

USING TEST EQUIPMENT

THE VOLTMETER

The **voltmeter** is an instrument used to measure voltage. It is a basic meter movement with a resistor in series, as shown in Figure 8-10. The current multiplied by the resistance will be the voltage drop across the resistance. An instrument used this way is calibrated in terms of the voltage drop across the resistor to read voltage.

A high value resistor provides a high impedance input for the meter. An ideal voltmeter would have an infinite input impedance. It would not affect the circuit under test in any way. Real-world voltmeters, however, have a finite value of input impedance. There is a possibility that the voltage you are measuring will change. This is because the meter adds some load to the circuit when you connect the voltmeter.

Use a voltmeter with a very high input impedance compared with the impedance of the circuit you are measuring. This prevents the voltmeter from drawing too much current from the circuit. Excessive current drawn from the circuit would significantly affect circuit operation. The input impedance of an ordinary voltmeter is about 20 kilohms per volt. Voltage measurements are made by placing the meter in parallel with the circuit voltage to be measured.

The range of a meter can be extended to measure a wide range of voltages. This is accomplished by adding resistance in series with it. Changing the multiplier resistor in Figure 8-10 changes the measuring range of the meter.

THE AMMETER

The **ammeter** depends on a current flowing through it to deflect the needle. The ammeter is placed in series with the circuit. That way, all the current flowing in the circuit must pass through the meter. Many times the meter cannot handle all the current. The range of the meter can be extended by placing resistors in parallel with the meter to provide a path for part of the current. A shunt resistor is shown in Figure 8-11. Selecting different shunt resistors changes the current-measuring range. The shunt-resistor values are calculated so the total circuit current can be read on the meter. Most ammeters have very low resistance, but low-impedance circuits require caution. There is a chance that the slight additional resistance of the series ammeter will disturb circuit operation.

Figure 8-10—When you use a voltmeter to measure voltage, the meter must be connected in parallel with the voltage you want to measure.

Figure 8-11—To measure current you must break the circuit at some point and connect the meter in series at the break. A shunt resistor expands the scale of the meter to measure higher currents than it could normally handle.

MULTIMETERS

A **multimeter** is a piece of test equipment that most amateurs should know how to use. The simplest kind of multimeter is the volt-ohm-milliammeter (VOM). As its name implies, a VOM measures voltage, resistance and current. VOMs use one basic meter movement for all functions. The movement requires a fixed amount of current (often 1 mA) for a full-scale reading. As shown in the two previous sections, resistors are connected in series or parallel to provide the proper voltage or current meter reading. In a VOM, a switch selects various ranges for voltage, resistance and current measurements. This switch places high-value dropping resistors in series with the meter movement for voltage measurement. It connects low-value shunt resistors in parallel with the movement for current measurement. These parallel and series resistors extend the range of the basic meter movement.

If you are going to purchase a VOM, buy one with the highest ohms-per-volt rating that you can find. Stay away from meters rated under 20,000 ohms per volt if you can.

Measuring resistance with a meter involves placing the meter leads across the component or circuit you wish to measure. Make sure to select the proper resistance scale. The full-scale-reading multipliers vary from 1 to 1000 and higher. The scale is usually compressed on the higher end of the range. See Figure 8-12. For best accuracy, keep the reading in the lower-resistance half of the scale. On most meters this is the right-hand side. Thus, if you want to measure a resistance of about 5000 ohms, select the R × 1000 scale. Then the meter will indicate 5.

A vacuum-tube voltmeter (VTVM) operates in the same manner as an ordinary VOM. There is one important difference. The meter in a VTVM is isolated from the circuit under test by a vacuum-tube dc amplifier. As a result, the only additional circuit loading is from the tube input impedance, which is very high. The standard VTVM input impedance is 11 MΩ. This is useful for measuring voltages in high-impedance circuits, such as vacuum-tube grid circuits and FET gate circuits.

Another type of meter is a field-effect transistor volt-ohm-milliammeter (FET VOM). This instrument uses a field-effect transistor (FET) to isolate the indicating meter from the circuit to be measured. FET VOMs have an input impedance of several megohms. They are the solid-state equivalent of a VTVM.

WATTMETERS

A **wattmeter** is a device in the transmission line that measures the power (in watts) coming out of a transmitter. Wattmeters are designed to operate at a certain line impedance, normally 50 ohms. Make sure the feed-line impedance is the same as the design impedance of the wattmeter. If impedances are different, any measurements will be inaccurate. Likewise, a power meter designed for use at 3-30 MHz (HF) will be inaccurate at VHF. For most accurate measure-

Figure 8-12—Resistance is measured across the component. For best accuracy the reading should be taken in the lower-resistance half of the scale.

ment, the wattmeter should be connected directly at the transceiver output (antenna) jack.

All wattmeters must contain some type of detection circuitry to enable you to measure power. The accuracy and upper frequency range of the wattmeter is limited by the detection device. Stray capacitance and coupling within the detection circuits can cause problems. A loss of sensitivity or accuracy occurs as the operating frequency is increased.

One type of wattmeter is a **directional wattmeter**. There are two kinds of directional wattmeters. One has a meter that reads forward power and another meter that reads reflected power. The other type has a single meter that can be switched to read either forward or reflected power.

You can use a directional wattmeter to measure the output power from your transmitter. Remember that the reflected power must be subtracted from the forward reading to give you the true forward power. The power reflected from the antenna will again be reflected by the transmitter. This power adds to the forward power reading on the meter. Let's say your transmitter power output is 100 watts and 10 watts of power is reflected from the antenna. Those 10 watts will be added into the forward reading on your meter when they are reflected from the transmitter. Your wattmeter reading would be incorrect. To find true forward power you must subtract the reflected measurement from the forward power measurement:

True forward power =

  Forward power reading – Reflected power reading

(Equation 8-1)

For example, suppose you have a wattmeter connected in the line from your transmitter. It gives a forward-power reading of 85 watts and a reflected-power reading of 15 watts. What is the true power out of your transmitter?

True forward power = 85 W − 15 W = 70 watts.

[Now turn to Chapter 13 and study all exam questions that begin with T4B as well as question T4C10. Review this section as needed.]

FREQUENCY ACCURACY

It's important to ensure that your transceiver frequency dial is accurate. If it isn't, you could transmit outside your allowed frequencies and be in violation of FCC rules. There are two good ways to ensure that your dial is accurate: use of a marker generator (crystal calibrator) and checking your dial by tuning in radio stations WWV or WWVH.

MARKER GENERATORS

The **marker generator** is a high-stability oscillator that produces an RF signal with a frequency that doesn't change. It is usually built into the receiver. This oscillator generates a series of signals that mark the exact edges of the amateur bands (and subbands, in some cases). It does this by oscillating at a low frequency that has harmonics falling on the desired frequencies. Most marker generators put out harmonics at 25, 50 or 100-kHz intervals. Since marker generators normally use crystal oscillators, they are often called **crystal calibrators**.

The marker generator is very useful. You can calibrate your receiver with it. You can then determine your precise transmitter frequency. First, turn on your marker generator (calibrator). This injects the calibrator signal into the receiver circuit. Then set your transceiver or receiver dial on the proper frequency marks. Calibration instructions are usually provided in your operator's manual. When the dial is calibrated, put your transmitter in the tune or spot position and read the frequency on the receiver dial.

The marker frequencies must be accurate. Now you know that the transmitter frequency is between the markers that show the ends of the band (or subband). In addition, the transmitter frequency must not be too close to the edge of the band (or subband). If it is, the sideband frequencies, especially in a voice transmission, will extend over the edge.

WWV AND WWVH

One easy way to calibrate your receiver frequency readout or dial is to tune in **WWV** (Fort Collins, Colorado) or **WWVH** (Kauai, Hawaii). These stations transmit reference signals on 1.5, 5, 10, 15 and 20 MHz. When you tune in WWV or WWVH, you'll hear a "tic" pulse every second and the precise time announced each minute.

SIGNAL GENERATORS

A **signal generator** produces a stable, low-level signal that can be set to a specific frequency. There are two different types of signal generators. *Audio-frequency signal generators* create signals in the audio range. *Radio-frequency signal generators* provide radio-frequency signals. These signals can be used to align or adjust tuned circuits for best performance. One common use of a signal generator is in the alignment of receivers. The generator can be adjusted to the desired signal frequency. Then the associated circuits can be adjusted for best operation. This is indicated by the appropriate output meter (maximum signal strength, for instance).

Sometimes a band of frequencies must be covered to chart the frequency response of a filter. For this application, a swept frequency generator is used. A swept frequency generator automatically sweeps back and forth over a selected range of frequencies.

You can use a signal generator to adjust the filter circuits in your transmitter. When you do, a dummy antenna must be connected to the transmitter output to avoid radiating a signal (dummy antennas are discussed in the "Station Accessories" section).

[Now turn to Chapter 13 and study questions T4C01 through T4C07. Review any material as necessary.]

ANTENNA SYSTEM MEASUREMENTS

A properly operating antenna system is essential for a top-quality amateur station. If you build your own antennas, you must be able to tune them for maximum operating efficiency. Even if you buy commercial antennas, there are tuning adjustments that must be made because of differences in mounting location. Height above ground and proximity to buildings and trees will have some effect on how your antenna operates. You should be able to monitor your antenna system periodically for signs of problems, and troubleshoot failures as necessary.

IMPEDANCE MATCH INDICATORS

A useful metering device is the **reflectometer (SWR meter)**. A reflectometer is a device used to measure *standing-wave ratio*, or *SWR*. SWR is a measure of the relationship between the amount of power traveling to the antenna and the power reflected back to the transmitter. The reflection is caused by *impedance mismatches* in the antenna system. The energy going from the transmitter to the antenna is represented by the forward, or incident, voltage. The energy reflected by the antenna is represented by the reflected voltage.

A reflectometer is placed in the transmission line between the transmitter and antenna. In the most common amateur application, the reflectometer is connected between the transmitter and Transmatch, as shown in Figure 8-13. A Transmatch is used to cancel out the capacitive or inductive component in the antenna impedance. With a Transmatch you can match the impedance of your antenna system to the transmitter output, normally 50 ohms. The reflectometer is used to indicate minimum reflected power as the Transmatch is adjusted. This indicates when the antenna system (including feed line) is matched to the transmitter output impedance.

To obtain a valid measure of the impedance match between an antenna and the feed line, place the SWR meter at

Figure 8-13—A reflectometer or SWR bridge should be connected between the transmitter and Transmatch in order to adjust the Transmatch properly.

the antenna feed point. If you put the meter at the transmitter end, feed line losses make SWR readings look lower than they really are. Most SWR meters are designed for HF use (from 3 to 30 MHz). These *may* also be useful at VHF, as long as the meter properly calibrates to full scale in the set position.

[Now turn to Chapter 13 and study questions T4C08, T4C09 and T4C11. Review any material as necessary.]

STATION ACCESSORIES

Until now, we have talked about the bare necessities—what you need to set up a basic Amateur Radio station. Few hams are content with the bare necessities, however. Most add accessories that make operating more convenient and enjoyable. Let's look at several popular ones: dummy antennas, antenna switches, keyers, station clocks and logbooks.

DUMMY ANTENNA

A **dummy antenna**, sometimes called a *dummy load*, is nothing more than a large resistor. It replaces your antenna when you want to operate your transmitter without radiating a signal. The dummy antenna safely converts the RF energy coming out of your transmitter into heat. The heat goes into the air or into the coolant, depending on the type of dummy antenna. It does all this while presenting your transmitter with a constant 50-ohm load. Dummy antennas are used to make test transmissions, or to make tuning adjustments after changing bands. Be sure the one you choose is rated for the type of operating (HF or VHF/UHF) and power level you'll be using.

Relatively inexpensive, a dummy antenna is one of the most useful accessories you can own. Every conscientious amateur should own one. You can build your own or buy one that's ready to use (see Figure 8-14). The container, often a gallon paint can, acts as a shield to keep RF energy from

being radiated. When filled with transformer oil, it allows the dummy antenna to dissipate RF power. You can use an antenna switch to connect the dummy antenna to the transmitter. See Figure 8-15.

Figure 8-14—A dummy antenna like this one uses a liquid coolant, often transformer oil, to dissipate heat. Dry dummy antennas dissipate heat into the air.

Figure 8-15—A dummy antenna is just a resistance that provides your transmitter with a proper load when you are tuning up and don't want to radiate a signal. Connect it to your antenna switch.

With a dummy antenna connected to the transmitter, you can make off-the-air tests—no signal goes out over the air. For transmitter tests, remember that a dummy antenna is a resistor. It must be capable of safely dissipating the entire (continuous) power output of the transmitter. A dummy antenna will often get warm during use, since the RF energy from the transmitter is turned into heat.

A dummy antenna should provide the transmitter with a perfect load. This usually means that it has a pure resistance (with no reactance) of approximately 50 ohms. The resistors used to construct dummy antennas must be noninductive. Composition resistors are usable, but wire-wound resistors are not. A single high-power resistor is best. Several lower-power resistors can be connected in parallel to obtain a 50-ohm load capable of dissipating high power.

ANTENNA SWITCH

An antenna switch enables you to choose between several antennas without disconnecting and reconnecting any cables. Figure 8-15 shows how an antenna switch connects to the rest of an amateur station. Antenna switches range from those that switch between only two antennas to those that switch between six or more. One position is usually used for the dummy antenna. Most antenna switches can be mounted easily on a wall, desk or shelf.

Be sure the one you buy is rated for the power level you are using. Also, the connectors on the switch should match the ones on your antenna feed lines and equipment. The most common coaxial connectors are PL-259, which mate to SO-239 connectors on a dummy antenna. (These connectors are also known, misleadingly, as UHF connectors.)

KEYERS

As you operate on the air and gain experience, your code speed will increase. A straight hand key works well at speeds up to about 15 words per minute. After that, you'll probably want to buy (or build) an electronic keyer to improve your high-speed sending.

An electronic keyer automatically produces dots and dashes of exactly the right length and spacing for the speed selected. Most are capable of sending at speeds ranging from 5 to 60 words per minute. The keyer uses a paddle that is moved to the left or right. For right-handed operation, left produces dashes; right produces dots. Some paddles provide a switch or other means to accommodate left-handed sending, in which case the dot and dash directions reverse.

Some keyers are called iambic, or squeeze, keyers. If both paddles are squeezed at the same time, the keyer will produce a string of alternating dots and dashes. Iambic keyers make it easier to send code at higher speeds because less hand movement is required. The iambic technique requires some practice, however. Figure 8-16 shows an electronic keyer. Some transceivers have built-in keyers.

Whether you use a straight key or an electronic keyer, make sure you have a good "fist" (how your CW sounds on the air). Practice until you can send clearly and confidently. Take care to space your words and letters properly. If you have a good fist, you'll communicate better and make a better impression on those you meet on the air.

LOGBOOK AND STATION CLOCK

Many hams keep a complete logbook to keep track of the stations they've contacted; you'll probably want to do so, too. A log is particularly handy for exchanging QSL cards.

If you keep a log, you'll need a clock to keep track of the time. Although you can use any clock that's handy, a 24-hour clock is most convenient. It eliminates the question of whether a contact was made in the AM or in the PM. In 24-hour time, 0300 is 3 AM, while 2200 is 10 PM. There can be only one time we call 16:15. But what about 4:15? Does it mean AM or PM?

Figure 8-16—Electronic keyers produce perfectly timed dots and dashes. This model is designed to be used with an iambic keyer.

UTC EXPLAINED

Ever hear of Greenwich Mean Time? How about Coordinated Universal Time? Do you know if it is light or dark at 0400 hours? Can you explain the International Date Line? If you answered no to any of these questions, you'd better read on!

Keeping track of time can be pretty confusing when you are talking to other hams around the world. Europe, for example, is anywhere from 4 to 11 hours ahead of us here in North America. Over the years, the time at Greenwich, England, has been universally recognized as the standard time in all international affairs, including ham radio. (We measure longitude on the surface of the earth in degrees east or west of the Prime Meridian. This imaginary line runs approximately through Greenwich, and is halfway around the world from the International Date Line.)

If you use UTC, you and the station you contact will be able to reference a common date and time, wherever you happen to be located. Mass confusion would occur if everyone used their own local time. Coordinated Universal Time (abbreviated UTC) is the name for what used to be called Greenwich Mean Time.

Twenty-four hour time lets you avoid the equally confusing question about AM and PM. If you hear someone say he made a contact at 0400 hours UTC, you will know immediately that this was 4 hours past midnight, UTC, since the new day always starts just after midnight. Likewise, a contact made at 1500 hours UTC was 15 hours past midnight, or 3 PM.

Maybe you have begun to figure it out: Each day starts at midnight, 0000 hours. Noon is 1200 hours, and the afternoon hours merely go on from there. You can think of it as adding 12 hours to the normal PM time—

3 PM is 1500 hours, 9:30 PM is 2130 hours, and so on. However you learn it, be sure to use the time everyone else does—UTC.

The photo shows a specially made clock, with an hour hand that goes around only once every day, instead of twice a day like a normal clock. Clocks with a digital readout that show time in a 24-hour format are quite popular as a station accessory.

Hams interested in communicating with other hams outside of the US keep track of the hour in terms of *Coordinated Universal Time*, or *UTC*. UTC is a 24-hour system, and requires a 24-hour clock. So you might as well invest in a good one. A conversion table that shows the UTC equivalent to your local time is easy to make. You can put together a dial-type device; see the accompanying sidebar about the N7BH World Time Finder.

THE S METER

S meters indicate the relative strength of a received signal. They are calibrated in S units from S1 to S9; above S9 they are calibrated in decibels (dB), as shown in Figure 8-17. In the 1940s at least one manufacturer made an attempt to establish some significant numbers for S meters. S9 was to be equal to a signal level of 50 microvolts, with each S unit equal to 6 dB. This means that for an increase of one S unit, the received signal power would have to increase by a factor of four. Such a scale is useful for a theoretical discussion. Real S-meter circuits fall far short of this ideal for a variety of reasons.

Suppose you were working someone who was using a 25-watt transmitter, and your S meter was reading S7. If the other operator increased power to 100 watts (an increase of four times), your S meter would read S8. If the signal was S9, the operator would have to multiply his power 10-fold to make your S meter read 10 dB over S9. It would take another

Figure 8-17—An S meter is calibrated in S units up to S9. Above S9 it is calibrated in decibels (dB) over S9.

THE N7BH WORLD TIME FINDER

You can eliminate errors in local-to-UTC time and date conversion by keeping a two-function clock set to local time and UTC. That solution, however, doesn't help determine the date and time in time zones other than local and UTC. The circular slide rule shown here works simultaneously for all time zones. (Within certain limits; many localities deviate from the system of Time Zones.)

The outer scale shows hours in the 24-hour format with plus and minus signs to show the gain or loss of a day at midnight. The inner scale is labeled with a letter and longitude to identify each Time Zone, UTC, the International Date Line and abbreviations for Time Zones in the US. For example, 2100T indicates 9 PM Mountain Standard Time or Pacific Daylight Time. (Note that adopting daylight time has the same effect as shifting one time zone to the east, and thus one hour closer to UTC.)

Time Zones

The system of Time Zones is based on a few facts about our planet. Because it takes 24 hours for the earth to rotate 360°, there is one hour of time change for every 15° of longitude. (There are 24 Time Zones, each 1 hour different from those adjacent.) The Prime Meridian (0° longitude) passes through Greenwich, England, and serves as a location for UTC. The International Date Line is 180° longitude and directly opposite the Prime Meridian. While this all forms a mathematical foundation for the Time Zone system, there are many deviations from a strict longitude/Time Zone correspondence for local convenience. For example, portions of Alaska and Hawaii share the same longitude, yet they are in different Time Zones.

Construction

You can assemble your own World Time Finder by making two photocopies of this page. Cut one copy on the line of the inner circle, and other on the line of the outer circle. Put each circle between two pieces of clear plastic, or laminate them between sheets of clear Con-Tact paper (a sticky-backed plastic material sold in many department stores). Punch a hole in the center of each disk and use a ⅛-inch rivet with washers for the center pin.

Use

Operate the World Time Finder by aligning the Time Zone letter for your areas with the current hour. In the example shown, it is 2000 hours (8 PM) at Zone U (PST). Zone U is 120° W longitude, or 8 hours earlier than UTC. If the zone-U date is July 1, then the UTC time and date is 0400 (4 AM) July 2. In Japan (Zone I), which is 135° E longitude and 9 hours later than UTC, it would be 1300 hours (1 PM) on July 2.

Note that this time conversion can be done by traveling either clockwise or counterclockwise around the chart. Going in a counterclockwise direction from Zone U to UTC, you pass midnight (in the direction of the arrow pointing to the plus sign) on the outer scale, thus gaining a day. If traveling in a clockwise direction, you pass the International Date Line (with the arrow to plus), again gaining a day. The gain or loss of date is established easily by the + or – signs next to the arrows.

---

10 times increase in signal strength to make the meter read 20 dB over S9.

S meters on modern transceivers may or may not respond in this manner. S-meter operation is based on the output of the automatic gain-control (AGC) circuitry; the S meter measures the AGC voltage. As a result, every S meter responds differently; no two S meters will give the same read-

ing. The S meter is useful for giving relative signal-strength indications, however. You can see changes in signal levels on an S meter that you may not be able to detect just by listening to the audio output level.

[Now turn to Chapter 13 and study questions T4D01 through T4D08. Review the material in this section if you have any difficulty with these questions.]

SAFETY WITH RF POWER

Amateur Radio is basically a safe activity but accidents can always occur if we don't use common sense. Most of us know enough not to place an antenna where it can fall on a

power line. We don't insert our hand into an energized linear amplifier, or climb a tower on a windy day. We also should not overexpose ourselves to RF energy. Large amounts of RF energy can be harmful to people because it heats body tissues. The effects depend on the wavelength, energy density of the incident RF field, and on other factors such as polarization.

The most susceptible parts of the body are the tissues of the eyes. They don't have heat-sensitive receptors to warn us of the danger before the damage occurs. Symptoms of over-exposure may not appear until after irreversible damage has been done. Though the problem should be taken seriously, with reasonable precaution, Amateur Radio operation can be safe.

SAFE EXPOSURE LEVELS

In recent years scientists have devoted a great deal of effort to determining safe RF-exposure limits. The problem is very complex. It's not surprising that some changes in the recommended levels have occurred as more information has become available. The American Radio Relay League believes that the 1982 **Radio Frequency Protection Guide of the American National Standards Institute (ANSI)** is a good protection standard; it took nearly five years to formulate and has undergone repeated critical review by the scientific community.

This guide recommends that special precautions be taken when transmitting with hand-held VHF and UHF radios. The antenna should be kept at least 1 to 2 inches away from the forehead, and power should be limited to no more than 7 watts.

The ANSI standards recognize the phenomenon of whole-body or geometric resonance and establishes a frequency-dependent maximum permissible RF exposure limit for humans. Exposure levels are expressed in terms of *power density*, measured in milliwatts per square centimeter. It is a measure of the radio-frequency power that strikes a person per square centimeter of body surface.

Whole-Body Resonance

Resonance occurs at frequencies for which the human body's length (height), if parallel to the antenna (vertical), is about 0.4 wavelength long. Because of the range of human heights, the resonant region spans a broad range of frequencies. This whole-body resonance establishes the frequency

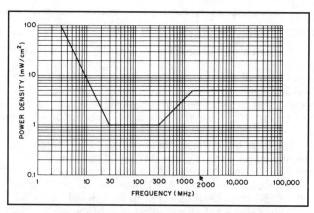

Figure 8-18—American National Standards Institute Radio Frequency Protection Guide for whole-body exposure of human beings.

range for the most stringent (lowest) permissible exposure level. Figure 8-18 shows that the lowest maximum exposure level is 1 mW/cm^2 for frequencies between 30 and 300 MHz. The frequencies in this range cause the greatest potential risk for human beings. RF energy at these frequencies is most easily absorbed by the human body. For this reason, it's important to reattach all shielding to a VHF amplifier after you've adjusted it. If it's turned on before all the shielding is in place, you could be exposed to dangerous amounts of RF energy.

On either side of those "corner" frequencies the rise is gradual. At 3 MHz the maximum permissible exposure level is 100 mW/cm^2; at 1500 MHz and above, 5 mW/cm^2. The valley region includes some active amateur bands (10, 6 and 2 meters) as well as all FM and TV broadcasting. The rationale for specifying a constant 5 mW/cm^2 above 1500 MHz takes into consideration that the extremely short wavelengths don't penetrate very deep into body tissue.

[This completes your study of Chapter 8. Now turn to Chapter 13 and study questions T4D09 through T4D16. Review the material in this section if you have any difficulty with these questions.]

KEY WORDS

Audio-frequency shift keying (AFSK)—A method of transmitting radioteletype information. Two switched audio tones are fed into the microphone input. AFSK RTTY is most often used on VHF. Also see **frequency-shift keying (FSK)**.

Autopatch—A device that allows repeater users to make telephone calls through a **repeater**.

Baudot—A code used in **radioteletype** communications. Each character is represented with a string of five bits of digital information. Each character has a different combination of bits.

Closed—A repeater that restricts access to those who know a special code. See **CTCSS**.

Connected—The condition in which two **packet-radio** stations are sending information to each other. Each is acknowledging when the data has been received correctly.

CQ—"Calling any station": the general call when requesting a conversation with anyone.

CTCSS—Continuous tone-coded squelch system; a subaudible tone system used on some repeaters. When added to a carrier, a CTCSS tone allows a receiver to accept a signal. Also called *PL*.

CW—A communications mode transmitted by on/off keying of a radio-frequency signal. Another name for international Morse code.

Data—Computer-based modes, such as **RTTY** and **packet**.

DE—The Morse code abbreviation for "from" or "this is."

Digipeater—A packet-radio station used to retransmit signals that are specifically addressed to be retransmitted by that station.

Digital communications—Computer-based communications modes. Also see **Data**.

Dummy load—A station accessory that allows you to test or adjust transmitting equipment without sending a signal out over the air. Also called *dummy antenna*.

Duplex operation—Receiving and transmitting on two different frequencies. Also see **simplex operation**.

DX—Distance, foreign countries.

Emergency—A situation where there is a danger to lives or property.

Emission types—Term for the different modes authorized for use on the Amateur Radio bands. Examples are **CW**, **SSB** and **FM**.

Frequency coordinators—Individuals or groups that recommend repeater frequencies.

Frequency-modulated (FM) voice—Mode of **voice (phone) communications** used on repeaters.

Frequency-shift keying (FSK)—A method of transmitting radioteletype information by switching an RF carrier between two separate frequencies. FSK RTTY is most often used on HF. Also see **audio-frequency shift keying**.

Ground-wave propagation—The method by which radio waves travel along the Earth's surface.

Input frequency—A repeater's receiving frequency. To use a repeater, transmit on the input frequency and receive on the **output frequency**.

Ionosphere—A region of charged (ionized) gases high in the atmosphere. The ionosphere bends radio waves as they travel through it, returning them to Earth. Also see **sky-wave propagation**.

K—The Morse code abbreviation for "any station respond."

Line of sight—The term used to describe VHF and UHF **propagation** in a straight line directly from one station to another.

Monitor mode—One type of **packet radio** receiving mode. In monitor mode, everything transmitted on a packet frequency is displayed by the monitoring TNC. This occurs whether or not the transmissions are addressed to the monitoring station.

Morse code—(see **CW**).

Narrow-band direct-printing telegraphy—The technical term for **radioteletype (RTTY)**.

Network—A term used to describe several packet stations linked together to transmit data over long distances.

Open—A repeater that can be used by all hams who have a license that authorizes operation on the repeater frequencies.

Output frequency—A repeater's transmitting frequency. To use a repeater, transmit on the **input frequency** and receive on the output frequency.

Overdeviating—Overloading an FM transceiver by speaking too loudly into the microphone or holding it too close to your mouth.

Packet radio—A system of digital communication whereby information is broken into short bursts. The bursts ("packets") also contain addressing and error-detection information.

Phone—Another name for **voice communications**.

Phonetic alphabet—Standard words used on voice modes to make it easier to understand letters of the alphabet, such as those in call signs. The call sign KA6LMN stated phonetically is *Kilo Alfa Six Lima Mike November*.

PL—(See **CTCSS**.)

Procedural signal (prosign)—One or two letters sent as a single character. Amateurs use prosigns in **CW** contacts as a short way to indicate the operator's intention. Some examples are κ̄ for "Go Ahead," or A̅R̅ for "End of Message." (The bar over the letters indicates that we send the prosign as one character.)

Propagation—The study of how radio waves travel.

Q signals—Three-letter symbols beginning with Q. Used on **CW** to save time and to improve communication. Some examples are QRS (send slower), QTH (location), QSO (ham conversation) and QSL (acknowledgment of receipt).

QRL—Ham radio **Q signal** meaning "Is this frequency in use?"

QSL card—A postcard that serves as a confirmation of communication between two hams.

QSO—A conversation between two radio amateurs.

Radioteletype (RTTY)—Radio signals sent from one teleprinter machine to another machine. Anything that one operator types on his teleprinter will be printed on the other machine. Also known as **narrow-band direct-printing telegraphy**.

Reflection—Signals that travel by **line-of-sight** propagation are reflected by large objects like buildings.

Repeater—An amateur station that receives a signal and retransmits it for greater range.

RST—A system of numbers used for signal reports: R is readability, S is strength and T is tone. (On **single-sideband** phone, only R and S reports are used.)

Sidebands—The sum or difference frequencies generated when an RF carrier is mixed with an audio signal. **Single-sideband** phone **(SSB)** signals have an upper sideband (USB—that part of the signal above the carrier) and a lower sideband (LSB—the part of the signal below the carrier). SSB transceivers allow operation on either USB or LSB.

Simplex operation—Receiving and transmitting on the same frequency. See **duplex operation**.

Single sideband (SSB)—Emission mode that describes the type of **voice** emission used most on the HF bands.

Skip zone—A "dead" area too distant for ground waves and too close for sky waves.

Sky-wave propagation—The method by which radio waves travel through the **ionosphere** and back to Earth. Sometimes called *skip*, sky-wave propagation has a far greater range than **line-of-sight** and **ground-wave propagation**.

Splatter—A type of interference to nearby stations. Occurs when a transmitter is overmodulated.

Sunspots—Dark spots on the surface of the sun. When there are few sunspots, long-distance radio propagation is poor on the higher-frequency bands. When there are many sunspots, long-distance HF propagation improves.

Sunspot cycle—The number of sunspots increases and decreases in a predictable cycle that lasts about 11 years.

Time-out timer—A device that limits the amount of time any one person can talk through a repeater.

Voice communications—Hams can use several voice modes, including **FM** and **SSB**.

73—Ham lingo for "best regards." Used on both **phone** and **CW** toward the end of a contact.

Chapter 9

You're on the Air!

I f you're like most people, as soon as your ticket arrives you'll want to get on the air and begin "working" other Amateur Radio operators. You'll be joining more than 2 million radio enthusiasts around the world.

The operating world is rich and varied. With your ham license, you can relax and enjoy a conversation (a *ragchew*) with another ham. You can send and receive messages anywhere in North America via the National Traffic System (handle *traffic*). You can contact large numbers of stations in a very short time in competition with other hams (enter *contests*). You can try to contact hams in as many different countries or distant places as possible (chase **DX**).

These are just a few of the activities that Amateur Radio operators enjoy. The more you operate, the more proficient you'll be on the air.

In this chapter, we'll give you some tips for operating with your new Novice or Technician license—on CW, phone, packet and through FM repeaters. Then we'll take a look at some other operating modes you may want to try. Later in the chapter we'll discuss how radio waves travel.

OPERATING SKILL

Poor operating procedure is ham radio's version of original sin. It is a curse that will not go away. Hams call these poor operators *lids*. No one wants to be a lid or to have a conversation with a lid. Calling someone a lid is the ultimate insult.

(Don't do it on the air. Just be sure your operating practices are such that no one *wants* to call *you* a lid.)

It is actually easier to be a good operator than to fall prey to sloppy habits. You can transmit a message without needless repetition and without unnecessary identification. On **CW (Morse code)**, you don't have to spell out each and every word. Good operating makes hamming more fun for everyone.

Clearly, the initial glamour of Amateur Radio is the opportunity to talk to people who share a common bond, the hobby of ham radio. It doesn't matter if they're across town or in another country. Novices and Technicians with Novice privileges have plenty of chances to work **DX** (contact distant stations) on the 10- and 15-meter HF ham bands. The 40- and 80-meter HF bands (7 and 3.7 MHz) and the VHF and UHF bands are useful primarily for shorter-range contacts.

Chapter 7 presents information on antennas for all types of locations. Once you choose the frequency bands, you'll want to install an antenna that will get you on the air. The success of your antenna installation will be evident once you actually begin making contacts: Either they'll hear you or they won't. You won't need a beam antenna to work DX. At times of high sunspot activity, many Novices report outstanding results using dipoles or converted CB antennas. Your goal is to put the best possible signal on the air. Of course, this will vary depending on your location and your budget. Don't worry about competing with the ham down the street who has a huge antenna atop a 70-foot tower.

In Amateur Radio, it's important to enjoy and take pride in your own accomplishment. You've heard it said "It's what you do with what you've got that really counts." That is especially true in Amateur Radio. To make up for any real or imagined lack of equipment, concentrate on improving your own operating ability.

Good operating skills can be like adding 10 dB to your signal. That's like increasing your power by 10 times! (More detailed operating information can be found in *The ARRL Operating Manual*. This book is available from your local radio dealer or directly from the ARRL.)

A good rule for Amateur Radio contacts (**QSO**s) is to talk as you would during a face-to-face conversation. When you meet someone for the first time, you introduce yourself once; you don't repeat your name over and over again. An exception would be if the other person is hard of hearing or you're meeting in a noisy place. Even then you would only repeat if the person couldn't hear you. When you are on the air, the other operator will tell you if she can't hear you clearly. Problem situations on HF include static (QRN), interference (QRM) or signal fading (QSB). Table 9-1 lists many common **Q signals**. Although hams use some Q signals in face-to-face conversations, when you're on the air, use them on CW only. On voice, say what you mean.

Now, let's take a look at some guidelines that will help you get accustomed to operating your new radio with your new privileges. We'll start with CW.

OPERATING CW

Even if you're anxious to key the mike, don't pass up the information in this section. There are many similarities between CW and voice operation. Non-CW modes such as voice

Table 9-1

Q Signals

These Q signals are the ones used most often on the air. (Q abbreviations take the form of questions only when they are sent followed by a question mark.)

QRG	Will you tell me my exact frequency (or that of ___)? Your exact frequency (or that of ___) is ___ kHz.
QRL	Are you busy? I am busy (or I am busy with ___). Please do not interfere.
QRM	Is my transmission being interfered with? Your transmission is being interfered with ___ (1. Nil; 2. Slightly; 3. Moderately; 4. Severely; 5. Extremely.)
QRN	Are you troubled by static? I am troubled by static ___. (1-5 as under QRM.)
QRO	Shall I increase power? Increase power.
QRP	Shall I decrease power? Decrease power.
QRQ	Shall I send faster? Send faster (___ WPM).
QRS	Shall I send more slowly? Send more slowly (___ WPM).
QRT	Shall I stop sending? Stop sending.
QRU	Have you anything for me? I have nothing for you.
QRV	Are you ready? I am ready.
QRX	When will you call me again? I will call you again at ___ hours (on ___ kHz).
QRZ	Who is calling me? You are being called by ___(on ___kHz).
QSB	Are my signals fading? Your signals are fading.
QSK	Can you hear me between your signals and if so can I break in on your transmission? I can hear you between signals; break in on my transmission.
QSL	Can you acknowledge receipt (of a message or transmission)? I am acknowledging receipt.
QSN	Did you hear me (or ___) on ___kHz? I did hear you (or ___) on ___kHz.
QSO	Can you communicate with ___direct or by relay? I can communicate with ___ direct (or relay through ___).
QSP	Will you relay to ___? I will relay to ___.
QST	General call preceding a message addressed to all amateurs and ARRL members. This is in effect "CQ ARRL."
QSX	Will you listen to ___on ___kHz? I am listening to ___ on ___kHz.
QSY	Shall I change to transmission on another frequency? Change to transmission on another frequency (or on ___ kHz).
QTB	Do you agree with my counting of words? I do not agree with your counting of words. I will repeat the first letter or digit of each word or group.
QTC	How many messages have you to send? I have ___ messages for you (or for ___).
QTH	What is your location? My location is ___.
QTR	What is the correct time? The time is ___.

(SSB and FM) and specialized digital modes are covered later in the chapter. The FCC refers to these types of radio signals as **emission types**.

An unmodulated carrier wave (a steady signal with no information included) is called a *test emission*. **CW**, also called international Morse code, is transmitted by on/off keying of a radio-frequency signal. This is demonstrated whenever you use a key to send CW. To send code, you press a lever. To stop sending, you let up on the lever. Code is either on or off. What could be simpler?

Calling CQ

There is no point in wasting words, and that includes trying to make a radio contact. Hams establish a contact when one station calls CQ and another replies. **CQ** literally means "Seek you: Calling any station." You can usually tell a good ham by the length of the CQ call. A good operator sends short calls separated by concentrated listening periods. Long CQs drive away more contacts than they attract! Generally, a "3 × 3" call is more than sufficient. Here's an example:

CQ CQ CQ DE KA7XYZ KA7XYZ KA7XYZ K

As you may have figured out, "3 × 3" refers to calling CQ three times, then **DE** ("from" or "this is") then your call sign three times. Perhaps the best way to get started is to listen for someone else's CQ. The Novice bands are usually alive with signals. You may choose to call CQ yourself, however, and that's okay. Always listen before you transmit, even if the frequency appears clear. To start, send **QRL?** ("Is this frequency in use?"). If you hear a C ("yes") in reply, then try another frequency. It is not uncommon to hear only one of the stations in a QSO. A frequency may seem clear even though there is a contact in progress. It's the worst of bad manners to jump on a frequency that's already busy. Admittedly, the Novice bands, especially 40 and 80 meters, are often crowded. Even so, if you listen first you'll avoid interfering with an in-progress QSO.

Whatever you do, don't send faster than you can reliably copy. That will surely invite disaster! It's best to *send at the speed you wish to receive*. This is especially true when you call CQ. If you answer a CQ, answer at a speed no faster than that of the sending station. Don't be ashamed to send PSE QRS (please send more slowly).

You will learn the meaning of many other **Q signals** as you gain operating experience. QTH means "location." Followed by a question mark, it asks, "What is your location?" QSL means "I acknowledge receipt" (of a message or information). A **QSL card** is a written confirmation of an amateur contact. QSY means "change frequency."

As you gain experience, you will develop the skill of sorting out the signal you want from those of other stations. Space in the ham bands is limited. Make sure your QSO takes up as little of it as possible to avoid interfering with other QSOs. When two stations are communicating, their transmitters should be on exactly the same frequency. We call this procedure *operating zero beat*. Both transmissions will have

the same pitch when you hear them in a receiver. When you tune in the other station, adjust your receiver for the strongest signal and the "proper" tone for your radio. This will usually be between 500 and 1000 Hz.

Zero-beat operating helps the other operator know where to listen for your signal. This is easy with a transceiver. The transmitter will automatically be close to zero beat when you tune in the received signal to a comfortable pitch. With a separate receiver and transmitter, you must be sure your transmitter is zero beat with that of the other station.

Correct CW Procedures

Establishing a contact—The best way to do this, especially at first, is to listen. When you hear someone calling CQ, answer them. If you hear a CQ, wait until the ham indicates she is listening, then call her. For example:

KA1WWP KA1WWP DE WL7AGA WL7AGA A̅R̅

(A̅R̅ is equivalent to over).

In answer to your call, the called station will reply WL7AGA DE KA1WWP R. . . . That R (roger) means she has received your call correctly. That's all it means—*received*. It *does not* mean (a) correct, (b) I agree, (c) I will comply, or anything else. It is not sent unless everything from the previous transmission was received correctly. Perhaps KA1WWP heard someone calling her but didn't quite catch the call because of interference (QRM) or static (QRN). In this case, she might come back with:

QRZ? DE KA1WWP A̅R̅ ("Who is calling me?").

Calling CQ—CQ means "I wish to contact any amateur station." Avoid calling CQ endlessly. It clutters up the air and drives off potential new friends. The typical CQ goes like this:

CQ CQ CQ DE KA1WWP KA1WWP KA1WWP K

The letter K is an invitation for any station to go ahead.

The QSO—During a contact, it is necessary to identify your station only once every 10 minutes. Keep the contact friendly and cordial. Remember, the conversation is not private. Many others, including nonamateurs, may be listening. Both on CW and phone, it is possible to be informal, friendly and conversational. This is what makes the Amateur Radio QSO enjoyable. During the contact, when you stand by, use K ("go") at the end of your transmission. If you don't want someone else to join the QSO, use K̅N̅. That says you want only the contacted station to come back to you. Most of the time K is sufficient (and shorter!).

Ending the QSO—When it's time to end the contact, don't keep talking. Briefly express your pleasure at having worked the other operator. Then, sign out:

SK WL7AGA DE KA1WWP

If you are leaving the air, add CL to the end, right after your call sign.

These procedures help hams communicate better on CW. They have no legal standing, however. FCC regulations say little about our internal procedures. They show we're not just hobbyists, but that we are an established, self-regulating communications service. Hams take pride in our distinctive procedures, and understandably so.

Table 9-2

Some Common Abbreviations Used on CW

Although abbreviations help to cut down unnecessary transmission, it's best not to abbreviate unnecessarily when working an operator of unknown experience.

AA	All after	GN	Good night	SASE	Self-addressed, stamped envelope
AB	All before	GND	Ground	SED	Said
ABT	About	GUD	Good	SIG	Signature; signal
ADR	Address	HI	The telegraphic laugh; high	SINE	Operator's personal initials
AGN	Again	HR	Here, hear		or nickname
ANT	Antenna	HV	Have	SKED	Schedule
BCI	Broadcast interference	HW	How	SRI	Sorry
BCL	Broadcast listener	LID	A poor operator	SSB	Single sideband
BK	Break; break me; break in	MA, MILS	Milliamperes	SVC	Service; prefix to service message
BN	All between; been	MSG	Message; prefix to radiogram	T	Zero
BUG	Semi-automatic key	N	No	TFC	Traffic
B4	Before	NCS	Net control station	TMW	Tomorrow
C	Yes	ND	Nothing doing	TNX-TKS	Thanks
CFM	Confirm; I confirm	NIL	Nothing; I have nothing for you	TT	That
CK	Check	NM	No more	TU	Thank you
CL	I am closing my station; call	NR	Number	TVI	Television interference
CLD-CLG	Called; calling	NW	Now; I resume transmission	TX	Transmitter
CQ	Calling any station	OB	Old boy	TXT	Text
CUD	Could	OC	Old chap	UR-URS	Your; you're; yours
CUL	See you later	OM	Old man	VFO	Variable-frequency oscillator
CW	Continuous wave	OP-OPR	Operator	VY	Very
	(that is, radiotelegraphy)	OT	Old-timer; old top	WA	Word after
DE	From, this is	PBL	Preamble	WB	Word before
DLD-DLVD	Delivered	PSE	Please	WD-WDS	Word; words
DR	Dear	PWR	Power	WKD-WKG	Worked; working
DX	Distance, foreign countries	PX	Press	WL	Well; will
ES	And, &	R	Received as transmitted; are	WUD	Would
FB	Fine business, excellent	RCD	Received	WX	Weather
FM	Frequency modulation	RCVR (RX)	Receiver	XCVR	Transceiver
GA	Go ahead (or resume sending)	REF	Refer to; referring to; reference	XMTR (TX)	Transmitter
GB	Good-by	RFI	Radio frequency interference	XTAL	Crystal
GBA	Give better address	RIG	Station equipment	XYL (YF)	Wife
GE	Good evening	RPT	Repeat; I repeat	YL	Young lady
GG	Going	RTTY	Radioteletype	73	Best regards
GM	Good morning	RX	Receiver	88	Love and kisses

Working "Split"

There is an exception to the zero-beat rule. Sometimes a DX station has a "pileup" of stations trying to make contact. If they are all calling on the DX station's frequency, no one (including the DX station!) can copy anything. Under these conditions, the DX station may choose to work *split*. We talked about split-frequency operating in Chapter 6. When you hear a DX station, but you never hear anyone else calling or working him, chances are he's working split. Tune higher in the band, and look for the pileup. It should be easy to find. DX stations work split when operating SSB, too. (For more helpful hints on working DX, see ARRL's *The DXCC Companion*.)

For amateurs to understand each other, we must standardize our communications. You'll find, for instance, that most hams use abbreviations on CW. Why? It's faster to send a couple of letters than it is to spell out a word. But there's no point in using an abbreviation if no one else understands you. Over the years, amateurs have developed a set of standard abbreviations (Table 9-2). If you use these abbreviations,

you'll find that everyone will understand you, and you'll understand them. In addition to these standards, we use a set of **procedural signals**, or *prosigns*, to help control a contact. The "Correct CW Procedures" sidebar explains these prosigns in more detail.

Answering a CQ

What about our friend KA7XYZ? In our last example he was calling CQ. What happened? Another ham, N2SN, heard him and answered the CQ:

KA7XYZ KA7XYZ DE N2SN N2SN $\overline{\text{AR}}$

Notice the "2 × 2" format. N2SN sent the call sign of the station he was calling twice, then sent DE, then sent his call sign twice. If N2SN thought he heard someone calling him, but wasn't quite sure, he would send:

QRZ? DE N2SN $\overline{\text{AR}}$

The Q signal QRZ means "who is calling me?" Use the prosign

$\overline{AR}$ (the letters A and R run together with no separating space) in an initial call to a specific station before officially establishing contact. When calling CQ, use the prosign **K**, because you are inviting any station to reply. KA7XYZ comes back to N2SN in a to-the-point manner (one you should copy).

N2SN DE KA7XYZ R GE UR RST 599 DENVER CO
$\overline{BT}$ NAME BOB HW BK

In this transmission, **RST** refers to the standard *readability*, *strength* and *tone* system of reporting signal reception. You'll exchange a signal report in nearly every Amateur Radio QSO. Don't spend a lot of time worrying about what signal report to give to a station you're in contact with. The scales are simply a general indication of how you are receiving the other station. As you gain experience with the descriptions given in Table 9-3, you'll be more comfortable estimating the proper signal report.

A report of RST 368 would be interpreted as "Your signal is readable with considerable difficulty, good strength, with a slight trace of modulation." The tone report is a useful indication of transmitter performance. When the RST system was developed, the tone of amateur transmitters varied widely. Today, a tone report of less than 9 is cause to ask a few other amateurs for their opinion of the transmitted signal. Consistently poor tone reports means your transmitter has problems.

A further word about signal reports, one that applies to all types of operating: If your signal report is *too* good, *reduce your power*. FCC rules say that amateurs must use the minimum power necessary to maintain communications. Whether you're operating through an FM repeater, on VHF or UHF simplex or on HF, if you don't have to use an amplifier or the highest power your transceiver is capable of, decrease your transmitting power. You'll be less likely to interfere with other stations and (if you're using a hand-held transceiver or other portable radio) you'll conserve your battery.

The basic information is only transmitted once. The other station will request repeats if necessary. Notice too that $\overline{BT}$ (B and T run together) is used to separate portions of the text. This character is really the double dash (=), and is usually written as a long dash or hyphen on your copy paper. HW means "how do you copy?" BK signifies that KA7XYZ is turning it over (back) to N2SN for his basic info. KA7XYZ does not sign both calls all over again. FCC rules require identification only at the end of a QSO, and once every 10 minutes.

At the end of the contact send 73 and sign off. **73** is a common abbreviation that means "Best regards."

Conversation is a two-way phenomenon. There is no reason that an Amateur Radio QSO can't be a back-and-forth process. No one wants to listen to a long, unnecessary monologue. Propagation conditions might change, hampering the QSO. KA7XYZ and N2SN will relate better if each contributes equally. Sometimes there is interference or marginal copy because of weak signals. It may help to transmit your call sign when you are turning it back to the other station.

Table 9-3
The RST System

READABILITY
1—Unreadable.
2—Barely readable, occasional words distinguishable.
3—Readable with considerable difficulty.
4—Readable with practically no difficulty.
5—Perfectly readable.

SIGNAL STRENGTH
1—Faint signals barely perceptible.
2—Very weak signals.
3—Weak signals.
4—Fair signals.
5—Fairly good signals.
6—Good signals.
7—Moderately strong signals.
8—Strong signals.
9—Extremely strong signals.

TONE
1—Sixty-cycle ac or less, very rough and broad.
2—Very rough ac, very harsh and broad.
3—Rough ac tone, rectified but not filtered.
4—Rough note, some trace of filtering.
5—Filtered rectified ac but strongly ripple-modulated.
6—Filtered tone, definite trace of ripple modulation.
7—Near pure tone, trace of ripple modulation.
8—Near perfect tone, slight trace of modulation.
9—Perfect tone, no trace of ripple or modulation of any kind.

The "tone" report refers only to the purity of the signal. It has no connection with its stability or freedom from clicks or chirps. Most of the signals you hear will be a T-9. Other tone reports occur mainly if the power supply filter capacitors are not doing a thorough job. If so, some trace of ac ripple finds its way onto the transmitted signal. If the signal has the characteristic steadiness of crystal control, add X to the report (for example, RST 469X). If it has a chirp or "tail" (either on "make" or "break") add C (for example, 469C). If it has clicks or noticeable other keying transients, add K (for example, 469K). Of course a signal could have both chirps and clicks, in which case both C and K could be used (for example, RST 469CK).

Tips for Better CW Operating

In summary, here are the points to keep in mind:
- *Listen before transmitting*. Send QRL? ("Is this frequency in use?") before transmitting. Listen again! It's worth repeating: Listen!
- *Send short CQs* and listen between each.
- Send no faster than you can reliably copy.
- *Use standard abbreviations* whenever possible—become familiar with them.
- Use prosigns and Q signals properly.
- Identify at the end of a QSO (the entire contact, not each

turnover) and every 10 minutes.
- Use R only if you've received 100 percent of what the other station sent.
- Be courteous.
- *Zero beat the other station's frequency* before calling (except when working "split").

TUNING UP

What is often the most exciting, most memorable and perhaps most terrifying moment in your entire ham experience? Your first on-the-air contact with another station! Before you have that experience, you'll want to know how to operate your station equipment.

The best place to start is the instruction manual. Before you even turn on your radio, read the instructions carefully so you'll be familiar with each control. Without turning on the equipment, try adjusting the controls. Nothing will happen, but you'll learn the location and feel of each important control.

After studying the manual and finding the important controls, you'll be ready to tune up. You must have your FCC license in hand before you can transmit or even tune up on the air!

If your transmitter requires tuning when you change bands, connect your transmitter output to a **dummy load** while you tune. This avoids on-the-air interference. (Chapter 8 has more information on this useful accessory.) Never tune up on the air because you could interfere with other hams.

Once you have tuned up the transceiver according to the instruction manual, disconnect the dummy load and connect the antenna (an antenna switch makes this easy). Now, you're

My First On-the-Air Contact

I slowly reach for the key, my other hand on the TR switch. All I have to do is throw the TR switch, which connects the antenna to the transmitter, and push down on the key. Then I'll be on the air and able to talk to the world. But I can't. My hand doesn't move, except to tremble slightly. It's not the rig, not the key. It's me! I can't find the nerve to do it. I can't send a signal to the world.

That's how it was for me when I tried to make my first contact. It had been weeks since I passed the Novice exam. Each day that passed without the arrival of my new ticket seemed endless. I knew that if that day would just end, the next would surely see the arrival of my license. Days, weeks passed. Then a month had gone by. The station was there waiting for me. All the equipment was in place and ready for operation. If only the license would arrive.

Finally, after what seemed to be several years, it did arrive. Can you believe it? It arrived on my birthday. What a present! The letter carrier had hardly made it to the sidewalk before I was at the rig. I turned on the transmitter and receiver. Nothing! What could be wrong? All the wires connected? Oh no! The power cords were just hanging there. I'd forgotten to plug them into the wall socket. The few seconds it took for the receiver to warm up seemed like hours. Then I heard the sound of radio. I was ready to enter the world of ham radio. Or so I thought.

After several minutes of total failure at my attempt to send CQ, I gave up. I needed help. But where to get it? Of course, the instructor of my Novice training class. He could, and would help me.

The next day I got in touch with him and explained my problem. Would he help me? Yes, he would be glad to stand behind me during that first contact. But, he was busy with something else. I'd have to wait a day or so. Another day without a contact? It might as well have been another year.

Although it didn't seem possible, the next day rolled around, and with it, my instructor. A few minutes of preparation and I was ready. I would give it my best, to sink or swim. Tuning the receiver to the frequency of my crystal-controlled transmitter, I found another station calling CQ. His sending was slow and steady. It had to be for me to copy it. "Go ahead, give him a try," advised my instructor. "What have you got to lose?"

. . . CQ CQ CQ CQ. His call seemed endless. Finally, he signed . . . CQ DE W3AOH. I flipped the TR switch and__ reached confidently for the key. W3AOH DE WN2VDN AR.

"That was a pretty short call you gave him. You should have sent your call sign more than once. He may have missed it." But he didn't. There it was, slow and steady. "WN2VDN DE W3AOH BT TNX CALL BT UR . . ." He had answered my call . . . my call. I'd done it! I made a contact!

"Pay attention and copy the code!" my instructor yelled at me. "He's still sending." Dahdididit dahdidah dah dahdahdah dididah dahdahdah . . . dahdididit . . . What on earth did that mean? I'd forgotten the code. Those strange sounds coming out of the speaker didn't make any sense at all. What was I going to do?

Fortunately, my instructor was copying it all. "He's just signed it over to you. Your turn now."

I flipped the TR switch again, and hit the key. This time there was no fear. The key was part of my hand, part of my mind. W3AOH DE WN2VDN . . . Until, that is, after I sent my call. I didn't know his name, his location, anything. So I sent the only thing I could think of. "Thanks for coming back to me OM. You're my first contact. I'm a bit nervous." Behind me, the instructor is saying, "You should have sent him your name and location, and his signal report." Too late, I was already signing it over to W3AOH.

Slow and steady came the reply. "Welcome aboard. Hope you enjoy ham radio as much as I do. . . ." This time I had no trouble copying him. After the contact was completed, I breathed a very large sigh of relief. But, I had done it. I had actually "talked" to another person via Morse code. I was a bona fide ham.

My instructor asked if he should hang around a little longer. But I didn't hear him. I had already begun looking around for another CQ to answer. I didn't need his help any longer; I was a ham.

ready to operate! If you use a Transmatch, you may have to transmit a brief low-power signal to adjust the circuit.

[Before you go on to the next section, turn to Chapter 12 and study questions N2A01, N2A02, N2A03 and N2A05 through N2A17. You should also study questions N8A01 and N8A03. Review this section if you have any difficulty with these questions.]

[**If you are studying for your Technician exam**, review questions T2A17, T2B05 through T2B10, T8A01 and T8A02 in Chapter 13 before moving on. Review this section if you have any difficulty.]

SSB VOICE OPERATING

To make the most of your voice privileges, you'll need to learn a different set of operating procedures. Those described here apply to voice operation on 10-meter **single sideband (SSB)** and on the "weak-signal" SSB frequencies on the VHF/UHF bands. Operating through FM repeaters is covered later in this chapter.

Any **voice communications** mode is known as **phone** under FCC rules. AM, SSB and FM voice are all phone **emission types**.

SSB is the most common voice mode on HF. You can use either the *lower* **sideband** or the *upper* **sideband** to transmit an SSB signal. Amateurs normally use the upper sideband for 10-meter phone operation.

Operating techniques and procedures vary from band to band. If you're accustomed to 80-meter CW, you may be uncertain about how to make a 10-meter SSB contact. A great way to become familiar with new techniques is to spend time listening. Be discriminating, though. Take a few moments to understand the techniques used by the proficient operators—the ones who are the most understandable and who sound the best. Don't simply mimic whatever you hear. This is especially true if you're going on the air for the first time.

Whatever band or mode you are using, there are three fundamental things to remember. These apply for any type of voice operating you might try. The first is that courtesy costs very little. It is often rewarded by bringing out the best in others. Second, the aim of each radio contact should be 100% effective communication. A good operator is never satisfied with anything less. Third, your "private" conversation with another station is actually open to the public. Many amateurs are uncomfortable discussing controversial subjects over the air. Also, never give any confidential information on the air. You never know who may be listening.

Keep it Plain and Simple

Even though it does not require the use of code or special abbreviations, proper voice procedure is very important. Voice operators say what they want to have understood. CW operators have to spell it out or abbreviate. One advantage of voice operation is speed: a typical voice QSO takes place at between 150 and 200 words per minute. Whether you're working a DX operator who may not fully understand our

Table 9-4
Voice Equivalents to Code Procedure

Voice	Code	Meaning
over	$\overline{AR}$	after call to a specific station
end of message	$\overline{AR}$	end of message
wait, stand by	$\overline{AS}$	please stand by
received	R	all received correctly (not a promise to take any specific action.)
go, go ahead	K	any station transmit
go only	$\overline{KN}$	addressed station only
clear	$\overline{SK}$	end of contact
closing station	CL	going off the air
break or back to you	$\overline{BK}$	the receiving station's turn to transmit
stroke, portable	$\overline{DN}$	operation away from primary station location

language, or talking to your friend down the street, speak slowly and clearly. That way, you'll have fewer requests to repeat information.

Avoid using CW abbreviations and prosigns such as "HI" and "K" for voice communications. Also, Q signals (QRX, QRV and so forth) are for CW, not voice, operation. QSL, QSO and QRZ have become accepted on voice, however. Abbreviations are used on CW to say more in less time. On voice, you have plenty of time to say what you mean. On CW, for example, it's convenient to send **K** at the end of a transmission. On voice, it takes less than a second to say "go ahead." Table 9-4 shows the voice equivalents for common CW prosigns.

Use plain language and keep jargon to a minimum. In particular, avoid the use of "we" when you mean "I" and "handle" or "personal" when you mean "name." Also, don't say "that's a roger" when you mean "that's correct." Taken individually, any of these sayings is almost harmless. Combined in a conversation, however, they give a false-sounding "radioese" that is actually less effective than plain language. "Roger" for example, means "I have received what you sent." It doesn't necessarily mean that what was sent was "correct."

Phonetic Alphabet

Sometimes you will have difficulty getting your call sign or other information across to the operator on the other end. This might be because conditions are poor, there is nearby interference, or the operator on the other end is not proficient in English. There is a standard International Telecommunication Union (ITU) **phonetic alphabet** to assist you. See Table 9-5. It is generally used when signing your call or passing information that must be spelled out. For

Table 9-5

Standard ITU Phonetics

A—Alfa (**AL** FAH)
B—Bravo (**BRAH** VOH)
C—Charlie (**CHAR** LEE) or (**SHAR** LEE)
D—Delta (**DELL** TAH)
E—Echo (**ECK** OH)
F—Foxtrot (**FOKS** TROT)
G—Golf (**GOLF**)
H—Hotel (HOH **TELL**)
I—India (**IN** DEE AH)
J—Juliett (**JEW** LEE ETT)
K—Kilo (**KEY** LOH)
L—Lima (**LEE** MAH)
M—Mike (**MIKE**)
N—November (NO **VEM** BER)
O—Oscar (**OSS** CAH)
P—Papa (PAH **PAH**)
Q—Quebec (KEH **BECK**)
R—Romeo (**ROW** ME OH)
S—Sierra (SEE **AIR** RAH)
T—Tango (**TANG** GO)
U—Uniform (**YOU** NEE FORM) or (OO NEE FORM)
V—Victor (**VIK** TAH)
W—Whiskey (**WISS** KEY)
X—X-RAY (**ECKS** RAY)
Y—Yankee (**YANG** KEY)
Z—Zulu (**ZOO** LOO)

Note: The **boldfaced** syllables are emphasized. The pronunciations shown in this table were designed for those who speak any of the international languages. The pronunciations given for "Oscar" and "Victor" may seem awkward to English-speaking people in the US.

example, KA5CHW signs his call *Kilo Alfa Five Charlie Hotel Whiskey*. WB2LZN signs *Whiskey Bravo Two Lima Zulu November*, and YU1EXY says *Yankee Uniform One Echo X-ray Yankee*.

The ITU phonetic alphabet is generally understood by hams in all countries. If you want people to understand you, stick to the standard phonetics. For example, WA2ECO should sign *Whisky Alfa Two Echo Charlie Oscar*, not something like *Worked All Two Electron Coupled Oscillators*. Use of nonstandard phonetics is often more confusing than using no phonetics at all.

Initiating a Contact

There are two ways to initiate an SSB voice contact: call **CQ** (a general call to any station) or answer a **CQ**. At first, you may want to tune around and find another station to answer. If activity on a band seems low, a CQ call may be worthwhile.

Before calling CQ, it is important to find a frequency that appears unoccupied by any other station. This may not be easy, particularly during crowded band conditions. Listen carefully—perhaps a weak DX station is on frequency. If you're using a beam antenna, rotate it to make sure the frequency is clear. If, after a reasonable time, the frequency seems clear, ask if the frequency is in use, then sign your call. "Is the frequency in use? This is KA1IFB." If, as far as you can determine, no one responds, you are ready to make your call.

As in CW operation, keep voice CQ calls short. Long calls are considered poor operating technique. You may interfere with stations already on frequency who didn't hear your initial frequency check. Also, stations intending to reply to the call may become impatient and move to another frequency. Call CQ three times, followed by "this is," followed by your call sign three times and then listen. If no one comes back, try again. If two or three calls produce no answer, it may be that there is too much noise or interference, or that atmospheric conditions are not favorable. At that point, change frequency and try again. If you still get no answer, try looking around and answer someone else's CQ.

An example of a good CQ call is:

"CQ CQ calling CQ. This is KA1WWP, Kilo Alfa One Whiskey Whiskey Papa, Kilo Alfa One Whiskey Whiskey Papa, calling CQ and standing by."

There is no need to say what band is being used, and certainly no need to add "tuning for any possible calls, dah-di-dah!" or "K someone please!" and the like.

When replying to a CQ, say both call signs clearly. It's not necessary to sign the other station's call phonetically. You should always sign yours with standard phonetics, however. Remember to keep calls short. Say the call sign of the station you are calling only once. Follow with your call sign, repeated phonetically, several times. For example, "W1AW, this is KA3HAM, Kilo Alfa Three Hotel Alfa Mike, KA3HAM, over." Depending on conditions, you may need to give your call phonetically several times. Repeat this calling procedure as required until you receive a reply or until the station you are calling has come back to someone else.

Listening is very important. If you're using *PTT* (push-to-talk), be sure to let up on the transmit button between calls so you can hear what is going on. With *VOX* (voice operated switch), you key the transmitter simply by talking into the microphone. VOX operation is helpful because, when properly adjusted, it enables you to listen between words. Remember: It is extremely poor practice to make a long call without listening. Also, don't continue to call after the station you are trying to contact replies to someone else. Wait for the contact to end before trying again, or try another spot on the band.

Conducting the QSO

Once you've established contact, it is no longer necessary to use the phonetic alphabet or sign the other station's call. FCC regulations say that you need to sign your call only every 10 minutes and at the conclusion of the contact. (The exception is when handling international third-party traffic. Then, you must sign both calls.) This allows you to enjoy a normal two-way conversation without the need for continual identification. Use "over" or "go ahead" at the end of a transmission to indicate that it's the other station's turn to transmit. (During FM repeater operation, it is obvious when you or the other station stops transmitting because you will hear the repeater carrier drop. In this case it may not be necessary to say "over.")

Signal reports on SSB are two-digit numbers using the RS portion of the RST system. No tone report is required. The maximum signal report would be "59"—that is, readability 5, strength 9. (If you get a report of, say, "59 plus 20 dB," it means your received signal reads 20 decibels higher than signal strength 9 on the transceiver S meter.) On FM repeaters, RS reports are not appropriate. FM signal reports are generally given as the amount of quieting. *Full quieting* means the received signal is strong enough to block all receiver noise.

Voice contacts are often similar in content to CW QSOs. Aside from signal strength, most hams exchange name, location and equipment information (especially antennas!). Once these routine details are out of the way, you can talk about your families, the weather, your occupation or any appropriate subject.

Working DX

During the years of maximum sunspot activity, 10-meter worldwide communication on a daily basis is commonplace. (The propagation section of this chapter discusses the effects of sunspots in detail.) Ten meters is an outstanding DX band when conditions are right. A particular advantage of 10 meters for DX work is that effective beam-type antennas tend to be small and light, making for relatively easy installation.

There are a few things to keep in mind when you contact amateurs from outside the United States. While many overseas amateurs have an exceptional command of English (which is especially remarkable since few US amateurs understand foreign languages), they may not be familiar with many of our local sayings. Because of the language differences, some DX stations are more comfortable with the "bare-bones" type contact, and you should be sensitive to their preferences.

During unsettled band conditions it may be necessary to keep the contact short in case fading or interference occurs. Take these factors into account when expanding on a basic contact. Also, during a band opening on 10 meters or on VHF, it is crucial to keep contacts brief. This allows many stations to work whatever DX is coming through.

When the time comes to end the contact, end it. Thank the other operator (once) for the pleasure of the contact and say goodbye: "This is KA9MAN, clear." This is all that is required. Unless the other amateur is a good friend, there is no need to start sending best wishes to everyone in the household including the family dog! Nor is this the time to start

digging up extra comments on the contact which will require a "final final" from the other station (there may be other stations waiting to call in). Hams usually end a contact by wishing the other station their best regards. Saying the CW abbreviation "**73**" is the common way to do that. (Notice this is not "73s" as is heard all too often on the bands.)

Tips for Better SSB Operating

• *Listen with care*. It is natural to answer the loudest station that calls, but you should answer the best signal. Not all amateurs can run high power, but there is no reason every amateur cannot have a signal of the highest quality. Do not reward the operator who has cranked up the transmitter gain to the point of being unintelligible, especially if a station with a nice sounding signal is also calling.

• *Use VOX or PTT*. If you use VOX, don't defeat its purpose by saying "aah" to keep the transmitter on the air. If you use PTT, let go of the mike button every so often to make sure you are not "doubling" with the other station. A QSO should be an interactive conversation. Don't do all the talking.

• *Take your time*. The speed of voice transmission (with perfect accuracy) depends almost entirely on the skill of the two operators concerned. Use a rate of speech that allows perfect understanding. The operator on the other end should have time to record important details of the contact. If you go too fast, you'll end up repeating a lot of information.

VHF AND UHF OPERATING

Operating SSB and CW on the VHF or UHF bands is called *weak-signal* operating. It's similar to operating on HF, but there are some differences you should know about. Most of these differences came about because of the way VHF and UHF signals travel. This is just a brief overview of VHF and UHF procedures. *The ARRL Operating Manual* has more information.

It's a good idea to use VHF or UHF for all local communications. The HF bands should be reserved for longer-distance contacts to reduce HF-band interference.

Grid Squares

One of the first things you'll notice when you tune the weak-signal frequencies is that most QSOs include an exchange of *grid squares*. Grid squares are a way of dividing up Earth's surface into 1° latitude × 2° longitude rectangles. The grid-square designator consists of two letters and two numbers. It's a shorthand way of describing your general location. For example, W1AW in Newington, Connecticut is in grid square FN31.

You can determine your grid square identifier from the *QST* article, "VHF/UHF Century Awards," published in January 1983, pp 49-51. (Reprints are available from the Awards Branch at ARRL HQ. Please include an SASE with your request.) ARRL also publishes a map (Figure 9-1) that

Figure 9-1—This map shows grid squares, used to specify your location for VHF and UHF SSB/CW operation.

---

shows the grid squares for the continental United States and most of Canada. This 12- × 18-inch map is available from ARRL HQ.

Operating Techniques

On HF, you never know who you will run into. On any given day, the HF bands are open for cross-country contacts. At least one band will be open for transcontinental QSOs. This makes for a vast number of potential contacts. Most HF QSOs are with operators you've never met before.

It's different on VHF and UHF. Transmission range is normally limited to your local area. A lot of the activity on these bands is relaxed ragchewing. Many groups meet regularly on VHF and UHF to discuss topics of mutual interest. You'll quickly identify one or more groups of "regulars." They'll probably be happy to have you join the group. These regular get-togethers are a wonderful way to meet people. The "regulars" can answer many of your questions about operating your Amateur Radio station.

When there is enhanced propagation, such as during a *tropo* opening, the VHF and UHF bands sound more like HF. They are teeming with operators eager to work stations not heard during normal conditions. Pileups occur during good VHF/UHF propagation conditions, just like on HF. In this case, the DX is a station from outside your local area, not from another country. Good VHF or UHF band openings are often short-lived. It's a good idea to limit the contact to the exchange of signal reports and locations. That way, everyone has a chance to work the DX. Other details can be left to the **QSL card**.

When conditions are good, call CQ sparingly, if at all. The old adage "if you can't hear 'em, you can't work 'em" is just as true at VHF as at HF. Stations can often be heard calling CQ DX on the same frequency as distant stations that they cannot hear.

One feature of VHF operating that is rarely found on the HF bands is the use of *calling frequencies* to provide a meeting place for operators using the same mode. Once a contact has been set up, a change to another frequency (a working frequency) is arranged. That lets others use the calling frequency.

The ARRL recommends band plans to coordinate various activities on each amateur band. For example, the 1.25-meter band plan specifies 222.1 MHz as a calling frequency. This frequency marks the bottom edge of the Novice subband,

OSCARs: Ham Satellites

OSCAR 1, the first Amateur Radio satellite, was launched in 1961. Since then, over a dozen more amateur satellites have flown. OSCAR stands for *O*rbiting *S*atellite *C*arrying *A*mateur *R*adio.

Home-Station Equipment and Modes

Amateur satellite communication takes place on the VHF/UHF bands, using voice, Morse code and data-communications modes. FM is rarely used on satellites for reasons explained later. You transmit to the satellite on its *uplink* frequency and listen on its *downlink* frequency. Uplink and downlink are on different bands. Typical satellite stations have a rig for each band, connected to separate antennas. Some manufacturers offer multiband VHF/UHF multimode rigs that can transmit and receive on separate bands at the same time. These rigs are ideal for satellite work. Linear transverters connected to your HF or VHF rig are an alternative.

Amateur satellites don't use fixed receive and transmit frequencies. Instead, they operate like flying, high-power mixers. A typical uplink frequency range is 435.601 to 435.637 MHz. Users can access the satellite anywhere in that range. Because the satellite is a mixer, where you transmit determines where your signal is retransmitted on the downlink. For the same satellite, the downlink range might be 2400.771 to 2400.747 MHz.

Satellites are battery powered and the batteries are recharged by solar panels. Because the satellite is a linear mixer, the more stations using the satellite and the stronger they are, the greater the drain on the satellite's batteries. Running excessive power overloads the satellite and makes all signals weaker! Because FM voice uses a continuous carrier, it imposes a large drain on a satellite's batteries. FM is not an efficient mode for satellite operation and is almost never heard.

Antennas

Amateur satellite stations use antenna systems that can be aimed at points in the sky as well as turned like terrestrial antennas. Unlike commercial TV-satellite "dishes," amateur satellite antennas are relatively small. Amateur satellite antennas need not be mounted high in the air like terrestrial VHF/UHF antennas, although higher antennas let you "see" the satellite when it is near your horizon.

Who Can You Talk To?

Amateur radio satellites offer hours of worldwide communications. DXing through the satellites is much easier than on the amateur HF bands where fading and interference can be very frustrating. Unlike commercial TV satellites that are parked over the equator, Amateur Radio satellites are constantly on the move, providing a wide variety of people and nations to work. Some DXpeditions to rare countries even bring along a satellite station. Another class of satellites in lower orbit offer packet-radio message forwarding. Who knows what the future will bring?

Many US Amateur Radio satellites are planned and built by the Radio Amateur Satellite Corporation (AMSAT). Other Amateur Radio satellites have been built by hams in Russia, Japan, France, Korea, Mexico and other countries. Amateur satellite experiments involving schools around the world have introduced Amateur Radio to thousands of children.

The ARRL Operating Manual has an entire chapter dedicated to satellite communications. It's the best place to start if you'd like more information. Two specialized ARRL books on amateur satellites are *The ARRL Satellite Anthology* and *The Satellite Experimenter's Handbook*.

---

however. You should not operate right on 222.1 MHz with a Novice license. You must always be aware of the *bandwidth* of your transmitted signal—how much frequency space it takes up. Also, your transmitted signal may vary from the operating frequency. You must keep your signal within your band limits.

Theoretically, by using the upper **sideband**—the portion of the SSB signal above the carrier—you could operate on a carrier frequency of 222.100 MHz and all of your transmitted signal would be within the Novice subband. In practice, however, this isn't a good idea. You should operate a few kilohertz above the bottom edge of the band, just to be safe. Check the frequencies near the bottom edge of the band for the other stations doing weak-signal work.

There are also **repeaters** operating in this frequency range. Be certain you are not doing weak-signal work on a repeater frequency in your area. Check the latest *ARRL Repeater Directory* for repeaters in your area. You can also ask some of the other local Amateur Radio operators what frequencies are used for weak-signal work in your area.

[Before going on to the next section, turn to Chapter 12 and study questions N2A18 through N2A20. Also study questions N8A06 and N8A11. Review this section if you have difficulty with any of these questions.]

[If you are studying for the Technician exam, also review questions T2A09, T2A10, T2A15, T2B11, T8A07 and T8A08 in Chapter 13. Reread this section if you have any difficulty.]

FM AND REPEATERS

More hams use **frequency-modulated (FM) voice** than any other communications mode. Most have an FM rig of some type. They use it to keep in touch with their local friends. Hams often pass the time during their morning and evening commute talking on the air. In most communities, amateurs interested in a specialized topic (such as chasing DX) have an FM frequency where they meet regularly to exchange information. At flea markets and conventions, hand-held FM units are in abundance as hams compare notes on the latest bargain.

VHF and UHF FM voice operation takes two forms: simplex and repeater. **Simplex operation** means the stations are talking to each other directly, on the same frequency. This is similar to making a contact on the HF bands.

The communications range for VHF and UHF FM simplex is usually limited to your local area. If you live high on a mountain and use a high-gain directional antenna, you may be able to extend your range considerably. Unfortunately, most of us do not have the luxury of ideal VHF/UHF operating conditions. Often, we want to make contacts even though we live in a valley, are driving in a car or are using a low-power, hand-held transceiver.

Enter repeaters. A **repeater** receives a signal on one frequency and simultaneously retransmits (repeats) it on another frequency. The frequency it receives on is called the **input frequency**, and the frequency it transmits on is called the **output frequency**. Often located atop a tall building or high mountain, VHF and UHF repeaters greatly extend the operating range of amateurs using mobile and hand-held transceivers. See Figure 9-2. If a repeater serves an area, it's not necessary for everyone to live on a hilltop. You only have to be able to hear the repeater's transmitter and reach the repeater's receiver with your transmitted signal.

EQUIPMENT CONSIDERATIONS

To use a repeater, you must have a transceiver that can transmit on the repeater's input and output frequencies. The input and output frequencies are separated by a predetermined amount that is different for each band. This separation is called the *offset*. Common repeater offsets for the VHF and UHF bands are listed in Table 9-6. For example, the offset on 1.25 meters is 1.6 MHz. A repeater on 1.25 meters might have its input frequency on 222.32 MHz and its output on 223.92 MHz.

Table 9-6

Repeater Input/Output Offsets

Band	Offset
6 meters	1 MHz
2 meters	600 kHz
1.25 meters	1.6 MHz
70 cm	5 MHz
33 cm	12 MHz
23 cm	20 MHz

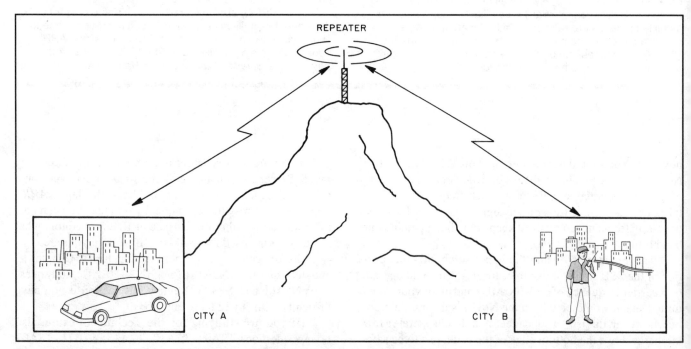

Figure 9-2—Stations in city A can easily communicate with each other, but the hill blocks their communications with city B. The hilltop repeater enables the groups to communicate with each other.

Most transceivers designed for FM repeater operation are set up for the correct offset. They usually have a switch to change between **simplex operation** (transmit and receive on the same frequency) and **duplex operation** (transmit and receive on different frequencies). So, if you wanted to use the repeater in the preceding example, you would switch your transceiver to the duplex mode and dial up 223.92 to listen to the repeater. When you transmit, your rig will automatically switch to 222.32 MHz (1.6 MHz away), the repeater input frequency.

When you have the correct frequency dialed in, just key your microphone button to transmit through ("access") the repeater. Most repeaters are **open**—that is, available for use by anyone in range. Some repeaters, however, have limited access. Their use is restricted to exclusive groups, such as members of a club. Such **closed** repeaters require the transmission of a continuous subaudible tone or a short "burst" of tones for access. These are called **CTCSS** (continuous tone-coded squelch system) or **PL** tones. There are also some repeaters available for use by everyone that require the use of special codes or subaudible tones to gain access. The reason for requiring access tones for "open" repeaters is to prevent interference from extraneous transmissions that might accidentally key the repeater. If you wish to join a group that sponsors a closed repeater, contact the repeater control operator.

FINDING A REPEATER

Most communities in the United States are served by repeaters. While the majority of repeaters (over 6000) are on 2 meters, there are more than 1600 repeaters on 222 MHz, more than 5000 on 440 MHz, over 70 on 902 MHz and more than 200 on 1270 MHz. More repeaters are being put into service all the time. Repeater frequencies are selected through consultation with **frequency coordinators**—individuals or groups that recommend repeater frequencies based on potential interference and other factors.

There are several ways to find the local repeater(s). Ask local amateurs or contact the nearest radio club. Each year, the ARRL publishes *The ARRL Repeater Directory*, a comprehensive listing of repeaters throughout the United States, Canada, Central and South America and the Caribbean. Besides finding out about local repeater activity, the *Directory* is handy for finding repeaters to use during vacations and business trips. See Figure 9-3.

Certain segments of each band are set aside for FM operation. For example, on 1.25 meters, repeater inputs are found between 222.32 and 223.28 MHz. The corresponding outputs are between 223.92 and 224.98 MHz. Frequencies between 223.42 and 223.9 MHz are set aside for simplex operation. On 70 cm, repeater inputs and outputs are located from 442 to 445 MHz and 447 to 450 MHz. On the 900-MHz band, repeater inputs are mostly located from 906 to 909 MHz, with outputs from 918 to 920 MHz. On 23 cm, repeater inputs run between 1270 and 1276 MHz, with corresponding outputs between 1282 and 1288 MHz. Simplex operation is between 1294 and 1295 MHz.

REPEATER OPERATING

Before you make your first FM repeater contact, you should learn some repeater operating techniques. It's worth a few minutes to listen and familiarize yourself with the procedures used by other hams in your area. Accepted procedures can vary slightly from repeater to repeater.

Your First Transmission

Making your first transmission on a repeater is as simple as signing your call. If the repeater is quiet, just say "KA4XYZ" or "KA4XYZ listening"—to attract someone's attention. After you stop transmitting, you will usually hear the unmodulated repeater carrier for a second or two. This *squelch tail* lets you know that the repeater is working. Someone interested in talking to you will call you after your initial transmission. Some repeaters have specific rules for making yourself heard. In general, however, your call sign is all you need.

Don't call CQ to initiate a conversation. It takes longer to complete a CQ than to transmit your call sign. (In some areas, a solitary "CQ" is permissible.) Efficient communication is the goal. You are not on HF, trying to attract the attention of someone who is casually tuning across the band. In the FM mode, stations are either monitoring their favorite frequency or not. Except for scanner operation, there is not much tuning across the repeater bands.

To join a conversation in progress, transmit your call sign during a break between transmissions. The station that transmits next will usually acknowledge you. Don't use the

ALABAMA					
ANNISTON					
Anniston	147.09	+	WB4GNA	o(ca)xz	CARA
Jacksonville	146.78	–	N4HOA	o	N4HOA
Jacksonville	147.04	+	WD4KTY	oae	AAERC
Lincoln	145.49	–	KC4OX	o	KC4OX
Munford	146.74	–	W4NTI	otalx	W4NTI
Talladega	147.36	+	W4GBQ	o	Tldga ARC
AUBURN/OPELIKA					
Auburn	147.06	+	KA4Y	o	KA4Y
Auburn	147.24	+	K4RY	o	AbrnU ARC
Opelika	146.94	–	W4ZBA	ox	E AL ARC
Opelika	147.12	+	W4ZBA	oaez	
BIRMINGHAM					
Bessemer	145.15	–	N4AHN	oaelxz	K4GTQ
Bessemer	145.39	–	N4LJE	o	N4LJE
Birmingham	145.19	–	N4IQT	ol	N4IQT
Birmingham	145.23	–	WB4TJX	oe	UAB ARC
Birmingham	146.76	–	K4TQR	oae	H & H RA
Birmingham	146.88	–	W4CUE	oaeWx	BARC
Birmingham	146.94	–	W4CUE	oae	BARC
Birmingham	147.28	+	KD4A	o	TelPnrRC
Birmingham	147.34	+	WA4MHO	o	WA4MHO
Centerpoint	146.62	–	K4ANU	o	K4ANU
Hueytown	147.14	+	WA4CYA	o(±Offset)	HueytwnRA
Leeds	146.68	–	N4IQT	o	Leeds RG
Oneonta	146.70	–	N4KGK	o	BlntCoARC
Palmerdale	145.45	–	WB4ZNQ	o	Cntrpt RC
Pelham	146.98	–	KC4AF	oe	ShlbyCoRC
Pell City	145.37	–	KB4LPR	oa	KB4LPR

Figure 9-3—Anyone who operates on VHF FM should have a copy of *The ARRL Repeater Directory*. Directory information includes the repeater call sign, location, frequency and sponsor.

word "break" to join a conversation—unless you want to use the repeater to help in an **emergency**. To make a distress call over a repeater, say *"break break"* and then your call sign to alert all stations to stand by while you deal with the emergency.

A further word about emergencies: Regardless of the band, mode or your class of license, FCC rules specify that, in case of emergency, the normal rules can be suspended. If you hear an emergency call for help, stop your conversation immediately and do whatever you can to establish contact with the station needing assistance. Immediately pass the information on to the proper authorities.

To call another station when the repeater is not in use, just give both calls. For example, "WA1MBK, this is N1BKE." If the repeater is in use, but the conversation sounds like it is about to end, wait before calling another station. If the conversation sounds like it is going to continue for a while, however, transmit only your call sign between their transmissions. After you are acknowledged, ask to make a quick call. Usually, the other stations will stand by. Make your call short. If your friend responds, try to meet on another repeater or on a simplex frequency. Otherwise, ask your friend to stand by until the present conversation ends.

Use plain language on a repeater. If you want to know someone's location, say "Where are you?" If you want to know whether someone you're talking with is using a mobile rig or a hand-held radio, just ask: "What kind of radio are you using?" You get the idea.

Courtesy Counts

If you are in the midst of a conversation and another station transmits his or her call sign between transmissions, the next station in line to transmit should acknowledge the new station and permit the new arrival to make a call or join the conversation. It is impolite not to acknowledge new stations, or to acknowledge them but not let them speak. The calling station may need to use the repeater immediately. He or she may have an emergency to handle, so let him or her make a transmission promptly.

A brief pause before you begin each transmission allows other stations to break in—there could be an emergency. Don't key your microphone as soon as someone else releases hers. If your exchanges are too quick, you can prevent other stations from getting in.

The *courtesy tones* found on some repeaters force users to leave a space between transmissions. The beeper sounds a second or two after each transmission to permit new stations to transmit their call signs in the intervening time period. The conversation may continue only after the beeper sounds. If a station is too quick and begins transmitting before the beeper sounds, the repeater may indicate the violation, sometimes by shutting down!

Keep transmissions as short as possible, so more people can use the repeater. Again, long transmissions could prevent someone with an emergency from getting the chance to call for help through the repeater. All repeaters encourage short

transmissions by "timing out" (shutting down for a few minutes) when someone gets long-winded. The **time-out timer** also prevents the repeater from transmitting continuously, due to distant signals or interference. Because it has such a wide coverage area, a continuously transmitting repeater could cause unnecessary interference. Continuous operation can also damage the repeater.

You must transmit your call sign at the end of a contact and at least every 10 minutes during the course of any communication. You do not have to transmit the call sign of the station to whom you are transmitting.

Never transmit without identifying. For example, keying your microphone to turn on the repeater without saying your station call sign is illegal. If you do not want to engage in conversation, but simply want to check if you are able to access a particular repeater, simply say "KJ4KB testing."

Fixed Stations and Prime Time

Repeaters were originally intended to enhance mobile communications. During commuter rush hours, mobile stations still have preference over fixed stations on some repeaters. During mobile prime time, fixed stations should generally yield to mobile stations. When you're operating as a fixed station, don't abandon the repeater completely, though. Monitor the mobiles: your assistance may be needed in an emergency. Use good judgment: Rush hours are not the time to test your radio extensively or to join a net that doesn't deal with the weather, highway conditions or other subjects related to commuting.

SIMPLEX OPERATION

After you have made a contact on a repeater, move the conversation to a *simplex* frequency if possible. The repeater is not a soapbox. You may like to listen to yourself, but others, who may need to use the repeater, will not appreciate your tieing up the repeater unnecessarily.

The function of a repeater is to provide communications between stations that can't otherwise communicate because of terrain, equipment limitations or both. It follows that stations able to communicate without a repeater should not use one. That way, the repeater is available for stations that need it. (Besides, communication on simplex offers a degree of privacy impossible to achieve on a repeater. On simplex you can usually have extensive conversations without interruption.)

Select a frequency designated for FM simplex operation. Otherwise, you may interfere with stations operating in other modes without realizing it. (The reason for this is simple: Changing to a simplex frequency is far easier than changing the frequencies a repeater uses.) Table 9-7 lists the established VHF/UHF simplex frequencies. Each band has a designated national FM simplex calling frequency, which is the center for most simplex operation. These frequencies are also indicated in Table 9-7. To see if you and the other station can communicate on a simplex frequency, listen on the

Table 9-7

Common VHF/UHF FM Simplex Frequencies

2-Meter Band	1.25-Meter Band	70-cm Band
146.52*	223.42	446.0*
146.535	223.44	
146.55	223.46	**33-cm Band**
146.565	223.48	906.5*
146.58	223.50*	
146.595	223.52	**23-cm Band**
147.42	223.54	1294.5
147.435	223.56	
147.45	223.58	
147.465	223.60	
147.48	223.62	
147.495	223.64	
147.51	223.66	
147.525	223.68	
147.54	223.70	
147.555	223.72	
147.57	223.74	
147.585	223.76	
	223.78	
	223.80	
	223.82	
	223.84	
	223.86	
	223.88	
	223.90	

*National simplex frequency

repeater input frequency. If you can clearly hear what's going into the repeater, you don't need the repeater to communicate.

Autopatch: Use it Wisely

An **autopatch** allows repeater users to make telephone calls through the repeater. To use most repeater autopatches, you generate the standard telephone company tones to access and dial through the system. The tones are usually generated with a telephone tone pad connected to the transceiver. Tone pads are available from equipment manufacturers as standard or optional equipment. They are often mounted on the front of a portable transceiver or on the back of a fixed or mobile transceiver's microphone. Whatever equipment you use, the same autopatch operating procedures apply.

There are strict guidelines for autopatch use. The first question you should ask is "Is the call necessary?" If it is an **emergency**, there is no problem—just do it! Other reasons may fall into a gray area. As a result, some repeater groups expressly forbid autopatch use, except for emergencies.

Don't use an autopatch where regular telephone service is available. One example of poor operating practice can be heard most evenings in any metropolitan area. Someone will call home to announce departure from the office. Why not make that call from work before leaving?

Never use the autopatch for anything that could be considered business communications. The FCC strictly forbids you to conduct communications in Amateur Radio for your business or for your employer. You may, however, use Amateur Radio to conduct your own personal communications. The rules no longer forbid you to use the autopatch to call your doctor or dentist to make an appointment, or to order food, for example.

Don't use an autopatch to avoid a toll call. Autopatch operation is a privilege granted by the FCC. Abuses of autopatch privileges may lead to their loss.

You have a legitimate reason to use the autopatch? Here's how most systems operate. First, you must access (turn on) the autopatch, usually by pressing a designated key on the tone pad. Ask the other hams on a repeater how to learn the access code. Many clubs provide this information only to club members. When you hear a dial tone, you know that you have successfully accessed the autopatch. Now, simply punch in the telephone number you wish to call.

Once a call is established, remember that you are still on the air. Unlike a normal telephone call, only one party at a time may speak. Both you and the other person should use the word "over" to indicate that you are finished talking and expect a reply. Keep the call short. Many repeaters shut off the autopatch after a certain time.

Turning off the autopatch is similar to accessing it. A key or combination of keys must be punched to return the repeater to normal operation. Ask the repeater group sponsoring the autopatch for specific information about access and turn-off codes, as well as timer specifics.

[Before you go on to the next section, turn to Chapter 12 and study questions N2A04 and N2B07 through N2B15. Review this section if you have difficulty with any of these questions.]

[If you are studying for the Technician exam, turn to Chapter 13 and look over questions T2A01 through T2A08, T2A11 through T2A14, T2A16, T2A18, T2B01 through T2B04, T2C03 and T8A06. Reread this section if you have any difficulty.]

INTERFERENCE AND SPLATTER

Chapter 10 covers various types of interference, and describes how to solve these problems. Voice modes require careful transmitter adjustment to produce the best signal quality. It's easy to avoid problems if you know what to do.

Most SSB transmitters and transceivers have a *microphone gain* control. This control may be on the front panel, or it may be inside the rig. Different microphones and different voices have different characteristics. You'll have to adjust the microphone gain control for your particular situation. Instructions for properly setting the mike gain are included in your instruction manual.

In FM transceivers, the mike-gain control, sometimes called the deviation control, is preset. It requires no adjustment unless there is a problem. Speaking too loudly or too close to the microphone can cause interference to those using nearby frequencies. If you're told that your signal is **overdeviating** (also called *overmodulating*), you're speaking too loudly or too close to the microphone.

Don't continuously adjust the mike gain. Set it according to the manual and leave it alone. Follow the manufacturer's instructions for use of the microphone. Some require close talking, while some need to be turned at an angle to the speaker's mouth. Always try to speak in an even amplitude, and at the same distance from the microphone. If you set the gain properly, you will minimize background room noise on your signal. You'll always hear one or two stations on the band with their mike gain set so high that you hear people talking in the background, music playing, dogs barking and the like. This is not good practice.

If you operate with the microphone gain set too high, you will most likely cause **splatter**—interference to stations on nearby frequencies. This applies to FM voice operation as well as SSB operation. If the mike gain or deviation control on your FM rig is too high, your signal will be too wide and will cause interference to stations on nearby frequencies. In addition, your signal may sound so awful that it is impossible to understand!

A *speech processor* (sometimes built into the transceiver, or available as an accessory) can be a mixed blessing. It can give your audio more "punch" to help you cut through interference and noise. If you use too much processing however, your signal quality suffers greatly. You'll sometimes hear operators with their speech processor set so high that they are difficult to understand. That's exactly the opposite of what speech processing is supposed to do.

If you set the speech processing level too high, your signal will sound bad. It will also cause splatter and interfere with stations on nearby frequencies. Make tests to determine the maximum speech processing level that can be used effectively. Make a note of the control settings. Turn your speech processor control down or off if it is not required during a contact.

[Before going on, turn to Chapter 12 and review questions N8B07 through N8B11. Reread this section if you have any questions.]

RADIOTELETYPE AND PACKET COMMUNICATIONS

Radioteletype, packet and AMTOR are terms used to describe amateur communications designed to be received and printed automatically. They are sometimes called **digital communications**, because they often involve direct transfer of information between computers. With the help of accessory equipment, the computer then processes the signal and sends it over the air from your transceiver. The station on the other end receives the signal, processes it and prints it out on a computer screen or printer. In this section, you will learn about two exciting forms of communication: **radioteletype (RTTY)** and **packet radio**.

EMISSION MODES

RTTY and packet communications are similar to CW. They both use two states to convey information. Instead of switching a carrier on and off as in CW operation, the carrier is left on continuously and switched between two different frequencies. The transmitter carrier frequency "shifts" between two frequencies, called MARK and SPACE. MARK is the ON state; SPACE is the OFF state. This is called **frequency-shift keying (FSK)**. On 10-meter RTTY, the MARK and SPACE frequencies are normally 170 Hz apart. This is known as 170-Hz shift RTTY.

On VHF, **audio-frequency shift keying (AFSK)** is used. AFSK is similar to FSK, except that an FM transmitter is used. Audio tones corresponding to MARK and SPACE are fed into the microphone jack and used to modulate the carrier. The most common VHF shift is 170 Hz, but you may hear 850-Hz shift as well.

It is interesting to note that you can also operate Morse code with an FM transceiver. You can feed an audio tone into the microphone jack, and then key the audio tone on and off to send Morse code. This type of signal is a *modulated CW* or *MCW* signal.

SENDING SPEEDS

Just as you can send CW at a variety of speeds, RTTY and packet transmissions are sent at a variety of speeds. The signaling speed depends on the type of transmission and on the frequency band in use. A *baud* is the unit used to describe transmission speeds for digital signals.

For a single-channel transmission, a baud is equivalent to one digital bit of information transmitted per second. A 300-baud signaling rate represents a transmission rate of 300 bits of digital information per second in a single-channel transmission. A digital bit of information has two possible conditions, ON or OFF. We often represent a bit with a 1 for ON or a 0 for OFF. The MARK and SPACE tones of a

RTTY Equipment

Years ago, all RTTY operation was done with mechanical teleprinters. You may still see some of these machines in old movies—they are big, slow and noisy. Today, amateurs use RTTY *terminal units (TUs)* or *multimode communications processors (MCPs)* for RTTY operation. The TU connects between a transceiver and a computer and decodes the RTTY signals for display by the computer. MCPs let you operate nearly any mode, so you won't have to buy new equipment if you decide to give a new mode a try. Newer versions use digital signal processing for added flexibility: if a new digital mode comes along, you won't need to replace the MCP—just the software!

When you shop for a communications terminal, there are a few things to keep in mind. Some are designed to work with only one type of computer. Some require additional software in the computer, and others require only a simple terminal program. As always, the best way to find out about a particular unit is to ask someone who has one. The next best way is to read the Product Review column in *QST*.

The following list is representative of what is available today. Some older used units may be available at hamfests and flea markets. If you plan to buy a used communications terminal, take an experienced RTTY operator along to the hamfest with you. Most hams love to talk about their favorite operating mode—ask around at your local club.

The trend in RTTY equipment today is toward the "all-mode" communications terminal. The AEA PK-232 (above) and MFJ-1278 (below) provide CW, Baudot, ASCII, AMTOR, FAX and packet-radio capability. They are used with any of a number of popular personal computers.

RTTY Communications Terminals

All models operate Baudot RTTY and ASCII in addition to the other modes listed.

Manufacturer	Model Number	Modes	Product Review
AEA	PK-900	CW, AMTOR, packet	
AEA	PK-232	CW, AMTOR, packet	January 1988 *QST*
AEA	DSP-2232/1232	CW, AMTOR, packet, PSK, fax, SSTV	
Kantronics	KAM	CW, AMTOR, packet	*June 1989 and January 1992 *QST*
Kantronics	KAM Plus	CW, AMTOR, PacTOR, G-TOR, packet	
MFJ	MFJ-1224	CW	*
MFJ	MFJ-1278	CW, AMTOR	September 1989 *QST*
PacComm	PacTor	PacTOR, CW, AMTOR (packet†)	January 1993 *QST*

*Equipment discontinued.
†Optional.

---

radioteletype signal also represent the two possible bit conditions.

On HF, we use signaling rates of up to 300 bauds. Sending speeds on VHF are generally faster than on HF. A common VHF sending speed is 1200 bauds. The most important thing to remember about sending speeds is that both stations use the same speed during a contact.

RTTY COMMUNICATIONS

Radioteletype (RTTY) is a popular form of communications. Its technical name is **narrow-band direct-printing telegraphy**. Novices can operate RTTY on 10 meters; Novices and Technicians can operate RTTY on the 222-MHz and 1270-MHz bands. On 10 meters, Novice RTTY operation is allowed from 28.1 to 28.3 MHz. On 222 and 1270 MHz, you are allowed to operate RTTY on all of the frequencies that the FCC authorizes you to operate on. You should, however, follow the band plans as described earlier. On VHF, most RTTY activity is on repeaters. Check with local amateurs to find out where the 222-MHz RTTY activity is in your area.

This section will tell you about three popular types of RTTY: *Baudot*, *AMTOR* and *ASCII*. Baudot and AMTOR are the most popular RTTY modes on HF. Until recently, high-speed ASCII was most often heard on VHF, but this mode has been almost completely replaced by packet radio.

Baudot Radioteletype

Radioteletype communication using the **Baudot** code (also known as the International Telegraph Alphabet number 2, or ITA2) is used on the Amateur Radio HF bands in most areas of the world.

The Baudot code represents each character with a string of five bits of digital information. Each character has a different combination of bits. There are only 32 possible Baudot code combinations. This limits the number of possible characters. All text is in upper-case characters. To provide numbers and punctuations, you shift between the letters case (LTRS) and the figures case (FIGS).

There are three common speeds for 10-meter Baudot RTTY communications: 60 WPM (45 bauds), 75 WPM (56 bauds) and 100 WPM (75 bauds). You'll often hear 60, 75 and 100 WPM RTTY referred to as "60 speed," "75 speed," and "100 speed," respectively. Both stations must use the same sending speed to make a RTTY contact.

Setting Up

To set up a RTTY station using Baudot, the first thing you will need is a *multimode communications processor (MCP)*, a RTTY terminal or a mechanical teleprinter. (Noisy and heavy, mechanical teleprinters are rarely used nowadays.) See Figure 9-4.

If you use an MCP, you'll also need a personal computer and the appropriate software (computer program). As its name implies, the MCP allows operation on several modes, often including AMTOR, CW and packet. Dedicated RTTY terminals (also called *RTTY modems* or *terminal units*) operate only RTTY. No personal computer is required although some RTTY modems do require a keyboard and display device (a terminal) or computer. In choosing a RTTY system, look for a modem with shift capabilities of 170 and 850 Hz. Check the *QST* Product Review column, and articles and ads in *QST* and other amateur magazines for information about these products.

Modem is a contraction of *modulator-demodulator*. A modem takes electrical signals from the computer or terminal and turns them into audio tones to *modulate* the RF signal. It also takes audio tones from the receiver and *demodulates* them into electrical signals for the computer. It is an interface between the computer or terminal and the radio.

What about a rig for RTTY? You can use almost any HF transceiver. Ideally, the receiver used for 170-Hz-shift RTTY should have the minimum practical bandwidth, preferably between 270 and 340 Hz. Many CW filters have bandwidths around 500 Hz and are good for 170-Hz RTTY reception. Some transceivers can use only the SSB filter when in the SSB or RTTY mode. Some RTTY operators have modified their transceivers to use the CW filter when the RTTY mode is selected. A switch can be added so it is possible to select

Figure 9-4—A RTTY modem connects between your transceiver and your computer.

the CW filter with the mode switch set for SSB operation. A transceiver with a frequency display of 10-Hz resolution is also helpful, although not necessary.

Many modern transceivers include an FSK mode. If yours does not, you can operate RTTY with the transceiver in the SSB mode. It is normal to use the lower-sideband mode for RTTY on HF SSB radio equipment. Most transceivers have two positions on the sideband-selection knob, labeled NORMAL and REVERSE. This labeling may cause some uncertainty when you use the rig for RTTY. On 10 meters, normal SSB operation uses upper sideband, so you will have to set the switch to the REVERSE position to select lower sideband for 10-meter RTTY.

If you are not using the correct sideband, your signal is "upside down," and other operators will have to change their normal operating setup to copy your signals. Be sure you select the lower sideband and transmit signals that are "right-side up." Consult your radio's manual for details.

On VHF, the most common practice is to use **AFSK**. Since most VHF RTTY work is done on local FM repeaters, almost any FM transceiver will work.

Your transmitter must be able to operate at full power for extended periods (called *100% duty cycle*) when transmitting conventional Baudot and ASCII radioteletype, as well as AMTOR Mode B. Some transceivers, designed for SSB and CW operation (which do not require constant transmission at full power), may overheat and possibly fail if subjected to a long RTTY transmission. Many operators reduce to half power during long RTTY transmissions to avoid overheating problems. AMTOR Mode A transmits data in shorter blocks and does not require constant transmission at full power.

Receiver-tuning accuracy is important. Thus, a tuning aid can be a great help in proper receiver adjustment. Most modems have some type of tuning indicator, possibly just flashing LEDs. Some have oscilloscopes that produce patterns such as those shown in Figure 9-5. The MARK signal is displayed as a horizontal line and the SPACE signal as a

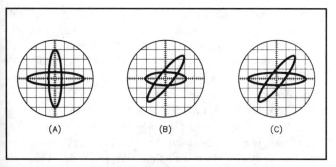

Figure 9-5—Oscilloscope RTTY tuning patterns. A shows the pattern produced by a properly tuned signal. The displays at B and C indicate improperly tuned signals.

vertical line. Theoretically this should appear as a "+" sign on the 'scope screen, but in practice it may look more like a pair of crossed bananas! To avoid the high cost of an oscilloscope, some manufacturers offer LED displays that imitate the 'scope display.

Typing the Message

No one can disagree that touch typing (typing without looking at the keys) is the best way to go. But, you can have enjoyable contacts even if you can only type with two fingers! It is a good practice to keep the number of characters per line to a maximum of 69. That line length can be handled by virtually any teleprinter designed for Baudot. Many US teleprinters will print 72 or 74 characters, but some foreign printers take only 69.

After a mistake, the "oops" signal is XXXXX (some operators may use EEEEEE). There is no need to overdo this, though, because generally the other operator will figure out what you meant to say anyway, and a page filled with XXXXX can be confusing.

Station Identification

When you are transmitting RTTY, simply send your call sign at least once every 10 minutes, just as you would on any other mode. FCC rules permit a station engaged in digital communications using Baudot, AMTOR or ASCII to identify in the digital code used for the communication. On the Novice 222- and 1270-MHz bands, identification may be given using one of the methods just mentioned or by voice. (Radiotelephone operation is permitted on the same frequency being used for the RTTY communication on these bands.) Identifying the station you are communicating with, in addition to your own station, is not legally required, but it will help those monitoring your QSO to determine band conditions.

Test Messages

A message consisting of repetitive RYs is useful for local and on-the-air testing of Baudot RTTY equipment:

RYRYRYRYRYRYRYRYRYRYRYRY DE W1AW

The use of the RY sequence goes back to the early Teletype machine days. To set the mechanical machines, it was necessary to send a string of Rs and Ys while the receiving operator adjusted a knob called the "rangefinder." The military and commercial services also used RYRYRYRYRY as a channel-holding signal during idle periods. The letters R and Y both alternate between MARK and SPACE tones in the Baudot code, so it was natural to use these letters for the range-setting operation. (If we represent the SPACE tone with a 0 and the MARK tone with a 1, then R is 01010 and Y is 10101. So you can see that a string of RY characters sends alternating SPACE and MARK tones.)

Use the RY sequence sparingly. RYs have a musical sound that is easily recognizable, and is useful for tuning a signal in properly. This makes it good for rousing attention when you are sending a CQ call, or when calling another station on schedule, but it is unnecessary during routine contacts. Don't use RY when you are working DX.

Making Contact

There are two ways of establishing contact with other amateur stations using radioteletype. You can answer someone else's call, or you can try calling CQ yourself. When you call CQ, you depend on someone else to tune your signal properly, and if you are a newcomer to the mode, that may be an advantage. By listening for other calls, you have a good chance of finding stations that you might not contact otherwise. You will want to use both methods at various times, but whichever one you choose, learn the proper procedure and follow it.

Calling CQ

Since it is a little easier to explain, we will start by describing the procedure for calling CQ on RTTY. You must first locate a clear frequency on which to make your call. Courtesy should be the rule. Ask if the frequency is busy by sending QRL?, or by asking in clear text. There may be a QSO on frequency, but you may not hear both sides of it.

The general CQ call, like RY, has a recognizable, musical sound. Take advantage of this fact, and send CQ in a pattern something like this:

CQ CQ CQ CQ CQ DE KA9HDQ KA9HDQ KA9HDQ K

Transmit "CQ" three to six times, followed by "DE," followed by your call sign three times. Many operators add their name and QTH (location) to the last line of the CQ call. This is a good idea, but keep it simple and short. As on voice, several short calls with pauses to listen are much better than one long call.

Calling Another Station

You have just heard 9H1EL in Malta calling CQ. He is booming in with an S9 signal, and this would be a new country for you. Make sure that he is tuned in properly, wait until he stops calling CQ and give him a call. Keep your calls short. Forget the RYs. Send 9H1EL's call sign only once, and your call sign no more than five times:

9H1EL DE KA1ION KA1ION KA1ION K

Then stand by and listen. If he doesn't answer, try another short call. Be sure to answer at the same speed you are receiving the other station.

Line Feeds and Other Things

When another station answers your call, gives you a report and then turns it back to you, what should you do? Always send a line feed (L/F) and a carriage return (C/R) first. This will put his computer and/or printer into the "letters" mode, and will eliminate the possibility of part of your message being garbled by numbers and/or punctuation marks.

Next, send his call and your call once each. On your first exchange, send your name and QTH message from the computer memory, if your system has such a message. If you have a type-ahead buffer on your system, you can compose some of your reply before it is actually transmitted. This saves time, especially if you are not a fast typist.

If you give another station a signal report of RST 599, it means you're copying solid. If you receive a similar report, it is unnecessary to repeat everything twice. Give honest reports—if the copy is solid, then the readability report is five. Adapt your operating technique to the reported conditions. If copy is marginal, repeats or extra spaces are in order. But with solid copy, you can zip right along.

Don't send a string of carriage returns to clear the screen. The other operator may be copying on a printer, and the carriage returns will waste a lot of paper. The last two characters of each transmission should be carriage returns, however.

Signing Off

You have now completed your QSO and are about to sign off. What **prosign** do you use? Well, the standard prosigns mean the same thing on RTTY as they do on CW, and should be adequate. Table 9-8 summarizes the meaning of the most commonly used prosigns. For more on RTTY, see the ARRL's *Your RTTY/AMTOR Companion*.

[Time to turn to Chapter 12 again. You should be able to answer questions N2B01, N2B02, N8A02 and N8A04. Review this section if you have any difficulty.]

[**If you are preparing for the Technician exam**, turn to Chapter 13 and study questions T8A03, T8A05 and T8A11.]

AMTOR: ERROR-FREE RTTY

Amateur Teleprinting Over Radio (AMTOR) is a more reliable transmission system than Baudot RTTY. AMTOR has the ability to detect errors, so you can be assured of getting perfect copy throughout the QSO. In many respects, AMTOR operation is the same as for Baudot RTTY. AMTOR characters are commonly sent at 100 bauds, and 170-Hz is the normal shift.

There are several modes of AMTOR operation: *Automatic Repeat Request (ARQ)* or *Mode A*, and *Forward Error Correction (FEC)* or *Mode B*. Mode B is further subdivided into Collective B-Mode and Selective B-Mode. Amateurs have added Mode L for monitoring AMTOR transmissions.

Stations in contact using AMTOR are classified either as Information Sending Stations (ISS) or Information Receiving Stations (IRS). The stations change identities as information is exchanged during a QSO. The station that originates the communication is also called the master station and the other station is called the slave station. The master/slave relationship does not change during the contact.

Mode A

Mode A is a synchronous system in which the ISS transmits blocks of three characters to the IRS. The ISS sends the block, listens for replies from the IRS, then sends three more characters. The transmitting and listening times are less than a quarter of a second, so your transceiver must be able to change from transmit to receive very quickly. The ISS keeps the three characters in memory until it gets a receipt from the IRS. When the IRS acknowledges, the ISS moves on to the next three characters. If the IRS does not acknowledge, the ISS keeps sending the three characters until they are acknowledged. AMTOR keeps trying until it gets it right!

Mode A is used only for contacts between two stations. It should not be used for calling CQ or in nets. This mode has

---

Table 9-8

Common Prosigns Used On RTTY

K—Invitation to transmit.

KN—Invitation to the addressed station only to transmit.

SK—Signing off. End of contact.

CL or CLEAR—I am shutting my station down.

SK QRZ—Signing off and listening on this frequency for any other calls. The idea is to indicate which station is remaining on the frequency.

Some operators use various other combinations of prosigns:

SK KN—Signing off, but listening for one last transmission from the other station.

SK SZ—Signing off and listening on this frequency for any other calls.

a characteristic "chirp" sound on the air. It has a 47% duty cycle, allowing transmitters to operate at full power output without worry about overheating. Mode L is used by stations to monitor (listen to) Mode A contacts.

Collective B-Mode

Collective B-Mode is basically a "broadcast" mode, although most hams don't use that word because FCC rules say that we are not allowed to "broadcast" in the entertainment sense of the word. In this mode, there is only one sending station, called the Collective B Sending station (CBSS) and any number of Collective B Receiving Stations (CBRSs). The CBSS sends each character twice: the first transmission of a character, called DX, is followed by four other characters, and then the repetition, called RX, is sent.

This is the mode that W1AW uses for bulletins and the one that should be used for calling CQ, for nets and for making multi-way contacts. It has a 100% duty cycle, so you may have to reduce your transmitter power output as explained before.

Selective B-Mode

Selective B-Mode is intended for transmissions to a single station or group of stations. It is similar to collective B-Mode. Only stations set up to accept this mode and recognize the specific selective-call identifier are intended to receive Selective B-Mode transmissions.

AMTOR Selective-Calling Identities

AMTOR is designed so that four-letter identities are used as the *selective-call identifier* (similar to a call sign). Amateurs can't simply use their call signs as their AMTOR selective-call identifier because amateur call signs consist of letters and numerals. Most AMTOR stations use the first letter and the last three letters of their call sign for the selective-call identifier. Using this algorithm, the selective-call identifier for W1AW would be WWAW. This scheme doesn't provide unique selective-call identifiers for every call sign (for example, WWAW would also be used for W2AW, W2WAW and others), so be aware that someone may ask you to choose another selective-call identifier if yours is already in use.

A recent version of AMTOR does permit numbers and up to six characters for the selective-call identifier. Not all AMTOR stations use this version, however, so you should start out by using a four-character selcall.

Making Contact on AMTOR

If you try to make your first contact using Mode A (ARQ) and there is a problem somewhere in the system, you will not make contact. It takes a two-way contact to make Mode A work at all.

Try your first contact using Collective B-Mode. Use this mode for calling CQ, and include your selective-call identifier. Use of Mode B should ensure that you can get the circuit working one way. In Mode B, you can check to confirm that you are using the correct frequency-shift polarity for both receiving and transmitting.

After making contact using Mode B, ask the other station to go into the monitor (listen) mode and try to receive your Mode A transmission. If there is a problem with Mode A, the most likely cause is that your transceiver is taking too long to change between receive and transmit. If that's the case, check with other AMTOR operators having the same equipment setup. If you are using commercial AMTOR gear, the manufacturer will probably know which transceivers will switch fast enough and which require specific modification.

If the other station (in monitor mode) is copying your Mode-A transmission, ask the station to make a Mode-A call to your selective-call identifier. The other station is acting as the master and your station the slave in this case. Finally, reverse the roles and try being the master station.

Because of the timing cycle specified in the AMTOR standards, it is not practical to contact AMTOR stations on the opposite side of the earth using Mode A. Mode B can be used for communication with stations too distant for Mode-A contacts. For more about AMTOR, see the ARRL's *Your RTTY/AMTOR Companion*.

ASCII OPERATION

ASCII (pronounced "askee") stands for American National Standard Code for Information Interchange. ASCII operation differs from Baudot RTTY operation in that it has a larger character set than ITA2. This is possible because ASCII characters consist of 7 digital information bits. Besides a number of control characters, ASCII also provides both upper- and lower-case letters and a more complete set of punctuation marks. ASCII is a character set that was designed for computer and **data** communications uses. For more information about the complete ASCII coded character set, and for the technical details about methods and transmission rates used for the ASCII code, see the *ARRL Handbook for Radio Amateurs*.

HF ASCII transmissions are normally sent at 110 or 300 bauds, using 170-Hz shift. Higher data rates are used on higher frequency bands. On 10 meters, the maximum permitted speed is 1200 bauds, while up to 19,600 bauds is permitted on 222 MHz and above. On VHF and UHF, 1200 bauds is the most common ASCII sending speed.

COMPUTER-BASED MESSAGE SYSTEMS

No discussion of modern radioteletype operation could be complete without some information about mailbox operation. A number of *Computer-Based Message Systems (CBMSs)* are operating on the ham bands. You may hear them called one of these terms: MSO (Message Storage Operation), bulletin board or mailbox.

All of these systems have some features in common.

They will automatically respond to calls on their operating frequency if the calling station uses the correct character sequence. You can send messages to be stored in the mailbox and retrieved later by another amateur. You may also request a listing of the messages on file and read any addressed to you, as well as any bulletin messages intended for general consumption. Some systems may allow messages to have password protection, restricting access to the information.

Message handling in this manner is third-party traffic, and the system operator (SYSOP) is required to observe appropriate rules concerning message content. Also the SYSOP is responsible for maintaining control of the transmitter and removing it from the air in the event of malfunction.

PACKET RADIO

Packet radio is communications for the computer age. More and more hams are adding computers to their ham shacks.

Early ham-oriented computer programs allowed computers to send and receive CW and RTTY. Some farsighted hams, however, developed a new mode of communications that unleashes the power of the computer. That mode is **packet radio**.

Being a child of the computer age, packet radio has the computer-age features that you would expect.

• It is **data** communications—high speed and error-free packet-radio communications lends itself to the transfer of large amounts of data.

• It is *fast*, much faster than the highest speed CW or RTTY.

• It is *error-free*—no "hits" or "misses" caused by propa-

PACKET RADIO EQUIPMENT

The following list is by no means complete, but it is a sample of what is available new and used. For more details on the equipment mentioned here (and others), see *QST*'s Product Review column, and a book published by the ARRL, *Your Packet Companion*.

Most of the packet controllers on the market today can trace their roots to two terminal node controller designs by the Tucson Amateur Packet Radio Corporation (TAPR), the TAPR TNC 1 and TNC 2. Equipment based on the TNC 2 is still available.

Packet Controllers

AEA PK-88. This version of the TNC 2 has beefed-up circuitry to suppress RFI. Its power requirements are relatively high—12 to 15 V dc at 400 mA.

Kantronics KPC3. The KPC3 has a standard RS-232-C serial port. Serial-port and radio-port data rates are software selectable. Six LEDs provide status information.

MFJ-1270. This unit provides a TTL interface (for Commodore 64 compatibility), as well as the standard RS-232-C serial port. (See the Product Review in September 1986 *QST*.)

Pac-Comm TNC-200. Another TNC 2 clone, the TNC-200 is available as either a kit or an assembled and tested unit. Other than cosmetic differences, Pac-Comm's versions of the TNC 2 are the same as the original.

Computer-Specific TNCs and Multimode Communications Processors

AEA PK-64. The PAKRATT Model PK-64 provides the Commodore 64 and 128 computers with packet-radio capability; in addition Morse, RTTY (Baudot and ASCII) and AMTOR operation are possible. The PK-64 plugs into the user port of a C64 or C128. (See Product Review, *QST*, June 1986. This unit has been discontinued but it is available used.)

AEA PK-900. The PAKRATT Model PK-900 provides Morse, Baudot, ASCII, AMTOR and packet radio capability

The Kantronics *KAM Plus* is a multimode communications processor that operates on AMTOR, packet, PacTOR and CW as well as RTTY. The KAM Plus now also includes G-TOR. PacTOR and G-TOR combine the best features of AMTOR and packet.

The MFJ 1270B is a popular and inexpensive TNC 2 clone for packet operation.

for any computer equipped with an RS-232-C serial port. The unit includes internal modems for HF (200 Hz frequency shift) and VHF operation.

Kantronics KAM Plus. The "Kantronics All Mode" does just that—Morse, Baudot, ASCII, AMTOR, PacTOR, G-TOR and packet radio. It may be connected to any computer or terminal with a standard RS-232-C serial port or TTL interface. Two radio ports are included for HF and VHF operation. (See the KAM Product Reviews in June 1989 and January 1992 *QST*.)

MFJ 1278. MFJ's all-mode data controller unit features

gation variations or electrical interference.

• It is *spectrum efficient*—several stations can share one frequency at the same time.

• It is a **network**—packet stations can be linked together to send data over long distances.

• It is *message storage*—packet-radio bulletin boards (PBBS) provide storage of messages for later retrieval.

How Does Packet Work?

Packet radio uses a *terminal node controller (TNC)* as the interface between computer and transceiver. A TNC is an enhanced modem. We discussed what a modem is and what it does in the section on RTTY. The TNC accepts information from your computer or ASCII terminal and breaks the data into small pieces called *packets*. In addition to the information from your computer, each packet contains addressing, error-checking and control information. The addressing information includes the call sign of the station that sent the packet, and the call sign of the station the packet is being sent to. The address may also include call signs of stations that are being used to relay the packet. The error-checking information allows the receiving station to determine whether the received packet contains any errors. If the received packet contains errors, the receiving station waits until the transmitting station sends it again.

Breaking up the data into small parts allows several users to share the channel. Packets from one user are transmitted in the spaces between packets from other users. The address section allows each user's TNC to separate packets intended for him from packets intended for other users. The addresses also allow packets to be relayed through several stations before they reach their ultimate destination. Having information in the packet that tells the receiving station if the packet has been received correctly assures perfect copy.

What Do I Need to Get on Packet Radio?

All you need to set up a VHF packet-radio station is a VHF or UHF FM transceiver (with an antenna), a computer

---

Packet, Morse, Baudot, ASCII, AMTOR, SSTV and FAX capability. TTL and RS-232 ports are standard. (See the Product Reviews in July and September 1989 *QST*.)

RADIOS FOR PACKET OPERATION
VHF

Just about any VHF FM transceiver can be used on VHF packet. Even a small hand-held transceiver can be used, although an outside antenna is a good idea. If there is very little packet activity in your area, you may need a bit more power to reach the nearest dedicated digipeater. Many manufacturers sell amplifiers that increase the power from a hand-held transceiver to 10 or 25 watts. Most mobile FM transceivers are rated at 10 to 50 watts, and these radios make great packet rigs.

Older crystal-controlled transceivers may be used on packet, but a few words of caution are in order. Many older rigs cannot switch from receive to transmit quickly enough for packet operation. Most TNCs have an adjustable "transmit delay" that causes the TNC to key the radio and then wait a second or so before actually sending the packet. Older radios may require this transmit delay. In addition, an older rig may have passed through the hands of several amateurs, and some of the internal settings may have been "adjusted." While this may not be a problem when you use the rig for voice communications, the rig may not work on packet. If you plan to use an older rig on packet, try to locate an experienced packet operator who can help you set up your packet station.

HF

Most SSB transceivers will work on HF packet. Again, your transceiver may require a bit of transmit delay from the TNC. Your TNC manual should cover wiring and operation guidelines in detail. On HF, packet operation uses lower sideband.

Antennas

Since many packet stations are put on the air by amateurs who have already been active on VHF FM, the same

The Kantronics KPC-3 packet-radio TNC features a 1200-baud modem.

In addition to the usual digital modes, this MCP operates on PacTOR, a mode that combines the best features of packet and AMTOR.

antennas are often used for both modes. As a result, vertically polarized antennas are the standard for VHF packet. Yagi antennas are usually not required, unless there is very little packet activity in your area or you have a particularly bad location.

Figure 9-6—A terminal node controller connects between your transceiver and your computer, like a RTTY modem.

or ASCII terminal and a TNC or multimode controller, which allows operation on several modes. (We'll use "TNC" to refer to both.)

The TNC connects between the transceiver microphone input and the computer. See Figure 9-6. For operation on 10 meters you'll need a 10-meter SSB transceiver in addition to the TNC and computer. The sidebar entitled "Packet Radio Equipment" lists some available equipment.

Your TNC manual should contain detailed instructions for wiring the TNC, radio and computer together. So many hams are on packet now that someone in the local radio club will probably be able to help you if you have problems, or ask around on the local voice repeater.

What Does It Look Like?

So far we've talked about the equipment you need and how packet works. But what does a packet contact look like? In the following examples, we'll look at the procedures used by most of the TNCs on the market today. Some TNCs may use different command formats—consult your operating manual if you're not sure.

First, you must tell the TNC your call sign. Most TNCs allow you to change your call sign at any time and have a way to remember it when the power is switched off. Before you can enter commands into the TNC, it must be in *command mode*. When the TNC is in command mode, you will see a prompt:

cmd:

This indicates that the TNC is waiting for input. To tell the TNC her call sign, KA1MJP types:

MYCALL KA1MJP <CR>

<CR> means "carriage return." On some computers this key may be labeled "ENTER" or have an arrow (←).

As in all other modes of Amateur Radio, packet allows you to "read the mail" or monitor channel activity. This is called the **monitor mode**, and looks like this:

> WA6JPR>WB6YMH: HELLO SKIP, WHEN IS THE NEXT OSCAR 13 PASS? K

> WB6YMH>WA6JPR: HANG ON WALLY, I'LL TAKE A LOOK.

To enable monitor mode, simply type: MON ON at the cmd: prompt. You may also need to type MFROM ALL. Consult the operating manual for your TNC to be sure. The call signs of the stations involved appear as "FROM>TO" and the contents of the packet appear after the ":." When it is in **monitor mode**, your computer will display everything that is transmitted on the packet frequency, whether or not it is addressed to you (see Figure 9-7). In addition, a packet station that is monitoring is not replying to any messages.

You can send a CQ by entering the converse mode of the TNC. You go to converse mode by typing:

CONV <CR>

(some TNCs allow you to type "K" instead of "CONV")
You can then type your CQ:

> MIKE IN SAN DIEGO LOOKING FOR ANYONE IN SIMI VALLEY

Your TNC adds your call sign as the FROM address and CQ as the TO address. The receiving station's TNC adds these addresses to the front of the text when it is displayed.

Making a Connection

You answer a CQ or establish a contact by using the CONNECT command. When two packet stations are **connected**, each station sends packets specifically addressed to the other station. When a station receives an error-free packet, it transmits a reply (acknowledgment) packet to let the sender know the packet has been received correctly.

To connect to another station, you type:

Connect WA1UQC <CR>

where WA1UQC is the call sign of the station you wish to contact. (Most TNCs let you use "C" as an abbreviation for "connect.")

If WA1UQC's packet-radio station is on the air and receives your connect request, your station and his will exchange packets to set up a connection. When the connection is completed, your terminal displays:

*** CONNECTED to WA1UQC

and your TNC automatically switches to the converse mode.

Now, everything you type into the terminal keyboard is sent to the other station. A packet is sent whenever you enter a carriage return. It's a good idea to use K, BK, O or > at the

```
W1AW-4>K1CE:You have new mail, please kill after reading:

W1AW-4>K1CE:Msg# TR  Size  To     From    @ BBS  Date      Title

W1AW-4>K1CE:5807  N   420 K1CE    K1MON   W1AW      870308 HI AGAIN

W1AW-4>K1CE:K1CE de W1AW: at 2101z on 870308, 142 active msgs, last msg #5839

W1AW-4>K1CE:(A,B,D,H,I,J,K,L,R,S,T,U,W,X) >

W1AW-4>K1CE:Enter title for message:

W1AW-4>K1CE:Enter message, ^Z (CTL-Z) to end, it will be message 5840

WA3VIL>KA1MJP:Hi Leslie, are you going to the club meeting? KK

W1AW-4>K1CE:K1CE de W1AW: at 2101z on 870308, 143 active msgs, last msg #5840

W1AW-4>K1CE:(A,B,D,H,I,J,K,L,R,S,T,U,W,X) >

W1AW-4>KE3Z:Max. path length: 2 digis

W1AW-4>KE3Z:KE3Z de W1AW: at 2123z on 870308, 143 active msgs, last msg #5840

KA1MJP>WA3VIL:No, Larry, I have too much work. I have to go

W1AW-4>KE3Z:(A,B,D,H,I,J,K,L,R,S,T,U,W,X) >

W1AW-4>KE3Z:Enter title for message:

KA1MJP>WA3VIL:back in tonight. Maybe next month. KK

W1AW-4>KE3Z:Enter message, ^Z (CTL-Z) to end, it will be message 5841
```

Figure 9-7—In monitor mode, your TNC displays all the packet activity on the frequency, whether or not the packets are addressed to your station.

end of a thought to say "Okay, I'm done. It's your turn to transmit."

When you are finished conversing with the other station, return to the command mode by typing <CTRL-C> (hold down the CONTROL key and press the C key). When the command prompt (cmd:) is displayed, type:

Disconne <CR>

and your station will exchange packets with the other station to break the connection. (Most TNCs let you use "D" as an abbreviation.) When the connection is broken, your terminal displays:

*** DISCONNECTED

If, for some reason, the other station does not respond to your initial connect request, your TNC will resend the request until the number of attempts equals the internal retry counter. When the number of attempts exceeds the retry counter, your TNC will stop sending connect requests and your terminal displays:

*** retry count exceeded
*** DISCONNECTED

A TNC can reject a connect request if it is busy or if the operator has set CONOK (short for CONnect OK) off. If this happens when you try to connect, your TNC displays:

*** WA1UQC busy
*** DISCONNECTED

Packet Radio Repeating

Sometimes terrain or propagation prevent your signal from being received by the other station. Packet radio gets around this problem by using other packet-radio stations to relay your signal to the intended station. All you need to know is which on-the-air packet-radio stations can relay signals between your station and the station you want to contact. Once you know of a station that can relay your signals, type:

Connect WA1UQC Via W1AW-5 <CR>

where WA1UQC is the call sign of the station you want to connect to and W1AW-5 is the call sign of the station that will relay your packets. The "-5" following W1AW is a secondary station identifier (SSID). The SSID permits up to 16 packet stations to operate with one call sign. For example, W1AW-5 is a 2-meter packet repeater and W1AW-6 is a 222-MHz packet repeater.

When W1AW-5 receives your connect request, it stores your request in memory until the frequency is quiet. It then retransmits your request to WA1UQC on the same frequency. This action is called *digipeating*, a contraction of "digital repeating." If WA1UQC's packet-radio station is on the air and receives the relayed connect request, your station and his will exchange packets through W1AW-5 to set up a connection. Once the connection is established, your terminal will display:

*** CONNECTED to WA1UQC VIA W1AW-5

W1AW-5 will continue to relay your packets until the connection is broken (see Figure 9-8).

Digital and voice repeaters both repeat, but the similarity ends there. Notice that digital repeaters differ from typical voice repeaters in a number of ways. A digital repeater (**digipeater**) usually receives and transmits on the same frequency (whereas a voice repeater receives and transmits on different frequencies). A digipeater does not receive and transmit at the same time (as compared to a voice repeater, which immediately transmits whatever it receives). Rather, a digipeater receives a packet, stores it temporarily until the frequency is clear, and then retransmits the packet. Also, a digipeater repeats only packets that are specifically marked to be repeated by that station (the address in the packet contains the call sign of the digipeater). A voice repeater repeats everything it receives on its input frequency.

If one digipeater is insufficient to establish a connection, you can specify as many as eight stations in your connect request. Additional digipeaters are added to the connect command separated by commas. For example, typing:

Connect WA1UQC Via W1AW-5, WA2FTC-1 <CR>

after the command prompt (cmd:), causes your TNC to send the WA1UQC connect request to W1AW-5 which relays it to WA2FTC-1. Then, WA2FTC-1 relays it to WA1UQC.

Don't use more than one or two digipeaters at any one time, especially during the prime time operating hours (evenings and weekends). Each time you use a digipeater, you are competing with other stations attempting to use the same digipeater. Each station that you compete with has the potential of generating a packet that may collide with your packet (which causes your TNC to resend the packet). The more digipeaters you use, the more stations you compete with, greatly increasing the chance of a packet collision. As a result, it may be difficult to get one packet through multiple digipeaters, and your TNC will quickly reach its retry limit and disconnect the link.

Any packet-radio station can act as a digipeater. Most TNCs are set up to digipeat automatically without any intervention by the operator of the station being used as a digipeater. You do not need his permission, only his cooperation, because he can disable his station's digipeater function. (In the spirit of Amateur Radio, most packet-radio operators leave the digipeater function on, disabling it only under special circumstances.)

Similar to VHF/UHF voice repeaters, some stations are set up as dedicated digipeaters. They are usually set up in good radio locations by packet-radio clubs. Besides location, the other advantage of a dedicated digipeater is that it is always there (barring a calamity). Stations do not have to depend on the whims of other packet-radio operators, who may or may not be on the air when their stations' digipeater functions are most needed.

Although you are not allowed to be the control operator of a voice repeater until you upgrade from the Novice class, you may leave your TNC's digipeater function enabled. The

Figure 9-8—Two packet stations connect to have a QSO. Everything you type is sent to the other station. Packets not addressed to your station are ignored by your TNC.

```
cmd:c aa2z v wlaw-5
*** CONNECTED to AA2Z via W1AW-5
Hi Bruce, what's up? KK
hi Mark. Just wondering if we can bring the desk over tonight. KK
No problem. Got the room all set up and everything. What time
do you think you'll be here? KK
Well, that's up to Rick. Probably around 6 or so. KK
OK, see you then. SK
See you at 6.
cmd:D
*** DISCONNECTED
```

FCC recognizes a distinction between digipeaters and voice repeaters in this case, and everyone realizes that an effective packet-radio system depends on having Novice digipeaters available.

The most common form of digipeater is the *node*. To reach a distant station, first connect to the node. Then, instruct the node to connect you to the distant station. The node acknowledges packets sent from either station, then relays them to the other station. This has a number of advantages over a simple digipeater.

VHF/UHF Packet Operating

Today, most amateur packet-radio activity occurs at VHF, on 2 meters, but activity on 222 MHz continues to grow. The most commonly used data rate on VHF is 1200 bauds, with frequency-modulated AFSK tones of 1200 and 2200 Hz. This is referred to as the "Bell 202" telephone modem standard.

Getting on the air is usually a simple matter of turning on your radio and tuning to your favorite packet-radio frequency. On 2 meters, common packet channels are 145.01, 145.03, 145.05, 145.07 and 145.09 MHz. On 222 MHz, Novice packet activity centers on 223.4 MHz. If there is a voice repeater on that frequency in your area, ask around at a club meeting or on the repeater. Someone is bound to know where the packet activity is.

If you are conducting a direct connect (a contact without using a digipeater), move your contact to an unused simplex frequency. It is very inefficient to use a frequency where other stations, especially digipeater stations, are operating. The competition slows down your packets and, in return, you are also slowing down the other stations. You should use a frequency occupied by a digital repeater only if you are using that digital repeater. Consult the latest edition of *The ARRL Repeater Directory* for more information on recommended packet operating frequencies.

VHF/UHF packet-radio operation is similar to VHF/UHF FM voice operation. Under normal propagation conditions, you can communicate only with stations that are in line of sight of your station. By using a repeater (voice or digital), this limitation is reduced. If a repeater is within range of your station, you may communicate with other stations that are also within range of the same repeater, even though those stations may be beyond your own line of sight. Packet expands on this by permitting the simultaneous use of several digipeaters to relay a packet from one station within range of the first digipeater in the link to a station within radio coverage of the last digipeater in the chain.

HF Packet Operating

HF packet radio is very different from VHF/UHF packet. An SSB transceiver is used to generate a 200-Hz-shift FSK signal, and 300 bauds is used rather than 1200 bauds.

Tuning is much more critical than it is on VHF. Tune your receiver very slowly, in as small an increment as pos-

sible (10-Hz increments are good) until your terminal begins displaying packets. Do not change frequency until a whole packet is received. If you shift frequency in midpacket, that packet will not be received properly and will not be displayed on your terminal even if you were on the correct frequency before or after the frequency shift.

Some TNCs and external modems have tuning indicators that make tuning a lot easier. Kits are also available to allow you to add a tuning indicator to a TNC without one.

The same rule of thumb that applies to VHF/UHF packet-radio operation also applies to HF operation. Move to an unused frequency if there is other packet-radio activity on the frequency you are presently using. Often, one frequency is used as a calling frequency where stations transmit packets to attract the attention of other stations who may wish to contact them. Once a contact/connection is established, clear the calling frequency by moving off to another unused frequency.

Packet Bulletin-Board Systems (PBBS)

We mentioned RTTY mailboxes in the section on radioteletype, and some of you may be familiar with landline telephone bulletin-board systems. The packet-radio equivalent is the *Packet Bulletin Board System (PBBS)*. The PBBS computer allows packet stations to store messages for other amateurs, download and upload computer files, and even to link one packet station through a "gateway" to another band (see Figure 9-9).

Some PBBS computers can automatically forward messages from one computer to another, so you can store a message at one PBBS that is ultimately meant for an amateur thousands of miles away. The message will be forwarded from one PBBS to another until it reaches its destination. A network is a system of packet stations that can interconnect to transmit data over long distances.

To use a PBBS, you must locate one. In any area with even a small amount of packet activity there is usually at least one PBBS. In addition, there are several HF PBBS stations, although many of the HF stations are set up for long-haul message traffic, not for individual-user connects. *The ARRL Operating Manual* contains more information about using a PBBS.

The Dreaded Beacon

All TNCs have a *beacon* function. This function allows a station to send an unconnected packet at regular intervals. These unconnected packets usually contain a message to the effect that the station originating the beacon is on the air and is ready, willing and able to carry on a packet-radio contact.

The purpose of the beacon function is to generate activity when there is none. This purpose was legitimate when there was little packet-radio activity. Back in the early 1980s, it was a rare occurrence when a new packet-radio station appeared on the air. Without beacons, that new operator might believe that his packet-radio station was the only one active

```
cmd:c klce v wlaw-4
*** CONNECTED to K1CE
K1CE     BBS - West Hartford, CT
Hello Bruce, Last on 0257/870307, new 369 - 368, active 15.
KB1MW   de K1CE    BBS: (B,D,H,I,J,K,L,R,S,T,U,W,X) >
ll 5
  Msg#  TR  Size  To      From    @ BBS  Date    Title
   367  N     78  K1BA    K1CE    WA1RAJ 870307  Greetings
   366  N    652  ALL     KA1KRP         870306  FLEA MARKET
   360  TY   271  K1CE    WB1ASH         870305  QTC 1 K1CE
   358  BN   679  ALL     WA1OCK         870304  novice enhancement
   244  BN   319  ALL     K1CE           870228  NTS Traffic Handling Basics
KB1MW   de K1CE    BBS: (B,D,H,I,J,K,L,R,S,T,U,W,X) >
r 244
  Msg#  TR  Size  To      From    @ BBS  Date    Title
   244  BN   319  ALL     K1CE           870228  NTS Traffic Handling Basics
If you are interested in NTS traffic handling via packet radio,
and would like to know the basics, download a new file NTS.DOC for
information.  Good luck, and thanks for your help in handling NTS
traffic in Connecticut. If I can be of any assistance, please
leave a message.   73,   Rick K1CE NTSCT Section Node

KB1MW   de K1CE    BBS: (B,D,H,I,J,K,L,R,S,T,U,W,X) >
x
KB1MW   de K1CE    BBS: Date 870307 Time 0258 Last msg # 368, 15 active msgs
Messages: L - List, R - Read, S - Send, K - Kill
Files:    W - What, D - Download from, U - Upload to
B - Bye, G - GateWay, H - Help, T - Talk to Rick
P - Path for callsign, N - Enter/Change your name
I - Information, J - Calls heard,  X - Short/Long Menu >
b
KB1MW   Bruce de K1CE  : 73, CUL
*** DISCONNECTED
```

Figure 9-9—A packet bulletin board (PBBS) lets amateurs store messages for other amateurs. The messages may be forwarded automatically to other stations in the PBBS network.

in the area. Similarly, packet-radio stations already on the air would not be aware of the new station's existence. It would be very discouraging to build a TNC (they were all kits in the early days), get on the air and find no one to contact. The beacon function was the solution to the problem. It let people know that a new packet station was on the air.

Today, beacons are usually unnecessary. There is absolutely no need to resort to beaconing in order to make your existence known. On HF, 2 meters and 222 MHz, there is plenty of activity in most areas. If you are getting on the air for the first time, monitor 145.01 or 223.4 MHz for a few minutes and you will quickly have a list of other stations that are on the air. When one of these stations disconnects, send a connect request to that station. After a few connections, your existence on the air will be known.

Instead of sending beacons, leave a message announcing your existence on your local PBBS. This is more effective than sending beacons because your message will be read even when your station is off the air.

Beacons only add congestion to already crowded packet-

radio channels, so do the packet-radio community a favor and disable your TNC's beacon function by typing:

Beacon Every 0 <CR>

What If I Still Have Questions?

The material presented here is only a very basic sketch of packet-radio operation. More material is presented in the following ARRL books: *Your Packet Companion*, *Your Gateway to Packet Radio* and *The ARRL Operating Manual*. *QST* and the other ham magazines often have articles about packet radio. Finally, there are sure to be a few "packeteers" in your local club. For a list of clubs in your area, write to the Educational Activities Department at ARRL HQ.

[Now turn to Chapter 12 and study questions N2B03 through N2B06 and N8A05. Review this section if you have any difficulty with these questions.]

[**If you are studying for the Technician exam**, turn to Chapter 13 and review questions T8A04 and T8B04. Review this section if you have any difficulty.]

PROPAGATION: HOW SIGNALS TRAVEL

Now that we've looked at several modes of operation, it's time to discuss how radio waves travel. As we've seen, amateurs enjoy the privilege of using many different frequency bands. Some of these are scattered throughout the high-frequency (HF) spectrum. Some of these HF bands work better during the day, and some work better at night. Some frequencies are good for long-distance communications, and others provide reliable short-range communications. Amateurs who want to work a certain part of the country or world need to know which frequency to use and when to be there. Experience combined with lots of careful listening is a good way to gain this knowledge.

Novices have privileges on four HF bands. To keep things simple, we'll refer to them as the "Novice bands," although Technicians who have passed a 5-WPM code test share these privileges. (All Technicians have other privileges on the VHF and UHF bands.)

Table 9-9 summarizes the HF privileges granted to radio amateurs. We often refer to amateur bands by approximate *wavelength* rather than frequency. When you hear someone refer to the 15-meter band, 15 meters is the approximate wavelength of a signal in that band. During a typical day, you can reach any part of the world using one of the Novice bands.

Radio waves travel to their destination in three ways:

• Directly from one point to another.
• Along the ground.
• Refracted, or bent, back to Earth by the **ionosphere**. (The ionosphere is a layer of charged particles, or ions, in Earth's outer atmosphere. These ionized gases make long-distance radio contacts possible on the HF bands.)

The study of how radio waves travel from one point to another is the science of **propagation.** Radio-wave propagation is a fascinating part of ham radio. Let's take a look at the different ways radio waves travel.

LINE-OF-SIGHT PROPAGATION

Line-of-sight propagation occurs when signals travel in a straight line from one antenna to another. These *direct waves* are useful mostly at VHF and higher frequencies. TV and FM radio broadcasts (those that are broadcast to your TV or radio antenna instead of through a cable) are examples of signals that are received as direct waves. When you transmit on a local repeater frequency, direct waves travel in a straight line to the repeater. The repeater sends the signals, in a straight line, to other fixed, portable and mobile stations. When you use a hand-held transceiver to communicate directly with another ham using a hand-held, your signals also travel in a straight line.

GROUND-WAVE PROPAGATION

In **ground-wave propagation**, radio waves travel along the Earth's surface, even over hills. They follow the curvature of the Earth for some distance. AM broadcasting signals

Table 9-9

High-Frequency Ham Bands, with Novice Privileges

HF Bands

For Full Amateur Privileges (MHz)	Novice Privileges (MHz)	Band (Meters)
1.8 - 2.0	None	160
3.5 - 4.0	3.675 - 3.725	80
7.0 - 7.3	7.10 - 7.15	40
10.10 - 10.15	None	30
14.00 - 14.35	None	20
18.068 - 18.168	None	17
21.00 - 21.45	21.1 - 21.2	15
24.89 - 24.99	None	12
28.00 - 29.70	28.1 - 28.5	10

travel by way of ground-wave propagation during the day. Ground wave works best at lower frequencies. You might have an 80-meter QSO with a station a few miles away, on the other side of a hill. For that contact, you are using ground-wave propagation.

Ground-wave propagation on the ham bands means relatively short-range communications. Stations at the top of the AM broadcast band (the 1600-kHz end) do not carry far during the day. Stations near the low end of the dial (540 kHz) can be heard at much greater distances. Amateur Radio frequencies are higher than the AM broadcast band, so the ground-wave range is shorter.

SKY-WAVE PROPAGATION (SKIP)

The **ionosphere** makes it possible for hams on opposite sides of the Earth to contact each other. The ions in the ionosphere refract (bend) some types of radio signals, bouncing them back to Earth. This is the way long-distance radio signals travel.

High in the atmosphere, 25 to 200 miles above Earth, the thin air is electrically charged (ionized) by radiation from the sun. When ionized by solar radiation, this region, the **ionosphere**, can refract radio waves. If the wave is bent enough, it returns to Earth. If the wave is not bent enough, it travels off into space. Ham radio contacts of up to 2500 miles are possible with one *skip* off the ionosphere. Worldwide communications using several skips (or hops) can take place if conditions are right.

Two factors determine skip-propagation possibilities between two points: the frequency in use and the level of *ionization*. The higher the frequency of the radio wave, the less it is bent by the ionosphere. The highest frequency at which the ionosphere bends radio waves back to a desired location on Earth is called the *maximum usable frequency*

Figure 9-10—Cool sunspots allow hot propagation on earth!

Ionization of the ionosphere results from the sun's radiation striking the upper atmosphere. It is greatest during the day and during the summer. The amount of radiation coming from the sun varies through the day, season and year. This radiation is closely related to visible **sunspots** (grayish-black blotches on the sun's surface). See Figure 9-10. Hams interested in long-distance communication always know where we are in the **sunspot cycle**. Sunspots vary in number and size over an 11-year cycle. More sunspots usually mean more ionization of the ionosphere. As a result, the MUF tends also to be higher.

By contrast, when sunspots are low, radiation, and thus the MUF, is lower. That is why HF communication is enhanced during times of greater sunspot activity.

Skip propagation has both a maximum range limit and a minimum range limit. That minimum is often greater than the ground-wave range. There is an area between the maximum ground-wave distance and the minimum skip distance where radio signals on a particular frequency will not reach. This "dead" area is called the **skip zone**. Figure 9-11 illustrates the difference between ground-wave and skip propagation. The drawing also shows the concept of skip zone and skip distance. Some radio signals may not be bent enough to bring them back to Earth.

(MUF). The MUF for communication between two points depends on solar radiation strength and the time of day. Radio waves that travel beyond the horizon by refraction in the ionosphere are called *sky waves*. **Sky-wave propagation** takes place when a signal is returned to Earth by the ionosphere.

PROPAGATION ON OUR HF BANDS

Operating on 10 meters (28 MHz)

At times the best of bands, 10 meters also can seem the worst of bands. The highest frequency amateur HF band, 10 is more subject to the whims of the sunspot cycle. During years of high sunspot activity, 10 meters remains open all day

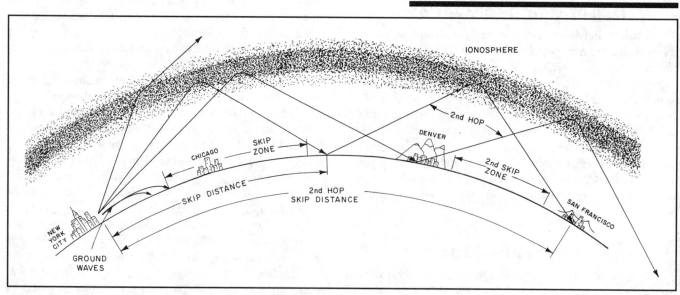

Figure 9-11—This drawing illustrates how skip zone, skip distance and ground waves are related. See text for details.

and most of the night. Strong signals arrive from all over the world. In years of low solar activity, you may hear nothing but local signals for days at a time. Even during the peak of the sunspot cycle, the MUF doesn't always reach 28 MHz.

When operating on 10 meters, remember that the part of the Novice band used for **CW** (Morse code) is a full 200 kHz wide. (The section of a band used for a certain mode is called a *subband*.) The voice subband is also 200 kHz wide. Some signals may be weak, but perfectly readable. Tune the entire Novice subband carefully. If there aren't any signals in the Novice portion, listen in the rest of the band. A good place to listen is around 28.5 MHz. If the band is open, you'll likely hear someone.

During the sunspot minimum, you'll have to work a little harder to see if 10 is open. Several automated *beacons* operate between 28.1 and 28.3 MHz. They transmit continuous signals to test propagation. If you can hear one of these beacons, the band is open. You'll find these beacons listed occasionally in *QST* and other ham publications. Even if 10 seems dead, it may just be a case of everyone listening and no one calling! Try calling CQ.

Ten meters offers many Novices their first taste of **DX**: The chance to communicate with hams in foreign lands. Familiarize yourself with some of the common DX call signs. That way, you won't be surprised the first time you hear one. In a typical week, you might hear a JA (Japan), an EA (Spain) or an LA (Norway). Some countries also use call signs beginning with a number. Don't be surprised if you work a 4Z (Israel) or a 9Y (Trinidad). Table 9-10 lists all of the currently assigned international prefixes.

In addition to DX potential, 10 meters also provides excellent short-range communications. Many clubs and local groups meet regularly on 10 meters. Check with a local radio club to find out if there are any nets in your area.

Ten meters is also the lowest-frequency Novice band that allows emissions other than CW. From 28.1 to 28.3 MHz, Novices are allowed to use CW and frequency-shift keying (FSK). FSK modes include packet, Baudot radioteletype (RTTY), ASCII and AMTOR. From 28.3 to 28.5 MHz, Novices are authorized to use CW and single-sideband (SSB) voice. The maximum power level that Novices may use for any mode on 10 meters is 200 watts PEP output.

Operating on 15 meters (21 MHz)

The 15-meter band shares several characteristics with the 10-meter band. It is not, however, as dependent on high sunspot activity. The 21-MHz frequency range is still high enough so that propagation conditions change dramatically with sunspot variations. Because of its lower frequency, 15 meters is open for longer periods and more often than 10. Also, the band is more stable in years of low sunspot activity.

DX is plentiful on 15 meters, even in the lean years of the solar cycle. During the day, stations in Europe, Africa and South America may be workable. You may hear signals from the Pacific and Japan in the late afternoon through early evening hours. Fifteen meters provides good communications to areas all over the US.

Many foreign stations operate within the Novice part of the band using **phone** transmissions. American hams are prohibited from doing this, but many other countries allow it. If you hear a foreign station on voice, give him a CW call at the end of his QSO. Many are glad to work Novices. This kind of operation won't give you much practice copying CW, but it will give you a chance to hear what phone DX stations sound like. You might even work a "new country."

Because 15 meters is so popular, you'll most likely hear stations if the band is open. Again, listen at the bottom of the CW and phone portions if you don't hear signals in the Novice band. This is where stations usually congregate.

Fifteen is one of the best bands around because of its DX possibilities. More stable than 10 meters, it is the most reliable Novice band for DX. It's usable even at the low point of the sunspot cycle. Higher in frequency than 40 and 80 meters, it provides good long-distance contacts.

Operating on 40 meters (7 MHz)

In many respects, the 40-meter band is one of the best US ham bands. It suffers from some of the worst interference you'll ever hear, however. International treaties allow the world outside of the Americas to use the 40-meter band above 7.1 MHz for high-power shortwave broadcasting. This causes quite a racket in North America.

During the morning and early afternoon hours, 40 meters provides good, reliable communications. Skip distance ranges from about 400 to 1200 miles. Later in the afternoon, however, the MUF drops. Skip distance then increases to several thousand miles. You'll have no trouble detecting the increase in range. You'll find yourself suddenly listening to those foreign broadcast stations.

Even though it has interference from the broadcasters, 40 meters is one of the most popular Novice bands. It is open to some part of the country 24 hours a day. During the morning and afternoon hours, you'll be able to chat with other hams to your heart's content. In the late afternoon, evening and at night, you'll be able to work more distant locations. Once you learn to live with the broadcast interference, you'll find yourself using 40 meters for many of your CW contacts.

Operating on 80 meters (3.5 MHz)

At 3.5 MHz, 80 meters is the lowest in frequency of all the Novice bands. It is very popular among Novices for stateside contacts. A half-wavelength dipole antenna for 80 is rather large (see Chapter 7). It need not be parallel to the ground—its shape is not critical. Most Novices manage to put up one that works fairly well.

During the daylight hours, 80 meters provides reliable communications out to about 350 miles. Although 80 is not as heavily populated as 40 during the day, you will probably find someone eager to talk with you.

Table 9-10

Allocation of International Call Signs

Call Sign Series	Allocated to	Call Sign Series	Allocated to	Call Sign Series	Allocated to
AAA-ALZ	United States of America	GAA-GZZ	United Kingdom of Great Britain and Northern Ireland	OFA-OJZ	Finland
AMA-AOZ	Spain			OKA-OLZ	Czech Republic
APA-ASZ	Pakistan	HAA-HAZ	Hungary	OMA-OMZ	Slovak Republic
ATA-AWZ	India	HBA-HBZ	Switzerland	ONA-OTZ	Belgium
AXA-AXZ	Australia	HCA-HDZ	Ecuador	OUA-OZZ	Denmark
AYA-AZZ	Argentina	HEA-HEZ	Switzerland	PAA-PIZ	Netherlands
A2A-A2Z	Botswana	HFA-HFZ	Poland	PJA-PJZ	Netherlands Antilles
A3A-A3Z	Tonga	HGA-HGZ	Hungary	PKA-POZ	Indonesia
A4A-A4Z	Oman	HHA-HHZ	Haiti	PPA-PYZ	Brazil
A5A-A5Z	Bhutan	HIA-HIZ	Dominican Republic	PZA-PZZ	Suriname
A6A-A6Z	United Arab Emirates	HJA-HKZ	Columbia	P2A-P2Z	Papua New Guinea
A7A-A7Z	Qatar	HLA-HLZ	South Korea	P3A-P3Z	Cyprus
A8A-A8Z	Liberia	HMA-HMZ	North Korea	† P4A-P4Z	Aruba
A9A-A9Z	Bahrain	HNA-HNZ	Iraq	P5A-P9Z	North Korea
BAA-BZZ	China	HOA-HPZ	Panama	RAA-RZZ	Russian Federation
CAA-CEZ	Chile	HQA-HRZ	Honduras	SAA-SMZ	Sweden
CFA-CKZ	Canada	HSA-HSZ	Thailand	SNA-SRZ	Poland
CLA-CMZ	Cuba	HTA-HTZ	Nicaragua	• SSA-SSM	Egypt
CNA-CNZ	Morocco	HUA-HUZ	El Salvador	• SSN-SSZ	Sudan
COA-COZ	Cuba	HVA-HVZ	Vatican City	STA-STZ	Sudan
CPA-CPZ	Bolivia	HWA-HYZ	France	SUA-SUZ	Egypt
CQA-CUZ	Portugal	HZA-HZZ	Saudi Arabia	SVA-SZZ	Greece
CVA-CXZ	Uruguay	H2A-H2Z	Cyprus	S2A-S3Z	Bangladesh
CYA-CZZ	Canada	H3A-H3Z	Panama	† S5A-S5Z	Slovenia
C2A-C2Z	Nauru	H4A-H4Z	Solomon Islands	S6A-S6Z	Singapore
C3A-C3Z	Andorra	H6A-H7Z	Nicaragua	S7A-S7Z	Seychelles
C4A-C4Z	Cyprus	H8A-H9Z	Panama	S9A-S9Z	Sao Tome and Principe
C5A-C5Z	Gambia	IAA-IZZ	Italy		
C6A-C6Z	Bahamas	JAA-JSZ	Japan	TAA-TCZ	Turkey
* C7A-C7Z	World Meteorological Organization	JTA-JVZ	Mongolia	TDA-TDZ	Guatemala
		JWA-JXZ	Norway	TEA-TEZ	Costa Rica
C8A-C9Z	Mozambique	JYA-JYZ	Jordan	TFA-TFZ	Iceland
DAA-DRZ	Germany	JZA-JZZ	Indonesia	TGA-TGZ	Guatemala
DSA-DTZ	South Korea	J2A-J2Z	Djibouti	THA-THZ	France
DUA-DZZ	Philippines	J3A-J3Z	Grenada	TIA-TIZ	Costa Rica
D2A-D3Z	Angola	J4A-J4Z	Greece	TJA-TJZ	Cameroon
D4A-D4Z	Cape Verde	J5A-J5Z	Guinea-Bissau	TKA-TKZ	France
D5A-D5Z	Liberia	J6A-J6Z	Saint Lucia	TLA-TLZ	Central Africa
D6A-D6Z	Comoros	J7A-J7Z	Dominica	TMA-TMZ	France
D7A-D9Z	South Korea	† J8A-J8Z	St. Vincent and the Grenadines	TNA-TNZ	Congo
EAA-EHZ	Spain			TOA-TQZ	France
EIA-EJZ	Ireland	KAA-KZZ	United States of America	TRA-TRZ	Gabon
† EKA-EKZ	Armenia			TSA-TSZ	Tunisia
ELA-ELZ	Liberia	LAA-LNZ	Norway	TTA-TTZ	Chad
† EMA-EOZ	Ukraine	LOA-LWZ	Argentina	TUA-TUZ	Ivory Coast
EPA-EQZ	Iran	LXA-LXZ	Luxembourg	TVA-TXZ	France
† ERA-ERZ	Moldova	† LYA-LYZ	Lithuania	TYA-TYZ	Benin
† ESA-ESZ	Estonia	LZA-LZZ	Bulgaria	TZA-TZZ	Mali
ETA-ETZ	Ethiopia	L2A-L9Z	Argentina	T2A-T2Z	Tuvalu
EUA-EWZ	Belarus	MAA-MZZ	United Kingdom of Great Britain and Northern Ireland	T3A-T3Z	Kiribati
EXA-EXZ	Russian Federation			T4A-T4Z	Cuba
† EYA-EYZ	Tajikistan			T5A-T5Z	Somalia
† EZA-EZZ	Turkmenistan	NAA-NZZ	United States of America	T6A-T6Z	Afghanistan
† E2A-E2Z	Thailand			† T7A-T7Z	San Marino
FAA-FZZ	France	OAA-OCZ	Peru	† T9A-T9Z	Bosnia and Herzegovina
		ODA-ODZ	Lebanon	UAA-UIZ	Russian Federation
		OEA-OEZ	Austria	† UJA-UMZ	Uzbekistan

Call Sign Series	Allocated to	Call Sign Series	Allocated to	Call Sign Series	Allocated to
UNA-UQZ	Russian Federation	ZNA-ZOZ	United Kingdom of Great Britain and Northern Ireland	5TA-5TZ	Mauritania
URA-UTZ	Ukraine			5UA-5UZ	Niger
† UUA-UZZ	Ukraine			5VA-5VZ	Togo
VAA-VGZ	Canada	ZPA-ZPZ	Paraguay	5WA-5WZ	Western Samoa
VHA-VNZ	Australia	ZQA-ZQZ	United Kingdom of Great Britain and Northern Ireland	5XA-5XZ	Uganda
VOA-VOZ	Canada			5YA-5ZZ	Kenya
VPA-VSZ	United Kingdom of Great Britain and Northern Ireland			6AA-6BZ	Egypt
		ZRA-ZUZ	South Africa	6CA-6CZ	Syria
		ZVA-ZZZ	Brazil	6DA-6JZ	Mexico
VTA-VWZ	India	† Z2A-Z2Z	Zimbabwe	6KA-6NZ	South Korea
VXA-VYZ	Canada	† Z3A-Z3Z	Macedonia (Former Yugoslav Republic)	6OA-6OZ	Somalia
VZA-VZZ	Australia			6PA-6SZ	Pakistan
† V2A-V2Z	Antigua and Barbuda	2AA-2ZZ	United Kingdom of Great Britain and Northern Ireland	6TA-6UZ	Sudan
† V3A-V3Z	Belize			6VA-6WZ	Senegal
† V4A-V4Z	Saint Kitts and Nevis			6XA-6XZ	Madagascar
† V5A-V5Z	Namibia	3AA-3AZ	Monaco	6YA-6YZ	Jamaica
† V6A-V6Z	Micronesia	3BA-3BZ	Mauritius	6ZA-6ZZ	Liberia
† V7A-V7Z	Marshall Islands	3CA-3CZ	Equatorial Guinea	7AA-7IZ	Indonesia
† V8A-V8Z	Brunei	• 3DA-3DM	Swaziland	7JA-7NZ	Japan
WAA-WZZ	United States of America	• 3DN-3DZ	Fiji	7OA-7OZ	Yemen
		3EA-3FZ	Panama	7PA-7PZ	Lesotho
XAA-XIZ	Mexico	3GA-3GZ	Chile	7QA-7QZ	Malawi
XJA-XOZ	Canada	3HA-3UZ	China	7RA-7RZ	Algeria
XPA-XPZ	Denmark	3VA-3VZ	Tunisia	7SA-7SZ	Sweden
XQA-XRZ	Chile	3WA-3WZ	Viet Nam	7TA-7YZ	Algeria
XSA-XSZ	China	3XA-3XZ	Guinea	7ZA-7ZZ	Saudi Arabia
XTA-XTZ	Burkina Faso	3YA-3YZ	Norway	8AA-8IZ	Indonesia
XUA-XUZ	Cambodia	3ZA-3ZZ	Poland	8JA-8NZ	Japan
XVA-XVZ	Viet Nam	4AA-4CZ	Mexico	8OA-8OZ	Botswana
XWA-XWZ	Laos	4DA-4IZ	Philippines	8PA-8PZ	Barbados
XXA-XXZ	Portugal	† 4JA-4KZ	Azerbaijan	8QA-8QZ	Maldives
XYA-XZZ	Myanmar	† 4LA-4LZ	Georgia	8RA-8RZ	Guyana
YAA-YAZ	Afghanistan	4MA-4MZ	Venezuela	8SA-8SZ	Sweden
YBA-YHZ	Indonesia	4NA-4OZ	Yugoslavia	8TA-8YZ	India
YIA-YIZ	Iraq	4PA-4SZ	Sri Lanka	8ZA-8ZZ	Saudi Arabia
YJA-YJZ	Vanuatu	4TA-4TZ	Peru	† 9AA-9AZ	Croatia
YKA-YKZ	Syria	* 4UA-4UZ	United Nations	9BA-9DZ	Iran
† YLA-YLZ	Latvia	4VA-4VZ	Haiti	9EA-9FZ	Ethiopia
YMA-YMZ	Turkey	4XA-4XZ	Israel	9GA-9GZ	Ghana
YNA-YNZ	Nicaragua	* 4YA-4YZ	International Civil Aviation Organization	9HA-9HZ	Malta
YOA-YRZ	Romania			9IA-9JZ	Zambia
YSA-YSZ	El Salvador	4ZA-4ZZ	Israel	9KA-9KZ	Kuwait
YTA-YUZ	Yugoslavia	5AA-5AZ	Libya	9LA-9LZ	Sierra Leone
YVA-YYZ	Venezuela	5BA-5BZ	Cyprus	9MA-9MZ	Malaysia
YZA-YZZ	Yugoslavia	5CA-5GZ	Morocco	9NA-9NZ	Nepal
Y2A-Y9Z	Germany	5HA-5IZ	Tanzania	9OA-9TZ	Zaire
ZAA-ZAZ	Albania	5JA-5KZ	Colombia	9UA-9UZ	Burundi
ZBA-ZJZ	United Kingdom of Great Britain and Northern Ireland	5LA-5MZ	Liberia	9VA-9VZ	Singapore
		5NA-5OZ	Nigeria	9WA-9WZ	Malaysia
		5PA-5QZ	Denmark	9XA-9XZ	Rwanda
ZKA-ZMZ	New Zealand	5RA-5SZ	Madagascar	9YZ-9ZZ	Trinidad and Tobago

• Half-series
* Series allocated to an international organization
† Provisional allocation in accordance with RR2088

Eighty comes into its own in the evening and nighttime hours. At that time, 10 and 15 meters shut down for the night and 40 opens up to foreign broadcasters. Many hams switch to 80 meters. The band provides strong, reliable local propagation. It has none of the uncertainty of 10 and 15, nor the interference of 40.

The long-range capabilities of the 80-meter band are subject to seasonal changes. During the summer months, nighttime communications may extend to 1000 miles. Atmospheric static levels, however, are often high. Contacts can be difficult to establish and maintain. The winter months are a different story. Static levels are usually very low, and cross-country and DX contacts are normal occurrences.

Like 40 meters, 80 is a 24-hour-a-day happening. It provides excellent local communications almost all the time. It offers long-distance possibilities at night during the winter months.

Eighty is often used to make and maintain schedules with friends. After operating on the band for a while, you will come to know many of your fellow 80-meter Novices. Plan to spend some of your evening operating time on 80 meters making new friends. You'll likely meet them face-to-face at local conventions or hamfests.

PROPAGATION ON OUR VHF AND UHF BANDS

Normally, you'll be contacting nearby stations—within 100 miles or so—on the VHF/UHF bands. The usual propagation mode on these bands is **line of sight**. The signal travels directly between stations, so if you're using a directional antenna, you normally point it toward the station you are trying to contact. VHF and UHF signals, however, are easily reflected by buildings, hills and even airplanes. Some of your signal reaches the other station by a direct path and some may be reflected. When such **reflections** occur, it is possible to contact other stations by pointing your antenna toward the reflecting object, rather than directly at the station you're trying to contact.

Reflections can cause problems for mobile operation, as the propagation path is constantly changing. The direct and reflected waves may first cancel and then reinforce each other. This causes a rapid fluttering sound called *picket fencing*.

Sometimes, especially during the spring, summer and fall months, it is possible to make VHF and UHF contacts over long distances—up to 1000 miles or more. This occurs during certain weather conditions that cause *tropospheric enhancement* and *tropospheric ducting*. When such "tropo" openings occur, the VHF and UHF bands are filled with excited operators eager to work DX. The troposphere is the layer of the atmosphere just below the stratosphere. It extends upward approximately 7 to 10 miles. In this region clouds form and temperature decreases rapidly with altitude. You can find out more about VHF and UHF propagation from *The ARRL Handbook*.

Popular modes on the VHF/UHF bands include frequency-modulated (FM) voice, SSB voice, CW, packet radio, and Baudot and ASCII RTTY. Unlike the HF bands, FCC rules and regulations do not usually set aside portions of these bands for specific modes. Any authorized mode may be used on any frequency within the band.

Band Plans

To avoid confusion and interference problems, amateurs have agreed to a voluntary *band plan* for VHF and UHF. The plan sets aside segments of the band for different uses. For example, there is a section set aside around 222.1 MHz for "weak signal" CW and SSB operation. Another segment is set aside for FM voice operation, and so on. For complete information on band plans, see the latest edition of *The ARRL Repeater Directory*. While Technicians have full frequency and mode privileges on the VHF/UHF bands, Novices are allowed to use only parts of two of these bands.

1¼ Meters (222 MHz)

The 222-MHz band (1¼ meters) is very popular for FM voice and packet radio operation, and is a good place to meet other local hams. The maximum power level that Novices may use for any mode on 222 MHz is 25 watts PEP output.

23 Centimeters (1270 MHz)

The highest-frequency Novice band, the 1270-MHz band lies in the UHF portion of the radio spectrum, and is the least-populated of all Novice frequencies. Novices are allowed to operate in the segment from 1270 to 1295 MHz. They may use any and all emission modes authorized for that band. Popular modes on 23 cm include FM voice, SSB, CW and amateur television (ATV).

The 1270-MHz Novice band is similar to 222 MHz. There are no FCC-mandated subbands for the different modes. Rather, frequency usage is determined by a voluntary band plan. The 23-cm band plan calls for FM voice and ATV operation from 1270 to 1295 MHz. For complete information on the 23-cm band plan, see the latest edition of *The ARRL Repeater Directory*. The maximum power level that Novices may use for any mode on 1270 MHz is 5 watts PEP output. Technicians are allowed to operate on the entire band, 1270 to 1300 MHz.

An FCC-allocated amateur-satellite band is located around 1265 MHz. Weak-signal CW and SSB operation takes place around 1296 MHz, including the exotic "moonbounce" mode, where signals are actually bounced off the moon's surface!

[Before moving on, turn to Chapter 12 and review questions beginning with N3A. Review this section if you have any difficulty.]

[If you're studying for the Technician exam, turn to Chapter 13 and look over questions T3A01 through T3A03. Review this section if you have any difficulty.]

OVER THE AIRWAVES . . . PAINLESSLY

Once your license is in hand, you'll discover that a whole new world of on-the-air operating activities awaits you. At first, you'll probably just make contacts. Maybe you'll *ragchew* (chat) with other hams. After a while, however, you'll want to try your hand at specialized activities. These include handling traffic, contesting or DXing. You will probably want to exchange QSL cards with the hams you work. You may want to go after some of the various awards available to hams.

You'll find that certain aspects of Amateur Radio seem more appealing than others. You should at least try some of the activities. You might find you enjoy them very much.

HANDLING TRAFFIC

Passing "radiogram" messages is a vital part of an amateur's emergency communications training. A scheduled sequence of *traffic nets*, under the banner of the ARRL National Traffic System, provides an orderly path for getting messages from their origins to their destinations. These can be anywhere in the US and Canada, and some points overseas. CW traffic nets furnish an opportunity to improve your code speed. They also serve as a training ground for amateurs who want to enhance their on-the-air operating abilities. Slow-speed CW nets are particularly popular among new amateurs.

Getting started in traffic handling is a simple matter. Familiarize yourself with net procedures, and then check in. The times and frequencies of most traffic nets are listed in the *ARRL Net Directory*. The *Directory* also includes background information and instructions for typical net operations. Emergency and traffic operating are addressed in more detail in the *Public Service Communications Manual* (PSCM). The *Net Directory* and *PSCM* are available from ARRL Headquarters for $1 each. Another source of information is *The ARRL Operating Manual*, which can be purchased from your local radio dealer or directly from ARRL.

Whether your interest is in emergency preparedness, handling messages for your friends in the community, or improving your code speed and operating skills, you will find traffic handling and net operation an enjoyable and educational aspect of Amateur Radio.

DX

When the DX bug bites, it bites hard, with little warning. One day as you call a casual CQ, a weak station answers. You are expecting to hear from someone in the next county or state. He sends his call sign. "Hey, that's a strange sounding call sign. Let's check the international call signs allocations list to see what country this guy might be in. Z2. That's Zim-

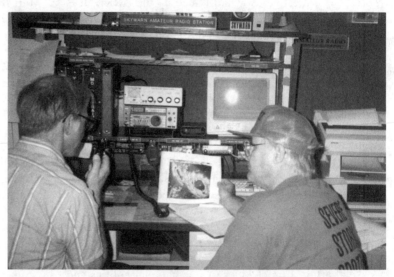

Emergency-preparedness paid off during the disaster that was Hurricane Andrew. Here, Jan Jubon, K2HJ (l), and Joe Peters, WB4WZZ, cover the night shift at the SKYWARN Amateur Radio station in the Washington, DC, forecast office of the National Weather Service. The time is early Monday morning, August 24, 1992, and Peters holds the latest printout of a GOES satellite image showing Hurricane Andrew as it begins to touch land in Florida. The amateur HF transceiver was used to monitor hurricane nets and 2 meters was used to coordinate local amateur staffing for the SKYWARN backup station. Other ham volunteers, on the scene in South Florida and elsewhere, worked around the clock for days and weeks to pass word of needed equipment and supplies to the outside world. *(photo courtesy of SKYWARN Net Manager Dan Gropper, KC4OCG).*

babwe in southeastern Africa! This guy is halfway around the world!" Before you know it, you have succumbed and become a DXer.

DX is one of those loosely defined terms that means different things to different people. The term **DX** (distance) is generally understood to include all stations that are not in your own country. It can be a two-mile contact across the border or a 10,000-mile QSO two continents away.

DX and DXing (working other hams in foreign lands) is a fascinating and absorbing aspect of Amateur Radio. You could easily spend an entire ham radio "career" solely in pursuit of DX. And indeed, some hams do.

It's a fact that any amateur station is capable of working DX. It is true that bigger antennas, linear amplifiers (when you get your General or higher class license) and well-honed operating techniques can make the process easier. But even the newest Novice with the most basic station is capable of snagging plenty of DX.

What is the ultimate goal of the DXer? The most common answer would probably be to "work 'em all." Contact and confirm by QSL card as many of the countries on the ARRL DXCC countries list as possible (presently about 325 in number). That will ensure your high standing among fellow DXers pursuing that same goal. The *DX Century Club (DXCC)* is a prestigious award issued by the ARRL. It is recognized worldwide as the premier award for DXing achievement in all of Amateur Radio. For the basic DXCC award, you'll need to work 100 different countries from the DXCC list. QSLs are required from each of them. The cards are then sent to ARRL for verification. An official application form is required. (The ARRL DXCC list of countries is available from ARRL Headquarters for $2. It's also included in *The ARRL Operating Manual.*) After you obtain the basic award, the challenge continues. A series of endorsements for working and confirming still more DXCC countries awaits you. DXCC membership is open to amateurs worldwide (although US and Canadian amateurs must be ARRL or CRRL members to participate). It is the basis for a friendly worldwide competition. Everyone tries to work more countries than their friends across town (or across the seas).

Working all the countries on the DXCC list is not a weekend affair. Complexities of international politics and the absence of amateur activity in some of the less-populated countries work against it. Some hams with 20 years or more of DX experience are still missing quite a few DXCC countries. Some relative newcomers, on the other hand, lack only a couple of countries.

There are many resources available to the avid DXer. At the top of the list is personal on-the-air experience. Spend lots of listening time on the bands. Get to know the peculiarities of radio-wave propagation. Get to know when and where DX stations are likely to hang out. Listen to successful DXers. Copy their good operating habits. You can develop effective strategies for working DX just by monitoring the bands.

Working DX requires a different set of operating skills. This is true from looking for DX QSOs, to sending and receiving QSL cards. Much has been printed on the art and science of working DX. Among the more useful printed materials are DX bulletins and newsletters. These bring subscribers current and helpful information on some of the "rarer" DX stations. The monthly "How's DX?" column in *QST* is chock full of useful DX news. For the beginner, ARRL publishes *The DXCC Companion, How To Work Your First 100 Countries.* This book covers every topic of interest to the DXer, including antennas, propagation, calling strategies, sending for QSL cards, coping with interference and applying for DXCC. A more general treatment of DXing techniques can be found in *The ARRL Operating Manual.* One last DX tip: Success comes only to the station that joins in the fray. After all, that's the name of the game. Good luck and good DXing.

CONTESTS AND CONTESTING

We all like to test our skills against others. In ham radio, contesting is an excellent way to learn good operating techniques. In contests, hams go "head-to-head"—and may the best operator win.

Contesting is a diverse facet of Amateur Radio. Most amateurs join in at one time or another to satisfy their competitive appetites. You might totally immerse yourself for the full 48 hours of a big contest. Those who do may come away with 2000 or more contest QSOs and the prize for first place! You can also spend but a few minutes to achieve a personal goal. Maybe you'd like to work a new DX country. Or you might try to make the most QSOs you ever made in a given time period.

Generally, a contest involves working as many different stations as possible in a given period of time. A premium is placed on working specific types of stations for "multipliers." Multipliers may be different states (or provinces) or counties in a North American contest. In international (DX) contests, multipliers may be different countries or International Telecommunication Union (ITU) Zones.

Contests come in all shapes and sizes, each tailored to one of the many different types of amateurs. There is a Novice/Technician Roundup (open to all, but awards are given only to Novice and Technician operators). There are DX contests for DXers. VHF contests exist for those who enjoy working the bands above 30 MHz. These are only a few of many types of contests. . . but you get the idea. "There is a contest for any taste."

"Winning" in Amateur Radio contesting is a relative term. One operator might turn in the number one score among thousands of participants. Another might significantly improve his or her operating techniques. All contests are learning experiences; both hams are "winners." Contesting sharpens operating skill and leads to greater station efficiency, for serious contenders and casual participants.

A good contest operator wastes neither words nor motions. Contesters are likely to have some of the best signals on the band. They may not have the most elaborate stations, however. Contests encourage operators to use their equipment and skills to the best advantage.

If you want to operate in a contest, you have to know the "exchange." You may have to send a signal report and a contact number, starting with 1. Other contests require a signal report and a zone number. You can find out by reading the rules in *QST* or by listening to what local hams are saying. The station you work will give you an exchange, too. Write it in your log. You'll notice right away that most signal reports are "59" or "599"! You can give an honest signal report, but don't elaborate.

Let's say you hear W6ISQ calling CQ in the ARRL 10-Meter Contest. You give only your call: KA1RRL (he knows his call!). W6ISQ responds: "KA1RRL 59 California" and you answer "Thanks, 59 Connecticut." The word "Thanks" indicates that you copied his exchange. W6ISQ may respond with "Thanks" or "**73**" (best regards) or may just start calling CQ again. He's not being rude, just efficient. Wasted words in a contest result in fewer contacts and a lower score. Staying on one frequency and calling CQ is called "running stations." Experienced operators can "run" over 150 stations (contacts) an hour, so every second counts! The number of contacts per hour is called the "rate."

Tuning the band looking for contacts is called "search and pounce," abbreviated "S and P." Running almost always generates more contacts, but sometimes you have to S and P to find multipliers. At first you should S and P. Later, find a clear frequency and try running for a while. Many times, the multipliers you need will find you. If your rate starts to drop, go back to S and P for a while.

Even if you've never entered a contest before, don't be afraid to try your hand! The "Contest Corral" column in *QST* and contest sections in most other Amateur Radio periodicals list hundreds of contests each year. The basic rules and entry requirements are also listed. Don't be gun-shy. Choose an event you wish to participate in. Read and understand the rules. Then, jump right in and give it a shot. Don't worry about letting your inexperience show. The high-scoring stations need us to keep their rates up!

FIELD DAY

Field Day is a special event. It's fun, challenging, frustrating, exciting, exhausting and satisfying. Field Day is Amateur Radio from A to Z.

Ask 10 amateurs "What is Field Day?" and you are likely to get 12 different answers. The premier operating event of the year is held on the fourth full weekend in June. It is sponsored by the American Radio Relay League. But that's where any and all agreement ends. Field Day is a chance to take to the hills and fields. It's a chance to test our ability to communicate in adverse conditions. Field Day stations try to contact as many other stations as possible during a 30-hour period. The multiplier on Field Day is your power source and transmitter power output. The biggest multiplier goes to 5-watt-or-less transmitters operating from "natural power" sources, like solar cells or water wheels. You can operate from home, too, but you'll have more fun in the field.

Field Day is, technically speaking, not a contest. Intended to test emergency preparedness and demonstrate the fun of Amateur Radio to nonhams, Field Day is a great chance to spend a weekend in the outdoors with family, friends and Amateur Radio.

One weekend spent at even a modest-sized Field Day setup is probably equal to a year's worth of "normal hamming." The experience draws heavily on ingenuity and resourcefulness. Even the newcomer is quickly pressed into service. Everyone must deal with the problems that occur in remote locations.

If Field Day sounds like your "cup of tea," write ARRL for the name of a radio club in your area. Although you can operate Field Day by yourself, you'll miss out on a lot of fun. Field Day is an experience not to be missed. Mark your calendar now.

QSL CARDS

Sending a **QSL card** to confirm a contact is a tradition almost as old as Amateur Radio itself. QSLs are sent more frequently for HF and VHF/UHF simplex QSOs than for VHF/UHF repeater and packet QSOs. Regardless of your operating interest, a QSL can serve as your ham radio "busi-

ness card" when you meet other hams at hamfests and the local radio store. If you want to chase DX or earn operating awards such as Worked All States you'll need QSLs from the stations you work. Unless you live in a less-populated state like Wyoming or Montana, you'll probably have to send a card first, and ask for a response. Including return postage will help bring a reply.

By the way, QSL sent as a **Q signal** on CW means "I acknowledge receipt," or "I will confirm this contact." QSL? asks "Can you acknowledge receipt?" or "Can you confirm this QSO for me?" QSL is an important "word" in amateur jargon. It is understood to mean a written or printed proof that you have communicated with another station.

Okay, you're just starting out in Amateur Radio. You're looking for a supply of QSL cards. You'll need to send QSLs to the stations whose call signs will soon fill your logbook. There are a few things to keep in mind as you select, fill out and send your QSLs.

QSLs come in all sizes, shapes, colors and textures. QSLs have been silk-screened on T-shirts, and sent as commercial telegrams. They've been molded in plastic and typed on tissue. Usually, however, they're printed on post-card sized card stock. (Remember, the US Postal Service will not accept cards smaller than 3½ by 5½ inches.)

Your QSLs can be printed professionally, or the cards can be homemade. (See the *Ham Ads* section of *QST* for suppliers.) The final design is left up to you. Keep the card's recipient in mind, however. He or she will form a lasting impression of you from your QSL card. Your QSL must also include certain information about each QSO.

The ultimate use of a QSL will be as an awards submission. There are many Amateur Radio awards. These include the DX Century Club (DXCC), Worked All States (WAS) and Worked All Continents (WAC). Most societies that issue the awards will honor your QSL as valid confirmation only if it contains the following essential information: your call sign, your specific location (QTH), confirmation that a two-way QSO has taken place, the call sign of the station worked, date of the QSO, time of the QSO (in UTC, please), frequency band used, mode (CW, SSB, RTTY, FM, and so on) used, and signal report given. When choosing the design for your QSL be sure the card has the proper layout for recording all QSO information.

There are conventions to follow when filling out QSLs. Always record times in *Coordinated Universal Time (UTC)*. It is frustrating to receive a QSL marked in local time. You'll have to convert some obscure "local" time to UTC. Otherwise, you won't be able to find the QSO in your logbook. Amateurs around the world use UTC. There's a good rule to follow when recording the QSO date on your QSL. The date is written in Arabic numerals. The month is written in Roman numerals. The year is written in Arabic numerals. Example: July 10, 1994 would be written as 10 VII 1994. Most DX (non-US or Canadian) amateurs record the date in this manner. Standardization makes life easier for all involved. An alternative is to abbreviate the month, but use the same format (10 JUL 1994). Do not use the method of just writing

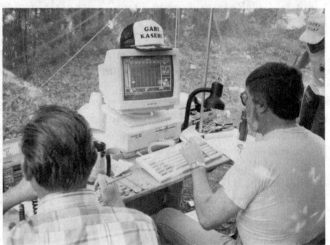

What do thousands of hams do on the fourth weekend in June every year? It's something they swear they'll never do again, but they usually do. You guessed it. It's Field Day!

Although points are awarded for various categories, it's not really a contest. Field Day is actually an emergency-preparedness exercise. Many groups of amateurs literally operate out in a field. Others try the nearest mountaintop, beach or backyard. The idea is to test your readiness to operate under adverse conditions. It's a group effort, and a great deal of fun. You may be in charge of untangling a longwire antenna at 3 AM. Or you may begin operating at sunrise after being up the entire night! It's an experience you're not likely to forget. It shows what the Amateur Service can do in emergencies.

Shortly after you begin your on-the-air operation, you will experience the excitement of collecting QSL cards from all over the world.

Arabic numerals, such as 10/7/94 or 7/10/94. (Is that July 10 or October 7?)

When filling out your QSL card, use regular ink or permanent marker. Pencil and most felt-tipped markers tend to fade. They smudge easily and deteriorate over the years. A rainy day is all it takes to "wipe out" a hard-earned contact. If you make a mistake when filling out a QSL card, throw it away. Do not try to correct it by writing over the mistake. Destroy the card and start again. Most organizations will not accept "altered" QSLs as valid awards confirmations. There is no way to determine who altered the card. It cannot be counted for credit.

Once the QSL is made out, you must get it to the operator of the station you've worked. There are many ways to send QSLs. They depend on its destination and how quickly you want it to arrive. Domestic QSLs (US, Canada or Mexico) are sent as postcards. They can also be mailed in envelopes, as First Class mail. Hams in rarer states or provinces receive many QSL requests. They appreciate receiving an SASE (self-addressed, stamped envelope) with your QSL. That helps defray postage costs and saves time in addressing many cards.

QSLs going to DX stations do not have to be sent directly to the DX station's address. They can be sent via a DX QSL bureau. A QSL bureau is a system developed to help amateurs exchange QSL cards with other amateurs in foreign countries. Hams send QSLs by regular mail to one central bureau location in their country. When enough QSLs for a foreign country accumulate, a bureau staff member sends them all to that country. Once the cards arrive, someone working in an incoming QSL bureau sorts and mails them to individual amateurs. Using a QSL bureau means that each amateur does not have to send single cards overseas.

The ARRL sponsors such a system. The outgoing DX QSL Bureau (for ARRL members only) sends your QSLs to bureaus around the world. The incoming DX QSL Bureau (for use by all US and Canadian amateurs) forwards DX QSLs to their final destinations in the US and Canada. Details on the use of the ARRL DX QSL Bureau system are printed in *QST*. When mailing DX QSLs directly overseas, remember an important rule. United States stamps on your SASE will not be accepted by foreign postal authorities. Instead of using a US stamp on the return envelope, enclose one or two International Reply Coupons. IRCs are available at US Post Offices. They can be exchanged for the proper stamps in another country. As an added bonus, you'll start receiving beautiful foreign stamps with your DX QSLs! Many successful stamp collectors trace their "collecting" roots back to Amateur Radio DX. *The ARRL Operating Manual* has more information about sending QSLs to, and receiving them from, DX stations.

Collecting QSL cards is a satisfying hobby within a hobby. You may collect QSLs for the fun of it, or in an attempt to earn an award.

QSLs, like money, are more fun to receive than to give. In receiving a QSL, there is an implied responsibility to return the courtesy. QSLing is a worthwhile chore—do it cheerfully and promptly.

AWARDS

Working toward operating awards is a popular aspect of Amateur Radio. The ARRL sponsors many coveted awards. These include Worked All States, DX Century Club and the A-1 Operator Club. In addition, the League also administers the Worked All Continents award. This is sponsored by the International Amateur Radio Union. They're all available to licensed Amateur Radio operators throughout the world. Hams in North America must be League members to apply for most ARRL awards. Foreign amateurs need not be League members.

Perhaps the first award you'll earn is the *Rag Chewers'*

Club award. To qualify, you need to maintain a contact for a solid half-hour. Your first contact will be quite exciting, and the first half hour will fly by. And, you'll have earned your RCC to boot! Just report your QSO to ARRL Headquarters, and we'll send you your award.

The *Worked All States (WAS)* award is probably the most popular achievement award, especially among US hams. Eligibility requirements are simple. You must establish two-way Amateur Radio contact with other hams in all 50 states. Each contact must be confirmed by a QSL card or other written verification. As a Novice, WAS is well within your reach. Many Novices qualify. Send an SASE along with your request for the application form to ARRL Headquarters.

The *DX Century Club (DXCC)* is a prestigious operating achievement award. You must work and collect QSL cards from licensed hams in at least 100 different countries to qualify for this award. ("Countries" as defined by the DXCC rules, that is. Alaska and Hawaii, for example, each count as separate "countries.") Although DXCC is somewhat difficult to achieve in the Novice bands, it is possible. Years of high solar activity make it easier. Your patience and perseverance will pay off in experience and satisfaction. All DX contacts made as a Novice will count toward DXCC. You don't lose anything when you upgrade.

The *Worked All Continents (WAC)* award is sponsored by the International Amateur Radio Union. It's the first DX award many new hams get. To qualify, you must work and collect QSL cards from one ham in each of the continents: North America, South America, Europe, Africa, Asia and Oceania. Write to ARRL Headquarters and request a WAC application form and complete rules. Please include a self-addressed, stamped envelope (SASE) with your request.

All amateurs should strive for membership in the *A-1 Operator Club*. Membership in this elite group attests to unusual competence and performance. The A-1 Operator Club recognizes many facets of Amateur Radio operation. These include keying, modulation, procedure, copying ability, and judgment and courtesy. Members of this club are smooth, courteous operators with clean signals. They set good examples for us all. You'll run into them only occasionally. The A-1 Operator club offers no list to aid those who seek membership. You must be recommended independently by two operators who already belong. If you ask to join, you may actually "fail" the test on the basis of courtesy and judgment!

Another ARRL-sponsored award is the *Certificate of Code Proficiency*. To qualify, you have to build your code speed up to at least 10 words per minute. Each month, W1AW transmits an official code practice qualifying run. This consists of successive five-minute periods of text. Speeds include 10, 15, 20, 25, 30 and 35 WPM. To apply for this award, underline one minute of text. Choose the period you believe you copied perfectly. Send it to ARRL Headquarters with an SASE. We'll tell you if you passed or failed. If you passed, you'll receive a free certificate. W1AW qualifying runs are announced in the "Contest Corral" section of *QST*. You can also request a W1AW operating schedule that includes a list of Qualifying Run times from League Headquarters.

Many other organizations besides the ARRL sponsor operating awards. Many Novices' first step toward WAS is the *Ten American Districts Award*. It's issued for working and confirming contacts in all 10 US call areas. Apply to the Lockheed Employees Recreation Club, 2814 Empire Blvd, Burbank, CA 91504. Enclose $1 with your application to cover postage and handling.

Other awards for working different US counties and international prefixes are available from CQ, 76 North Broadway, Hicksville, NY 11801. Write them for information on their awards program. You'll find that there are a lot of certificates available. You could even wallpaper your entire shack. Decide which awards you want, tune up your rig and go get 'em!

LOCAL RADIO CLUBS

Your local Amateur Radio club is a good place to meet other hams. At club meetings, you'll meet others with similar interests. They can help you get on the air. They can introduce you to new aspects of Amateur Radio. They can also amuse you with tales of the "good old days." Through membership in your local club, you'll make fast, lifelong friendships.

Most radio clubs sponsor a wide range of activities for their members. These range from banquets to auctions, from flea markets to community service projects, and from training classes to Field Day efforts. By participating, you'll add enjoyment and dimension to your amateur experience.

Special-interest clubs may interest you as well. Some examples of these are contest clubs, DX clubs and VHF clubs. Other groups, such as Explorer posts, canoers, motorcyclists and marathoners use Amateur Radio as an integral part of their activities. If you develop a special interest within Amateur Radio, you'll want to investigate—or even help start—

FIND YOURSELF A CLUB

There are 2000-plus Amateur Radio clubs scattered throughout the US and Canada. The Sheboygan (Wisconsin) ARC is one of the most active. One popular activity is helping American Field Service foreign exchange students from Latin America. The club helps the students contact their families, thousands of miles away. It's a thrill for all involved. It's rewarding to bring together homesick students and their anxious families.

The club also sponsors licensing classes and a Novice buddy program. That's where senior members of the club assist newcomers. They help in setting up stations and antennas. Big or small, problems are solved on a one-to-one basis. Long-lasting friendships are begun.

Whether you're looking for your first transceiver or just want to learn more about Amateur Radio, a local club is a great way to go. Write to Clubs, ARRL, 225 Main St, Newington, CT 06111, and ask for the name of one nearby.

a local special-interest club. Often, special-interest club members are the most knowledgeable hams in your area. Many are glad to help a beginner learn the ropes.

How do you find out about these clubs? More than 2000 are affiliated with the ARRL. To find the club (or clubs) in your area, write to the Educational Activities Department at League Headquarters. We'll send you the information you need.

If there is no club in your area, why not start your own? If you supply the members, ARRL will supply the information. Hobbies are more fun if you have someone to share them with.

THE ARRL

In addition to participating locally, you should also take part on the national level. The American Radio Relay League (ARRL) is the membership organization of Amateur Radio operators in the United States. There are many reasons why you, as a ham, should join the League.

Each month, the ARRL sends its official journal, *QST*, to every member. In *QST* you'll find articles dealing with nearly every aspect of Amateur Radio. As a new ham, you'll appreciate the beginners' articles. They explain how transmitters, receivers, antennas and accessories work. There are easy-to-build construction projects. In addition, *QST* carries information on contests, hamfests (where exams are often given) and the latest news from the world of ham radio. In the back of *QST*, you'll find page after page of informative ads.

These can help you choose the right equipment and accessories.

QST isn't the only reason you should join the ARRL, though. The League represents the interests of Amateur Radio in Washington. ARRL makes sure the FCC and other government agencies know where US hams stand on current issues. ARRL also answers many questions from members. The staff at League Headquarters exists to help its members. DXers can take advantage of the *ARRL Outgoing QSL Service*. It saves considerable money in forwarding your cards to the right place. These are only a few of the services available to ARRL members.

The ARRL is much more than an organization in Newington, Connecticut. The ARRL is each local member

Whether you're into DX, simple construction projects or late-breaking news that affects all amateurs, the ARRL's monthly journal, *QST*, is indispensable.

across the country. The ARRL Field Organization depends on local volunteers to carry out various programs. The work of Amateur Radio begins at the local level by amateurs working in official elected or appointed capacities. There are jobs for those interested in many aspects of public service. These include emergency communications, communications for community events and handling routine messages. There are jobs for public relations people, government liaisons and those who will provide technical assistance. Share some of your talent and enthusiasm by becoming involved with The American Radio Relay League.

By supporting your League, you're supporting Amateur Radio. Write to us today.

Propagation: A Closer Look
Bandwidth

KEY WORDS

Critical frequency—The highest frequency at which a vertically incident radio wave will return from the ionosphere. Above the critical frequency, radio signals pass through the ionosphere instead of returning to Earth.

D region—The lowest region of the ionosphere. The D region contributes very little to short-wave radio propagation. It acts mainly to absorb energy from radio waves as they pass through it This absorption has a significant effect on signals below about 7.5 MHz during daylight.

Direct wave—A (usually) VHF or UHF radio wave that travels in a straight line-of-sight path between the transmitter and receiver.

E region—The second lowest ionospheric region. Exists only during the day. Under certain conditions, may refract radio waves enough to return them to earth.

Emission type—The different kinds of radio signals, such as SSB, FM and CW and RTTY.

F region—A combination of the two highest ionospheric regions, the F1 and F2 regions. Refracts radio waves and returns them to earth. Its height varies greatly depending on the time of day, season of the year and amount of sunspot activity.

Frequency deviation—The amount the carrier frequency in an FM transmitter changes as it is modulated.

Maximum usable frequency (MUF)—The highest-frequency radio signal that will reach a particular destination using E- or F-region skip propagation. The MUF may vary for radio signals sent to different destinations.

Scatter—A small amount of signal that is reflected from the Earth's surface after traveling through the ionosphere. The reflected signals may go back into the ionosphere along several paths and be refracted to Earth again. Scatter can help provide communications into a station's skip zone.

Splatter—A type of interference to nearby stations. Occurs when a transmitter is overmodulated.

Temperature inversion—A condition in the atmosphere in which a region of cool air is trapped beneath warmer air.

Troposphere—The region in Earth's atmosphere just above the Earth's surface and below the ionosphere.

Tropospheric bending—When radio waves are bent in the troposphere, they return to earth farther away than the **visible horizon**.

Tropospheric ducting—A type of VHF propagation that can occur when warm air overruns cold air (a **temperature inversion**).

Visible horizon—The most distant point one can see by line of sight.

IONOSPHERIC PROPAGATION

The earth's upper atmosphere consists mainly of oxygen and nitrogen. There are traces of hydrogen, helium and several other gases. The atoms making up these gases are electrically neutral: They have no charge and exhibit no electrical force outside their own structure. The gas atoms absorb ultraviolet radiation from the sun, which knocks electrons out of the atom. In this process, the atoms become positively charged. A positively charged atom is an ion. The process by which ions are formed is called *ionization*. Several ionized regions appear at different heights in the atmosphere. Each region has a central region where the ionization is greatest. The intensity of the ionization decreases above and below this central region in each region. See Figure 9-12.

The ionosphere consists of several regions of charged particles. These regions have been given letter designations, as shown in Figure 9-13. Scientists started with the letter D just in case there were any undiscovered lower regions. None have been found, so there is no A, B or C region.

The **D region**: The lowest region of the ionosphere affecting propagation is the D region. This region is in a relatively dense part of the atmosphere about 30 to 55 miles above the Earth. When the atoms in this region absorb sunlight and form ions, they don't last very long. They quickly combine with free electrons to form neutral atoms again. The amount of ionization in this region varies widely. It depends on how much sunlight hits the region. At noon, D-region ionization is maximum or very close to it. By sunset, this ionization disappears. The D region is ineffective in *refracting* or bending high-frequency signals back to earth. The D region's major effect is to absorb energy from radio waves. As radio waves pass through the ionosphere, they give up energy. This sets some of the ionized particles into motion. Effects of absorption on lower frequencies (longer wavelengths) are greater than on higher frequencies. Absorption

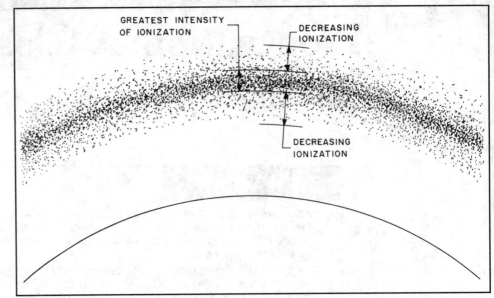

Figure 9-12—A cross section of a region of the ionosphere. The intensity of the ionization is greatest in the central region and decreases above and below the central region.

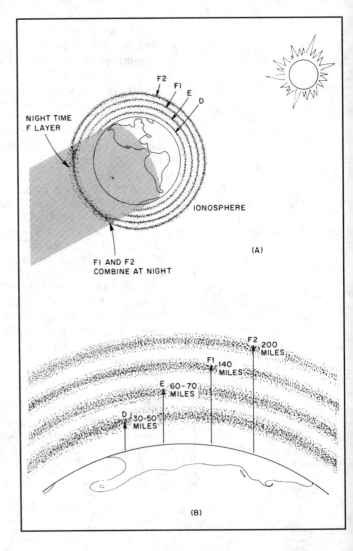

Figure 9-13—The ionosphere consists of several regions of ionized particles at different heights above Earth. At night, the D and E regions disappear and the F1 and F2 regions combine to form a single F region.

also increases proportionally with the amount of ionization. The more ionization, the more energy the radio waves lose passing through the ionosphere. Absorption is most pronounced at midday. It is responsible for the short daytime communications ranges on the lower-frequency amateur bands (160, 80 and 40 meters).

The next region of the ionosphere is the **E region**, at an altitude of about 60 to 70 miles. At this height, ionization produced by sunlight does not last very long. This makes the E region useful for bending radio waves only when it is in sunlight. Like the D region, the E region reaches maximum ionization around midday. By early evening the ionization level is very low. The ionization level reaches a minimum just before sunrise, local time. Using the E region, a radio signal can travel a maximum distance of about 1250 miles in one hop. *Sporadic E*, or E skip, is a type of sky-wave propagation that allows long-distance communications on the VHF bands (6 meters, 2 meters and 222 MHz). Although, as its name implies, sporadic E occurs only during certain times of the year, it is the most common type of VHF sky-wave propagation.

The **F region**: The region of the ionosphere most responsible for long-distance amateur communication is the F region. This region is a very large region. It ranges from about 100 to 260 miles above the Earth. The height depends on season, latitude, time of day and solar activity. Ionization reaches a maximum shortly after noon local standard time. It tapers off very gradually toward sunset. At this altitude, the ions and electrons recombine very slowly. The F region remains ionized during the night, reaching a minimum just before sunrise. After sunrise, ionization increases rapidly for the first few hours. Then it increases slowly to its noontime maximum.

During the day, the F region splits into two parts, F1 and F2. The central region of the F1 region forms at an altitude of about 140 miles. For the F2 region, the central region forms at about 200 miles above the Earth. These altitudes vary with the season of the year and other factors. At noon in the summer the F2 region can reach an altitude of 300 miles. At night, the two regions recombine to form a single F region slightly below the higher altitude. The F1 region does not have much to do with long-distance communications. Its effects are similar to those caused by the E region. The F2 region is responsible for almost all long-distance communication on the amateur HF bands. A one-hop radio transmission travels a maximum of about 2500 miles using the F2 region.

[Now turn to Chapter 13 and study questions T3A04 through T3A11 and T3B01 through T3B08. Review this section as needed.]

Virtual Height

An ionospheric region is a region of considerable depth. For practical purposes, think of each region as having a definite height. The height from which a simple reflection from the region would give the same effects (observed from the ground) as the effects of the gradual bending that actually takes place is called the *virtual height*. See Figure 9-14.

The virtual height of an ionospheric region for various frequencies is determined with an ionosonde. An ionosonde is a variable-frequency transmitter and receiver. It directs radio energy vertically and measures the time required for the signal to make a round-trip. As the frequency increases, there is a point where no energy will return. The highest frequency at which vertically incident waves return to earth is the **critical frequency**. Signals at higher frequencies pass through the ionosphere into space.

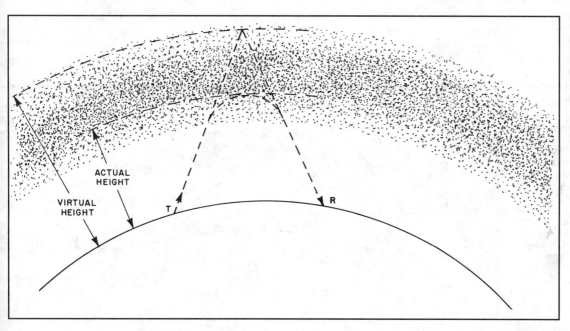

Figure 9-14—The virtual height of a region in the ionosphere is the height at which a simple reflection would return the wave to the same point as the gradual bending that actually takes place.

Radiation Angle and Skip Distance

A radio wave at a low angle above the horizon requires less refraction to bring the wave back to earth. This is why we want antennas with low radiation angles to work DX.

Figure 9-15 illustrates some of the effects of radiation angle. The high-angle sky waves are bent only slightly in the ionosphere and pass through it. The wave at the somewhat lower angle is just able to return. In daylight it might be returned from the E region. The point of return for high angle waves is relatively close to the transmitting station. The lowest angle wave returns farther away, to point B.

The highest radiation angle that will return a radio wave to earth under specific ionospheric conditions is the *critical angle*. Waves meeting the ionosphere at greater than the critical angle will pass through the ionosphere into space.

Maximum Usable Frequency

There is little doubt that the critical frequency is important to amateur communication. More interesting to radio amateurs is the frequency range over which communication can be carried via the ionosphere. Most amateurs want to know the **maximum usable frequency** (abbreviated MUF). The MUF applies to a particular direction and distance at the time of day when you want to communicate. The MUF is the highest frequency that allows a radio wave to reach the desired destination using E- or F-region propagation. The MUF is subject to seasonal variations as well as changes throughout the day, based on the amount of ultraviolet solar radiation.

Why is the MUF so important? If we know the MUF, we can make accurate predictions about which bands give the best chance for communication over a particular path. Predictions concerning amateur bands affording the best chance for communication via particular paths can be made. For example, we wish to contact a distant station. We know that the MUF for that path is 17 MHz at the time we wish to make our contact. The closest amateur band below that frequency is the 20-meter, or 14-MHz, band. The 20-meter band offers our best chance to contact that long distance location, then. There is no single MUF for a given location at any one time. The MUF will vary depending on the direction and distance to the station we wish to contact. Understanding and using the MUF is one way you can change your study of propagation from guesswork into a science.

[Now turn to Chapter 13 and study exam questions T3B09 through T3B11. Review this material as needed.]

Solar Activity

It should be clear that the sun plays the major role in the ionization of different regions in the ionosphere. The sun is the dominant factor in sky-wave communication. To communicate effectively over long distances, you must understand how solar conditions will affect your radio signals.

You know about the day-to-day sun-related cycles on earth, such as the time of day and the season of the year. Conditions affecting radio communication vary with these same cycles. Long-term and short-term solar cycles also influence propagation in ways that are not so obvious. The condition of the sun at any given moment has a very large effect on long-distance radio communication. The sun is what makes propagation prediction an inexact science.

Man's interest in the sun is older than recorded history. Sunspots are dark regions that appear on the surface of the sun. They were observed and described thousands of years ago. Observers noted that the number of sunspots increased and decreased in cycles. The solar observatory in Zurich, Switzerland has been recording solar data on a regular basis since 1749. The solar cycle that began in 1755 was designated Cycle 1. The low sunspot numbers marking the end of Cycle 21 and the beginning of Cycle 22 occurred in September of 1986. Sunspot cycles also vary a great deal. For example, Cycle 19 peaked with a smoothed mean sunspot number of over 200. Cycle 14 peaked at only 60.

Maximum ionization occurs during a sunspot cycle

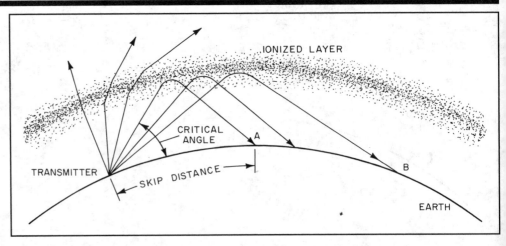

Figure 9-15—Ionospheric propagation. The waves at left leave the transmitter above the critical angle—they are refracted-in the ionosphere, but not enough to return to earth. The wave at the critical angle will return to earth. The lowest angle wave will return to earth farther away than the wave at the critical angle. This explains the emphasis on low radiation angles for DX work.

peak. High sunspot numbers usually mean good worldwide radio communications. During sunspot cycle peaks, the 20-meter amateur band is open to distant parts of the world almost continuously. The Cycle 19 peak in 1957 and 1958 was responsible for the best propagation conditions in the history of radio. Sunspot cycles do not follow consistent patterns. There can be periods of high activity that seem to come from nowhere during periods of low sunspot activity. There is an important clue to anticipating variations in solar radiation levels. Radio propagation changes resulting from these variations can be predicted. This clue is the time it takes the sun to rotate on its axis, approximately 27 days. Active areas capable of influencing propagation can recur at four-week intervals. These active areas may last for four or five solar rotations. High MUF and good propagation for several days indicates similar conditions may develop approximately 27 days later.

Another useful indication of solar activity is solar flux, or radio energy coming from the sun. Increased solar activity produces higher levels of solar energy. More solar energy produces greater ionization in the ionosphere. Scientists use sophisticated receiving equipment and large antennas pointed at the sun to measure solar flux. A number called the *solar-flux index* is given to represent the amount of solar flux. The solar-flux index is gradually replacing the sunspot number as a means of predicting radio-wave propagation.

The solar-flux measurement is taken at 1700 UTC daily in Ottawa, Canada, on 2800 MHz (10.7 centimeters). The information is then transmitted by the National Institute of Science and Technology station WWV in Fort Collins, Colorado. Both the sunspot number and the solar-flux measurements tell us similar things about solar activity. To get the sunspot number, the sun must be visible. The solar-flux measurement may be taken under any weather conditions.

Using the Solar Flux

We can use the solar-flux numbers to make general daily predictions about band conditions. Figure 9-16 shows a relationship between average sunspot numbers and the 2800-MHz solar-flux measurement. The solar flux varies directly with the activity on the sun. Values range from around 60 to 250 or so. Flux values in the 60s and 70s mean fair to poor propagation conditions on the 14-MHz (20-meter) band and higher frequencies. Values from 90 to 110 or so indicate good conditions up to about 24 MHz (12 meters). Solar-flux values over 120 indicate very good conditions on 28 MHz (10 meters) and even up to 50 MHz (6 meters) at times. On 6 meters, the flux values must be in the 200s for reliable long-range communications. The higher the frequency, the higher the flux value required for good propagation.

Sky-wave or *skip propagation* is primarily an HF propagation mode; it is rare at VHF. When the solar-flux values are very high, however, you may find enhanced propagation on 6 meters and even on 2 meters sometimes. If you notice strong

signals on 10 meters from stations 500 to 600 miles away, this is a good indication you should look for skip on 6 and 2 meters.

The Scatter Modes

All electromagnetic-wave propagation is subject to scattering influences. These alter idealized patterns to a great degree. The earth's atmosphere, ionospheric regions and any objects in the path of radio signals scatter the energy. Understanding how scattering takes place helps us use this propagation mode to our advantage.

Forward Scatter

There is an area between the outer limit of ground-wave propagation and the point where the first signals return from the ionosphere. We studied this area, called the *skip zone*, earlier in this chapter. See Figure 9-11. The skip zone is often described as if communications between stations in each other's skip zone were impossible. Actually, some of the transmitted signal is scattered in the atmosphere, so the signal can be heard over much of the skip zone. You usually need a very sensitive receiver and good operating techniques to hear these signals, as they are usually weak and distorted.

The troposphere is a region of the atmosphere below the ionosphere. VHF "tropo" scatter is usable out to about 500 miles from the transmitting station.

Ionospheric scatter, mostly from the height of the E region, is most apparent at frequencies up to about 60 or 70 MHz. This type of forward scatter may be usable out to about 1200 miles. Ionospheric scatter propagation is most noticeable on frequencies above the MUF.

Figure 9-16—The relationship between smoothed Zurich sunspot numbers and 2800-MHz solar flux values.

Figure 9-17—Meteors passing through the atmosphere create trails of ionized gas. These ionized trails can be used for short-duration communications.

Another means of ionospheric scatter is provided by meteors entering Earth's atmosphere. As the meteor passes through the ionosphere, a trail of ionized particles forms. These particles can scatter radio energy. See Figure 9-17. This ionization is short-lived and can show up as short bursts of signals with little communication value. There may be longer periods of usable signal levels, lasting up to a minute or more. Meteor scatter is most common between midnight and dawn. It peaks between 5 and 7 AM local time. Meteor scatter is an interesting method of amateur communication at 21 MHz or higher, especially during periods of low solar activity.

Scatter

You can observe a complex form of scatter when you are working very near the MUF. The transmitted wave is refracted back to earth at some distant point. This may be an ocean area or land mass. A portion of the transmitted signal reflects back into the ionosphere. Some of this signal comes toward the transmitting station. The reflected wave helps fill in the skip zone, as shown in Figure 9-18.

Scatter signals are generally weak and subject to distortion, because the signal may arrive at the receiver from many different directions. Scatter (sometimes called *backscatter*) is usable from just beyond the local range out to several hundred miles. Under ideal conditions, scatter is possible over 3000 miles or more. The term "sidescatter" is more descriptive of what probably happens on such long paths, however.

[Now study exam questions T3C01, T3C02, T3C09 and T3C11 in Chapter 13. Review the material in this section as needed.]

LINE-OF-SIGHT PROPAGATION

In the VHF and UHF range (above 144 MHz), most propagation is by *space wave*. Figure 9-19 shows how space waves travel within visible range. Space waves are made up of a **direct wave**, and usually one or more *reflected waves*. The direct wave and reflected wave arrive at the receiving station slightly out of phase. They may cancel or reinforce each other to varying degrees. This depends on the actual path of the reflected wave(s). Effects are most pronounced when either, or both, stations are mobile. When one or both stations are moving, the paths are constantly changing. The waves may alternately reinforce and cancel each other. This causes a rapid fluttering sound (called "picket fencing").

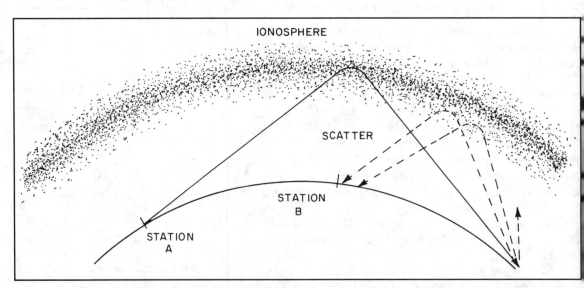

Figure 9-18—When radio waves strike the ground after passing through the ionosphere, some of the signal may reflect back into the ionosphere. If some of the signal is reflected from the ionosphere toward the transmitting station, the energy may be scattered back into the skip zone.

Figure 9-19—Line-of-sight propagation is accomplished by means of the space wave, a combination of a direct ray and one or more reflected rays.

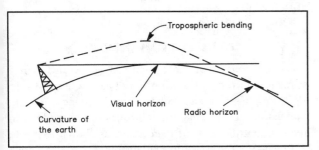

Figure 9-20—Under normal conditions, tropospheric bending causes radio waves to be returned to Earth beyond the visible horizon.

Figure 9-21—When a cool air mass is overrun by a mass of warmer air, a "duct" is formed, allowing VHF radio signals to travel great distances with little attenuation.

TROPOSPHERIC BENDING AND DUCTING

The **troposphere** consists of atmospheric regions close to the earth's surface. **Tropospheric bending** is evident over a wide range of frequencies. It is most useful in the VHF/UHF region, especially at 144 MHz and above. Instead of gradual changes in air temperature, pressure and humidity, distinct regions may form in the troposphere. Adjacent regions having significantly different densities will bend radio waves passing between regions. (In the same way, light is bent when it passes from air into water. That's why a spoon in a glass of water seems to bend when it enters the water.)

The **visible horizon** is the most distant point one can see. It is limited by the height of the observer above ground. Slight bending of radio waves occurs in the troposphere. This will cause signals to return to earth somewhat beyond the geometric horizon. See Figure 9-20.

Under normal conditions, the temperature of the air gradually decreases with increasing height above ground. When there's a stable high-pressure system, a mass of warm air may overrun cold air. When warm air covers cold air, we have a **temperature inversion**. Radio waves can be trapped below the warm air mass. They can travel great distances with little loss. The area between the earth and the warm air mass is known as a *duct*. See Figure 9-21.

The term for radio-wave propagation through a duct is **tropospheric ducting**. As with tropospheric bending, we find ducting primarily at VHF, 144 MHz and higher. Ducts usually form over water, though they can form over land as well. The lowest usable frequency for duct propagation depends on two factors: the depth of the duct and the amount the refractive index changes at the air-mass boundary.

[At this point, turn to Chapter 13 and study questions T3C03 through T3C08 and question T3C10. Review any material necessary.]

BANDWIDTH

The amount of space in the radio-frequency spectrum that a signal occupies is called its *bandwidth*. The bandwidth of a transmission is determined by the information rate. Thus, a pure, continuous, unmodulated carrier has a very small bandwidth with no sidebands. A television transmission, which contains a great deal of information, is several megahertz wide.

RECEIVER BANDWIDTH

Receiver bandwidth determines how well you can receive one signal in the presence of another signal that is very close in frequency. The enjoyment you'll experience will depend greatly on how well you can isolate the signal you are receiving from all the others nearby.

Bandwidth is a measure of *selectivity*: how wide a range of frequencies is received with the receiver tuned to one frequency. For example, if you can hear signals as much as 3 kHz above and 3 kHz below the frequency to which you are tuned, your receiver has a bandwidth of at least 6 kHz. If you cannot hear signals more than 200 Hz above or below the frequency to which you are tuned, the bandwidth is only 400 Hz. The narrower the bandwidth, the greater the selectivity and the easier it is to copy one signal when there is another one close by in frequency.

Selectivity is determined by special *intermediate-*

frequency (IF) filters built into the receiver. Some receivers have several filters so you can choose different bandwidths. They are necessary because some **emission types** occupy a wider frequency range than others. A 250-Hz-bandwidth filter is excellent for separating CW signals on a crowded band, but it's useless for listening to SSB, AM or FM transmissions. A wider filter is needed to allow all the transmitted information to reach the detector.

Most receivers designed for single-sideband voice operation come standard with a filter selectivity of around 2.8 kHz. This is ideal for SSB, which usually has a bandwidth between 2 and 3 kHz. Although it is also usable on CW, a narrower bandwidth is needed to prevent adjacent CW signals from getting through at the same time.

As we have seen in these examples, radioteletype emissions are wider than CW signals, and SSB signals are even wider than that. Next, we will consider FM and PM, which can occupy even more bandwidth.

BANDWIDTH IN FM AND PM

Frequency deviation is the instantaneous change in frequency for a given signal. The frequency swings just as far in both directions, so the total frequency swing is equal to twice the deviation. In addition, there are sidebands that increase the bandwidth still further. A good estimate of the bandwidth is twice the maximum frequency deviation plus the maximum modulating audio frequency:

$$Bw = 2 \times (D + M)$$

where
  Bw = bandwidth
  D = maximum frequency deviation
  M = maximum modulating audio frequency

An FM transmitter using 5-kHz deviation and a maximum audio frequency of 3 kHz uses a total bandwidth of about 16 kHz. The actual bandwidth of a typical FM signal is somewhat greater than this, but it is a good approximation. The relatively large bandwidth of an FM signal results in good-quality audio that can be understood even if a signal is weak. This is why FM is used for local VHF and UHF voice communications.

FREQUENCY DEVIATION

With direct FM, frequency deviation is proportional to the amplitude of the modulating signal. With phase modulation, the frequency deviation is proportional to both the amplitude and the frequency of the modulating signal. Frequency deviation is greater for

higher audio frequencies. This is actually a benefit, rather than a drawback, of phase modulation. In radio communications, speech frequencies between 300 and 3000 Hz need to be reproduced for good intelligibility. In the human voice, however, speech sounds between 2000 and 3000 Hz are not as strong as at lower frequencies. Something must be done to increase the amplitude of these frequencies when a direct FM transmitter is used. A circuit called a preemphasis network amplifies the sounds between 2000 and 3000 Hz. With phase modulation, the preemphasis network is not required, because the deviation already increases with increasing audio input frequency.

MODULATION INDEX

The *sidebands* that occur from FM or PM differ from those resulting from AM. With AM, only a single set of sidebands is produced for each modulating frequency. FM and PM sidebands occur at integral multiples of the modulating frequency on either side of the carrier, as shown in Figure 9-22. Because of these multiple sidebands, FM or PM signals inherently occupy a greater bandwidth. The additional sidebands depend on the relationship between the modulating frequency and the frequency deviation. The ratio between the peak carrier frequency deviation and the audio modulating frequency is called the *modulation index*.

Given a constant input level to the modulator, in phase modulation, the modulation index is constant regardless of the modulating frequency. For an FM signal, the modulation index varies with the modulating frequency. In an FM system, the ratio of the maximum carrier-frequency deviation to the highest modulating frequency is called the *deviation ratio*. The modulation index is a variable that depends on a set of operating conditions, but the deviation ratio is a constant.

Figure 9-22—When an RF carrier is frequency-modulated by an audio signal, multiple sidebands are produced. FM sidebands are spaced above and below the center frequency of the carrier. This drawing shows a group of sidebands resulting from the application of a 2-kHz tone to the transmitter modulator. The sidebands farthest from the center frequency are the lowest in amplitude. The sidebands nearest to center frequency can be considered the significant sidebands, their amplitudes being fairly large.

The deviation ratio for narrow-band FM is 5000 Hz (maximum deviation) divided by 3000 Hz (maximum modulation frequency) or 1.67.

Frequency multiplication offers a means for obtaining any desired amount of frequency deviation, whether or not the modulator is capable of that much deviation. Over-deviation (overmodulation) of an FM transmitter causes **splatter**, out-of-channel emissions that can cause interference to adjacent frequencies. A frequency-multiplier stage is an amplifier that produces harmonics of the input signal. By using a filter to select the desired harmonic, the proper output frequency is obtained. A frequency multiplier is not the same as a mixer stage, which is used to obtain the designed output frequency in an SSB or CW rig. A mixer shifts, or translates, an oscillator frequency to some new frequency.

[Now turn to Chapter 13 and look at questions T8B05 through T8B11. Review this section if you have difficulty.]

KEY WORDS

Chirp—A slight shift in transmitter frequency each time you key the transmitter.

Front-end overload—Interference to a receiver caused by a strong signal that overpowers the receiver RF amplifier ("front end"). See also **receiver overload**.

Harmonics—Signals from a transmitter or oscillator occurring on whole-number multiples of the desired operating frequency.

High-pass filter—A filter designed to pass high-frequency signals, while blocking lower-frequency signals.

Key click—A click or thump at the beginning or end of a CW signal.

Key-click filter—A circuit in a transmitter that reduces or eliminates **key clicks**.

Low-pass filter—A filter designed to pass low-frequency signals, while blocking higher-frequency signals.

Radio-frequency interference (RFI)—Disturbance to electronic equipment caused by radio-frequency signals.

Receiver overload—Interference to a receiver caused by a strong RF signal that forces its way into the equipment. A signal that overloads the receiver RF amplifier (front end) causes **front-end overload**. Receiver overload is sometimes called **RF overload**.

RF overload—Another term for **receiver overload**.

Spurious emissions—Signals from a transmitter on frequencies other than the operating frequency.

Superimposed hum—A low-pitched buzz or hum on a radio signal.

Television interference (TVI)—Interruption of television reception caused by another signal.

But What if I Have Trouble?

T his chapter describes some common problems you may encounter when you get on the air. Although you may never have any trouble, you should be aware of possible problems and know how to cure them. We show you how to identify interference to consumer electronics equipment. We also suggest ways to cure interference problems caused by overload and harmonics. Finally, we discuss some common transmitter problems: key clicks, chirp and superimposed hum. When you finish this chapter you'll be able to identify these problems and describe their cures.

SPURIOUS SIGNALS

An ideal transmitter emits a signal only on the operating frequency and nowhere else. Real-world transmitters radiate undesired signals, or **spurious emissions**, as well. Using good design and construction practices, manufacturers and home builders can reduce spurious emissions so they cause no problems. These undesired signals can, however, cause you some problems. Spurious emissions fall into two general categories: harmonics and parasitic oscillations. On the HF bands, you are more likely to have harmonic problems. Parasitic oscillations can occur in equipment designed for HF or VHF/ UHF operation.

HARMONICS

Harmonics are whole-number multiples of a given frequency. For example, the second harmonic of 100 Hz is

Figure 10-1—Harmonics are signals that appear at whole number multiples of the resonant, or fundamental, frequency. This drawing represents a spectrum analyzer display screen, and shows a 2-MHz signal and some of its harmonics. A spectrum analyzer is an instrument that allows you to look at energy radiated over a wide range of frequencies. Here, the analyzer is adjusted to display an RF energy between 1 MHz and 11 MHz. Each vertical line in the background grid denotes an increment of 1 MHz. The first thick black vertical "pip" represents energy from the fundamental signal of a 2-MHz oscillator. The next pip, two vertical divisions later, is the second harmonic at 4 MHz (twice the fundamental frequency). The pip at the center of the photo is the third harmonic at 6 MHz (three times the fundamental frequency). This figure shows the second, third, fourth and fifth harmonics.

harmonic energy. Fortunately, that energy is so small that it is not likely to cause problems.

Good transmitter engineering calls for tuned circuits between stages in amateur transmitters. These circuits reduce or eliminate spurious signals such as harmonics. The tuned circuits allow signals at the desired frequency to pass, but they *attenuate* (reduce) harmonics.

Transmitters with vacuum tubes in the final amplifier often use a *pi-network* impedance-matching circuit at the final amplifier output. The plate-tuning and antenna-loading capacitors in the pi network are adjustable. When properly tuned, the pi network allows maximum power to pass at the frequency of operation but reduces harmonics.

Transmitters with solid-state final amplifiers usually operate over a wide range of frequencies with no external tuning controls. In these transmitters, the band switch activates a separate tuned *band-pass filter* for each band. As the name implies, this filter allows energy to pass at the desired frequency and attenuates spurious signals above and below.

How can you be sure that your transmitter does not generate excessive harmonics? The FCC requires all transmitter manufacturers to prove that their equipment complies with its regulations. This means that commercially manufactured equipment usually produces clean signals. If you build a transmitter from a magazine or book article, check for information on harmonic radiation. ARRL requires that all transmitter projects published in *QST* and our technical books meet the FCC specifications for commercially manufactured equipment.

PARASITIC OSCILLATIONS

Parasitic oscillations in the final amplifier occur on frequencies unrelated to those for which the amplifier is designed. These spurious signals, called *parasitics*, are not harmonics. They result from an unintentional tuned feedback path in the final-amplifier circuit. Parasitics happen because an amplifier tube or semiconductor often works well at frequencies much higher than those it is used on. Parasitic oscillations usually take place in the VHF/UHF range. Most transmitters have special tuned circuits built into the final-amplifier circuit to prevent parasitic oscillations. They are called *parasitic suppressors*. Figure 10-2 shows examples of parasitic suppressors connected in amplifier circuits.

A defective component in your VHF or UHF transceiver may cause it to radiate spurious signals. These signals may interfere with other hams on the same bands, or even users of other bands (including public safety and commercial bands).

NEUTRALIZATION

Spurious signals may result from improper *neutralization* of the transmitter. An oscillator is an amplifier that has some of the output signal fed back to the input. This *positive feedback* is fine for oscillators, but we don't want positive feedback in an amplifier. Positive feedback in an audio amplifier for example, causes a howling noise. You may have

200 Hz. The fifth harmonic of 100 Hz is 500 Hz. Every oscillator generates harmonics in addition to a signal at its *fundamental frequency*. For example, consider an oscillator tuned to 7.125 MHz in the 40-meter band. It also generates signals at 14.250 MHz (second harmonic), 21.375 MHz (third harmonic), 28.500 MHz (fourth harmonic), and so on. Figure 10-1 shows a spectrum-analyzer display of the output of an oscillator that has many harmonics.

Harmonics can interfere with other amateurs or other users of the radio spectrum. The second through fourth harmonics of a 40-meter transmitter fall in the 20, 15 and 10-meter amateur bands. Imagine the interference (QRM) that would result if everyone transmitted two, four, six or more harmonics in addition to the desired signal!

To prevent this chaos, FCC regulations specify limits for harmonic and other spurious radiation. For example, if you are operating a 100-W-output transmitter, your harmonic signals can total no more than 10 milliwatts. As you can see, a transmitter that complies with the rules still generates some

Figure 10-2—A shows the schematic diagram of a vacuum tube amplifier. The coil and resistor combination Z1 is a parasitic suppressor. B shows the schematic diagram of a transistor power amplifier. Z2 and Z3 are ferrite beads that also function as parasitic suppressors.

heard this noise from a public-address system with the gain set too high.

We don't want positive feedback in radio amplifiers, either. In an improperly neutralized amplifier, some of the output feeds back to the input and causes the amplifier to oscillate. The signal generated may be transmitted. If the signal is strong enough, it may generate more power than the transmitter was designed to handle, and may damage the transmitter. Neutralization eliminates or neutralizes this positive feedback by applying negative feedback. If you suspect that you need to neutralize your transmitter, you may have to adjust an internal control. Figure 10-3 shows a vacuum tube amplifier circuit with a neutralizing capacitor. The capacitor sometimes is nothing more than a piece of wire running from under the chassis, next to the tube. The tube itself makes up half the capacitor.

An amplifier that requires neutralization oscillates on the frequency band it is designed to amplify. Compare this with parasitics, which happen at much higher frequencies. You may need to operate your transmitter to check for proper neutralization, and for other transmitter adjustments and tests. You should do this without transmitting a signal on the air. The best way to do so is to use a *dummy load*, described in Chapter 8.

Modern solid-state transmitters don't normally require neutralization. If your transmitter or transceiver has a vacuum-tube final amplifier, you may need to adjust the neutralization if you change the tubes. Often, even replacing

Figure 10-3—Schematic diagram of a vacuum tube amplifier. C1 is a neutralizing capacitor. It applies negative feedback to the amplifier input and prevents oscillation.

tubes doesn't upset the adjustment. The procedure is not the same for all transmitters. Follow the instructions in your transceiver or transmitter manual, if you have it. Otherwise, ask your instructor or a local ham to help you.

Remember that you have exposed the high-voltage components of your rig when you are neutralizing it. Be extra careful to avoid the possibility of electric shock.

[Now turn to Chapter 12 and study questions N8B01, N8B02, N8B03, N8B04 and N8B05. Review this section if you have any difficulty with those questions.]

INTERFERENCE WITH OTHER SERVICES

Radio frequency interference (RFI) has given radio amateurs headaches for years. It can occur whenever an electronic device is surrounded by a field of RF energy. Your rig

emits RF energy each time you transmit. This RF energy may interfere with your own or your neighbor's television set (causing **television interference—TVI**). You may also have

Figure 10-4—The boom in home electronic devices continues. Many amateur antennas are close to a growing number of home-entertainment devices. The result: an RFI problem that shows little sign of disappearing.

Figure 10-5—A high-pass filter can prevent fundamental energy from an amateur signal from entering a television set. This type of high-pass filter goes in the 300-ohm feed line that connects the television with the antenna.

problems with a stereo system, electronic organ, video cassette recorder, telephone or any other piece of consumer electronic equipment.

If you have a very visible antenna in your yard, your neighbors may blame you for any interference they experience, even when you're not on the air! If you have any problems with interference to your own equipment, it's a good bet that your neighbors do too. On the other hand, if you can show your neighbors that you don't interfere with *your* television, they may be more open to your suggestions for curing problems.

So what should you do if someone complains of interference? First, make sure that your equipment is operating properly. If the complaint is TVI, check for interference to your own TV. If you see it, stop operating and cure the problem before you go back on the air.

Even if you don't interfere with your own TV, don't stop there. Simply telling your neighbors you're not at fault can cause even more problems. Try to work with your neighbors to determine if your rig is actually causing the interference. If so, try to help solve the problem. A more-experienced ham can be a great help. If you don't know any other hams in your area, write to ARRL HQ. We'll try to help you find a knowledgeable local ham.

[Now go back to Chapter 12 and study question N4D11. Review this section if you have any problems.]

RECEIVER OVERLOAD

Receiver overload is a common type of TV and FM-broadcast interference. It happens most often to consumer electronic equipment near an amateur station or other transmitter. When the RF signal (at the fundamental frequency) enters the receiver, it overloads one or more circuits. The receiver front end (first RF amplifier stage after the antenna) is most commonly affected. For this reason, we sometimes call this interference **front-end overload** or **RF overload**.

A strong enough RF field may produce spurious signals in the receiver, which cause the interference. Receiver overload interference may occur in your neighbor's house or just your own. Receiver overload can be caused by transmitters operating on any frequency. It is the most-common interference problem caused by VHF and UHF transmitters.

Receiver overload usually has a dramatic effect on the television picture. Whenever you key your transmitter, the picture may be completely wiped out. The screen may go black, or it might just become light with traces of color. The sound (audio) will probably be affected also. In an FM receiver, the audio may be blocked each time you transmit. More often than not, overload affects only TV channels 2 through 13. In cases of severe interference, however, it may also affect the UHF channels.

The objective in curing receiver overload is to prevent the amateur signal from entering the front end of the entertainment receiver. The first step is to have the equipment owner or a qualified service technician install a **high-pass filter**. See Figure 10-5.

Install the filter at the TV or FM receiver input. The best location is where the antenna feed line connects to the TV or

FM tuner. It is *not* a good idea for an Amateur Radio operator to install a filter on a neighbor's entertainment equipment. Only the owner or a qualified technician should install the filter. If you install the filter, you might later be blamed for other problems with the TV set. A high-pass filter is a tuned circuit that passes high frequencies (TV channels start at 54 MHz). The filter blocks low frequencies (the HF amateur bands are in the range of 1.8-30 MHz).

[Now study questions N4D01, N4D02, N4D03 and N4D05 in Chapter 12. Review this section as needed.]

HARMONIC INTERFERENCE

Another problem for hams is harmonic interference to entertainment equipment. As we learned before, harmonics are multiples of a given frequency. Your HF transmitter radiates undesired harmonics along with your signal. Your transmitting frequency is much lower than the TV or FM channels. Some harmonics will fall within the home entertainment bands, however.

The entertainment receiver cannot distinguish between the TV or FM signals (desired signals) and your harmonics (undesirable intruders). If your harmonics are strong enough, they can seriously interfere with the received signal. Harmonic interference shows up as a crosshatch or a herringbone pattern on the TV screen. See Figure 10-6.

Unlike receiver overload, harmonic interference seldom affects all channels. Rather, it may bother the one channel that has a harmonic relationship to the band you're on. Generally, harmonics from amateur transmitters operating below 30 MHz affect the lower TV channels (2 through 6). Ten-meter transmitters usually bother channels 2 and 6, and channels 3 and 6 experience trouble from 15-meter transmitters.

Harmonic interference must be cured at your transmitter. As a licensed amateur, you must take steps to see that harmonics from your transmitter do not interfere with other services. All harmonics generated by your transmitter must be attenuated well below the strength of the fundamental frequency. If harmonics from your transmitting equipment exceed these limits, you are at fault.

In this section we will discuss some of the several possible cures for harmonic interference. Try each step in order and the chances are good that your problem will be solved quickly.

The first step you should take is to install a **low-pass filter** like the one shown in Figure 10-7. The filter goes in the transmission line between your transmitter and antenna or antenna tuner. As the name implies, a low-pass filter is the opposite of a high-pass filter. A low-pass filter allows RF energy in the amateur bands to pass freely. It blocks very high frequency harmonics that can fall in the TV and FM bands. Low-pass filters usually have a specified cutoff frequency, often 40 MHz, above which they severely attenuate the passage of RF energy.

Even if your transmitter is working well within FCC specifications, you may need additional attenuation to reduce harmonics. Remember, your goal is to eliminate inter-

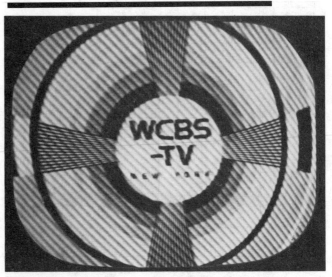

Figure 10-6—Harmonics radiated from an amateur transmitter may cause "crosshatching."

Figure 10-7—A low-pass filter. When connected in the coaxial cable feed line between an amateur transmitter and the antenna, a low-pass filter can reduce the strength of transmitted harmonics.

ference. Good-quality low-pass filters often attenuate signals falling in the entertainment bands by 70 or 80 dB. This is significantly better than the 40 to 50 dB typical of amateur transmitters. A decibel (dB) is a number (the logarithm of a ratio) used to describe how effective the filter is. Larger numbers indicate better filtering. Figure 10-8 shows the output of a transmitter before and after filtering.

Another source of interference is RF energy from your transmitter that enters the ac power lines. The *ac power-line filter* is another kind of low-pass filter. It prevents RF energy from entering the ac line and radiating from power lines inside and near your house.

Figure 10-8—A shows a spectrum-analyzer display of the signals emitted from an amateur transmitter. The pip at the left of the display (the one that extends to the top horizontal line) is the fundamental. All other pips represent harmonics. This particular transmitter generates several harmonics. On an analyzer display, stronger signals create taller pips. Here, the harmonic signals are quite strong. In fact, the third harmonic is about one-tenth as strong as the fundamental. The fundamental signal is 100 W, so the transmitter is radiating a potent 10-W signal at the third harmonic. Most of these harmonics will cause interference to other services. B shows the output from the same transmitter, operating at the same power level at the same frequency, after installation of a low-pass filter between the transmitter and the analyzer. The harmonics have all but disappeared from the display. The stronger of the two remaining harmonics is only about 100 microwatts—weak enough that it is unlikely to cause interference.

Multiband Antennas

You can also run into trouble if you use a multiband antenna. If your antenna works on two or three different bands, it will radiate any harmonics present on those frequencies. After all, we *want* the antenna to radiate energy at a given frequency. It cannot tell the difference between desired signal energy and unwanted harmonic energy. This problem does not usually affect home entertainment equipment. It may cause interference to other amateurs or to other radio services operating near the amateur bands, however.

For example, a multiband dipole antenna that covers 80 and 40 meters may radiate 40-meter energy while you operate on 80 meters. The second harmonic of a 3.7-MHz signal falls above the 40-meter amateur band, at 7.4 MHz. If your transmitter is free of excessive harmonic output, you will probably not have a problem. (Note: Older, tube-type equipment can sometimes radiate excessive harmonics if it isn't "tuned up" properly. If you use a rig with a "plate tuning"

control, be sure to adjust it according to the manufacturer's instructions.)

Proper shielding and grounding are essential to reduce harmonic and other spurious radiation. The only place you want RF to leave your transmitter is through the antenna connection. Your transmitter must be fully enclosed in a metal cabinet. The various shields that make up the metal cabinet should be securely screwed or welded together at the seams. You must also connect the transmitter to a good earth ground connection.

Remember: A low-pass filter will only block harmonics from reaching your antenna. It will do nothing for a poorly shielded transmitter that leaks stray RF from places other than the antenna connector. Figure 10-9 summarizes the steps you can take to reduce harmonic radiation from your station.

[Now turn to Chapter 12 and study questions N4D04, N4D06, N4D07, N4D08, N4D09, N4D10 and N8B06. Review this section as necessary.]

SIGNAL PURITY

As a licensed Amateur Radio operator, you're responsible for the quality of the signal transmitted from your station. The rules require your transmitted signal to be stable in

frequency and pure in tone or modulation. If your signal is not "clean," it is unpleasant to listen to. It may also cause interference to others using the band or other services.

Figure 10-9—Suggested techniques for filtering harmonic energy from the leads of an amateur transmitter.

Unfortunately, some problems can creep up and give you a dirty signal. Three of the most common problems with CW signals are **key clicks**, **chirp** and **superimposed hum**. Superimposed hum is sometimes a problem with voice signals, also.

KEY CLICKS

As you listen on the air, you will notice that some signals have a click or thump on *make* (the instant the key contacts close) or *break* (the instant they open) or both. This **key click** is more than just annoying; it can interfere with other stations.

If you use an oscilloscope (a type of test equipment resembling a TV set) to monitor a CW signal with key clicks, you will see excessively square CW keyed waveforms. Imagine a good, stable transmitter sending a CW signal on 3.720 MHz. What happens when we turn the transmitter on and off with a telegraph key? You'd think that the only frequency the output energy could have would be 3.720 MHz. If you turn the transmitter output on and off rapidly, however, something else happens. Unwanted energy in the form of key clicks appears for several kilohertz on either side of the operating frequency. The transmitter creates these clicks during the instant that it is turned on and off.

When we say "keyed rapidly," we don't mean that we're sending fast (like 35 WPM CW). We mean that the transmit-ter goes from zero power to full power and back again very quickly and abruptly. We call the time it takes to go from no power to full power the *rise time*. We call the return trip from full power to zero power the *fall time*.

An oscilloscope display of a CW signal with short rise and fall times in shown in Figure 10-10A. This signal has key clicks. A click-free signal is shown in part B. Notice how the beginning and ending of this CW pulse are soft and round. Compare this with the square shoulders of the signal with clicks. Part C shows a string of dots and dashes that will generate interference from key clicks.

We use a **key-click filter** or *shaping filter* to eliminate this unwanted and bothersome energy. The filter makes the transmitter output increase to maximum more slowly on make, and fall from full output more slowly on break. The keying waveform is softer, limiting the output energy to a few hundred hertz on either side of the operating frequency.

CHIRP

Chirp is another common transmitter problem. Chirp occurs when the oscillator in your transmitter shifts frequency slightly whenever you close your telegraph key. The result is that other stations receive your transmitted signal as a chirping sound rather than as a pure tone. Your "dahdidahdit"

(A)

(B)

(C)

Figure 10-10—CW signals viewed on an oscilloscope. A shows a CW dot with no intentional wave shaping. This signal has key clicks. B shows a similar dot, shaped to eliminate key clicks. Notice how the signal at B builds and decays gradually. Compare this with the abrupt on-and-off characteristics at A. C shows a string of unshaped CW characters. Notice the sharp *spikes* at the beginning of each pulse. These spikes are audible in a receiver as key clicks. Such spikes can appear at either end of a CW character.

will sound like "whoopwhiwhoopwhip." It isn't very much fun to copy a chirpy CW signal!

Chirp usually happens when the oscillator power supply voltage changes as you transmit. Your transmitter may also chirp if the load on the oscillator changes when you transmit. If the supply voltage changes, you must improve the *voltage regulation*. With better regulation, the voltage won't shift when you key your transmitter.

If it's not a voltage problem, then what? Amplifier stages after the oscillator may be loading it down and pulling its frequency. You may need a better buffer (isolation) or driver stage between the oscillator and the next stage in your transmitter. Some oscillators are sensitive to temperature changes. If there is too much current through the frequency-determining components, their temperature may increase and the resonant frequency will change.

SUPERIMPOSED HUM

The last kind of signal problem we discuss is **superim-** **posed hum**. Power supplies contain filters that remove the ac present in the rectifier output. The result is a pure, filtered dc output. If a filter capacitor fails, then the filtering action will be incomplete.

If the filter doesn't work properly, ac will be present in the power supply output. This ac will also work its way into the transmitter output. Instead of hearing a pure tone or your voice, stations receiving your signal will hear a low-pitched hum as well. If enough ac is present, the signal will have a raspy tone. Your CW signal may even buzz as if you were keying an electric razor! To cure hum, check the power-supply filter circuit. A bad filter can cause superimposed hum.

[Now study questions N8A07, N8A08, N8A09 and N8A10 in Chapter 12. Review this section as necessary.]

HOW TO GET HELP

Transmitting a signal with chirp, key clicks, hum or spurious emissions is a violation of both the letter and spirit of the regulations governing Amateur Radio. The best way to find out if you have a bad-sounding signal is from the stations you work. If another operator tells you about a problem with your signal, don't be offended. He cares about the image of the Amateur Service and only wants to help you. You might never be aware of a problem otherwise.

Don't let this chapter scare you. Spurious signals, unwanted harmonics, poor neutralization, key clicks, parasitic oscillations, television interference (TVI) or other equipment difficulties can all be solved. Two ARRL publications will be helpful: *The ARRL Handbook for Radio Amateurs* and *Radio Frequency Interference*. Check with your Amateur Radio instructor or any experienced ham for help. There's no substitute for experience. Another ham may know just how to solve your problem and return you to the air as soon as possible. If you'd like to set up a station in an area where RFI may be a real problem, you'll want to read *Low Profile Amateur Radio*. This ARRL book tells you how to enjoy Amateur Radio with few problems from an apartment, condo, dorm room or car, or in the field.

KEY WORDS

A1A emission—The FCC emission type designator that describes international Morse code telegraphy (CW) communications without audio modulation of the carrier.

Bandwidth—A range of associated frequencies (measured in hertz). For example, bandwidth is used to describe the range of frequencies in the radio spectrum that a radio transmission occupies.

Code key—A device used as a switch to generate Morse code. Also called a **hand key**, a **straight key** or a **telegraph key**.

Code-practice oscillator—A device that produces an audio tone, used for learning the code.

CW (continuous wave)—The FCC emission type that describes international Morse code telegraphy communication without audio modulation of the carrier. Hams usually produce Morse code signals by interrupting the continuous-wave signal from a transmitter to form the dots and dashes.

Dash—The long sound used in Morse code. Pronounce this as "dah" when verbally sounding Morse code characters. Dashes are three times longer than dots.

Dot—The short sound used in Morse code. Pronounce this as "dit" when verbally sounding Morse code characters if the dot comes at the end of the character. If the dot comes at the beginning or in the middle of the character, pronounce it as "di."

Emission—RF signals transmitted from a radio station.

Emission privileges—Permissible types of transmitted signals.

Fist—The unique rhythm of an individual amateur's Morse code sending.

Hand key—Another name for a **code key**.

MCW—The FCC emission type that describes international Morse code telegraphy communication with audio modulation of the carrier.

Phone—Voice communications.

Q signals—Three-letter symbols beginning with "Q," used in amateur CW work to save time and for better comprehension.

Straight key—Another name for a **code key**.

Telegraph key—Another name for a **code key**.

Traffic net—An on-the-air meeting of amateurs, for the purpose of relaying messages.

W1AW—The Headquarters station of the American Radio Relay League. This station is a memorial to the League's cofounder, Hiram Percy Maxim. The station provides daily on-the-air code practice and bulletins of interest to hams.

Chapter 11

Morse Code—The Amateur's International Language

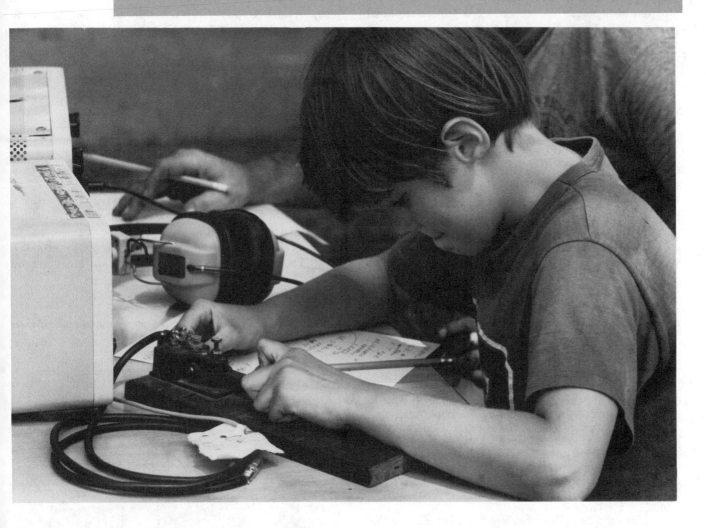

To qualify for a Novice license, you must pass a written exam about FCC Rules, electronics theory and basic amateur operating practices. In addition, you will have to show your ability to communicate using the international Morse code. The code exam is given at 5 words per minute (WPM).

To earn the Technician license, you do not have to pass a Morse code exam. That license does not include privileges on the amateur high-frequency bands, which provide direct worldwide communications. As a Technician class licensee, you can earn those privileges, however. To join the excitement of communicating with other Amateur Radio operators around the world, you simply pass the 5-WPM Morse code exam.

This book won't teach you the Morse code. In fact, you won't even find a copy of the Morse code printed in this book. That's because Morse code is best learned by sound, not by sight. We don't want to confuse you with printed dots and dashes. That will only slow you down later.

As we said, the code is best learned by listening to sounds, and for that you'll need either cassette tapes or a computer program (or both). The ARRL offers a code-teaching package of cassette tapes, called *Your Introduction to Morse Code*. That package contains two 90-minute cassette tapes to teach you all of the code characters required by the FCC. Practice text at 5 WPM ensures that you are prepared for the 5-WPM code exam. We also have sets of cassette tapes to help you increase your code speed right up to

22 WPM, to help you prepare for the 20-WPM Amateur Extra code exam!

The ARRL also offers two computer programs to teach Morse code. For the IBM PC and compatible computers, the ARRL sells the GGTE *Morse Tutor* program, which is an excellent Morse-code training program. You can tell the computer what code speed you want it to use for the characters and words it sends. *Morse Tutor* drills you on individual characters and combinations of characters. You can practice with a mixture of random characters or with words. The program also includes a random QSO generator, which produces an unlimited variety of Amateur-Radio-type con-

versations. These practice QSOs are very similar to the code exams.

The ARRL also sells the GGTE *Advanced Morse Tutor*. The advanced version of the program includes all the features of the regular *Morse Tutor* and more. The main difference is that the advanced version will send code from an ASCII text file you create. This version also includes some additional text analysis and reporting features.

The cassette tapes and both programs are available directly from ARRL Headquarters or from the many dealers who sell ARRL publications. See the advertising section at the back of this book for an ARRL publications order form.

CODE IS FUN!

Using the code is an exciting way to communicate. Many long-time hams beam with pride when they proclaim that they don't even own a microphone! You are fluent in another complete language when you know the code. You can chat with hams from all around the world using this common language. With the practice you will gain by making on-the-air contacts, your speed will increase quickly. With more advanced licenses, you'll be operating on more portions of the

bands, increasing your realm of contacts.

Amateur Radio operators must know the international Morse code if their license permits them to transmit on frequencies below 30 MHz. An international treaty sets this rule, and the United States follows the treaty. Because of this, the Federal Communications Commission requires you to pass a code test to earn your Novice license.

There's a lot more to the code than just satisfying the

A FAMILY OF HAMS

The Dalton family of Independence, Missouri is a special group of hams. David (NØLOG) and Elaine (NØNLF) began to study for Amateur Radio licenses in 1989. Two of their children, Sarah (10 at the time) and Daniel (then 8) also became interested when they saw mom and dad studying. Both children passed their Novice exams in January 1990.

Encouraged by the excitement of the Novice bands and the interest of other hams in their local radio club, both children upgraded to Technician in May. Less than a year after first being licensed, Daniel (NØMAT) and Sarah (NØMAS) upgraded to General. David has upgraded to Amateur Extra and Elaine to General.

Sarah and Daniel have become very excited about ham radio. They have taken their hand-held 2-meter radios to school and demonstrated them to their classes, they attend club meetings and enjoy participating in nets. Daniel took his hand-held radio along when he went out selling popcorn for his Cub Scout Pack and on other Webelos Scout activities. Now he takes it on Boy Scout campouts. Both children carry their radios while on their newspaper routes. This may be Elaine's favorite aspect of ham radio—keeping in touch with her busy family. The family has helped their local radio club provide public-service communications at several local events.

James, another member of the Dalton family, began to study Morse code at the age of 6, and learned over half the Morse code characters after only a few weeks of studying. James (KBØJPG) passed his Novice exam before he completed the first grade!

terms of a treaty, however. Morse code goes back to the very beginning of radio, and is still one of the most effective radio-communication methods. We send Morse code by interrupting the **continuous-wave** signal generated by a transmitter, and so we call it **CW** for short.

For one thing, it takes far less power to establish reliable communications with CW than it does with voice (**phone**). On phone, we sometimes need high power and elaborate antennas to communicate with distant stations, or *DX* in the ham's lingo. On CW, less power and more modest stations will provide the same contacts. Finally, there is great satisfaction in being able to communicate using Morse code. This is similar to the satisfaction you might feel from using any acquired skill.

CONSERVE TIME AND SPECTRUM SPACE

Another advantage of CW over phone is its very narrow **bandwidth**. Morse code makes efficient use of spectrum space. The group of frequencies where hams operate, the *ham bands*, are narrow portions of the whole spectrum. Many stations use the bands, and because they are so crowded, interference is sometimes a problem. A CW signal occupies only about one-tenth the bandwidth of a phone signal. This means 10 CW signals can fit into the space taken up by one phone signal.

Over the years, radiotelegraph operators have developed a vocabulary of three-letter **Q signals**, which other radio-telegraphers throughout the world understand. For example, the Q signal *QRM* means "you are being interfered with." Just imagine how hard it would be to communicate that thought to someone who didn't understand one word of English. (Chapter 9 includes a list of common Q signals.)

Another advantage to using Q signals is speed. It's much faster to send three letters than to spell out each word. That's why you'll use these Q signals even when you're chatting by code with another English-speaking ham. Speed of transmission is also the reason radiotelegraphers use a code "shorthand." For example, to acknowledge that you heard what was transmitted to you, send the letter R. This means "I have received your transmission okay."

Standard Q signals and shorthand abbreviations reduce the total time necessary to send a message. When radio conditions are poor and signals are marginal, a short message is much more likely to get through.

Many hams prefer to use CW in **traffic nets**. (A net is a regular on-the-air meeting of a group of hams.) Using these nets, hams send messages across the country for just about anyone. When they send messages using CW, there is no confusion about the spelling of names such as Lee, Lea or Leigh.

CW: SOMETIMES THE ONLY CHOICE

When WA6INJ's jeep went over a cliff in a February snowstorm, he was able to call for help using his mobile rig. This worked well at first, but as the search for him continued, he became unable to speak. The nearly frozen man managed to tap out Morse code signals with his microphone push-to-talk button. That was all his rescuers had to work with to locate him. Morse code saved this ham's life.

For some types of transmissions, CW (also called type **A1A emission** by the FCC) is the only available choice. As a Novice, CW is the only **emission privilege** you may use on the 80, 40 and 15-meter amateur bands. (One exception to this rule is if you are in Alaska, and involved in some emergency communications. In that case you may use single sideband [J3E or R3E] on 5167.5 kHz.) CW is the only mode that may be used on *all* amateur frequencies.

Some hams like to bounce VHF and UHF signals off the surface of the moon to another ham station on the earth. Because of its efficiency, hams use CW in most of this *moonbounce* work. They could use voice signals, but this increases the power and antenna-gain requirements quite a bit.

On some frequencies, amateurs communicate by bouncing their signals off an auroral curtain in the northern sky. (Stations in the Southern Hemisphere would use an auroral curtain in the southern sky.) Phone signals become so distorted in the process of reflecting off an auroral curtain that they are difficult or impossible to understand. CW is the most effective way to communicate using signals bounced off an aurora.

Figure 11-1—Even the moon isn't immune from ever-ambitious hams, and code is the most efficient type of signal to bounce off its surface and be reflected back to earth.

Morse Code From the Heart

The power of modern medicine kept one ham alive; the power of Morse code kept him in touch with loved ones.

Reprinted from July 1990 QST.

Have you ever felt Morse code rather than heard it?

Although I've been a licensed amateur for 34 years and have had many wonderful experiences in this hobby, none can compare to the one I'll never forget. It happened on Monday, August 26, 1985.

My husband Ralph, W8LCU, went to the hospital the Friday before for an ECG and heart catheterization. He was told that an immediate quadruple bypass was needed and he could not leave the hospital. Arrangements were made for the following Monday. Ralph had always been healthy and never showed an inkling of heart problems, so this came as quite a shock to us. The doctors told us that after his surgery he would not be speaking or doing much else, until perhaps the next day or until he was off the respirator, heart pump and everything else that goes with this type of surgery.

Over the weekend we had many things to discuss and Amateur Radio never entered my mind.

Monday evening, in the intensive-care unit (ICU) after surgery, I held Ralph's hand. He began tapping on my palm. I didn't think much of it, but his eyes opened and seemed to be telling me something. He moved his fingers to my wrist and suddenly, as clear as if I was hearing CW on the radio, came the letters P A (tears coming down my face) I N (pause) Is HELL. "Pain is hell."

Before I could say or do anything, four nurses were in the room. I told them what Ralph had just said to me in Morse code. I think they were ready to have me taken to the mental ward, but they gave him a shot. Then they stood by in disbelief as he tapped out a short message to our son and daughter and a final I LUV U before drifting off to sleep.

Meanwhile, in another hospital across the state, a friend of ours, Vanessa, KA8TRH, was back in her room after major surgery. Her first words to her husband, Chuck, N8EOJ, were "How is Ralph doing?" Sure enough, through the wonders of CW, hand-held transceivers, repeaters and many hams along the way, relays of Ralph's progress were sent to N8EOJ all day long. When Ralph tapped CW to me, within moments hams all across the state knew all was well.

The ICU and other units were buzzing for days as doctors came into Ralph's room to find out about this "new" form of communication. They told Ralph that they thought more people should know Morse code, themselves included. Of course, Ralph told them that with a little study and determination anyone can do it.

Needless to say, CW will always be important to me, either in an emergency or just for plain fun.

Today, Ralph is doing fine and I, while I was an Advanced at the time, have gone on to earn my Amateur Extra Class license.—*Donna Burch, W8QOY*

You will feel a special thrill and a warm satisfaction when you use the Morse code to communicate with someone. This feeling comes partly from sending messages to another part of the world without regard to language barriers. Sending and receiving Morse code is a skill that helps set amateurs apart from other people. It provides a common bond between amateurs worldwide.

GETTING ACQUAINTED WITH THE CODE

The basic element of a Morse code character is a **dot**. The length of the dot determines how long a **dash** should be. The dot length also determines the length of the spaces between elements, characters and words. Figure 11-2 shows the proper timing of each piece. Notice that a dash is three times as long as a dot. The time between dots and dashes in a character is equal to the length of a dot. The time between letters in a word is equal to three dot lengths and the space between words is seven dot lengths.

The lengths of Morse code characters are not all the same, of course. Samuel Finley Breese Morse (1791-1872) developed the system of dots and dashes in 1838. He assigned the shortest combinations to the most-used letters in plain-language text. The letter E has the shortest sound, because it is the most-used letter. T and I are the next most-used characters, and they are also short. Character lengths get longer for letters used less often.

An analysis of English plain-language text shows that

Figure 11-2—Whether you are a beginner or an expert, good sending depends on maintaining proper time ratios or "weight" between the dots, dashes and spaces, as shown.

the average word (including the space after the word) is 50 units long. By a unit we mean the time of a single dot or space between the parts of a character. The word PARIS is 50 units long, so we use it as a standard word to check code speed accurately. For example, to transmit at 5 words per minute (WPM), adjust your code-speed timing to send PARIS five times in one minute. To transmit at 10 WPM, adjust your timing to send PARIS 10 times in one minute.

As you can see, the correct dot length (and the length of dashes and spaces) changes for each code speed. As a result, the characters sound different when the speed changes. This leads to problems for a person learning the code. Also, at slower speeds, the characters seem long and drawn out. The slow pace encourages students to count dots and dashes, and to learn the code through this counting method.

Learning the code by counting dots and dashes introduces an extra translation in your brain. (Learning the code by memorizing Morse-code-character dot/dash patterns from a printed copy introduces a similar extra translation.) That extra translation may seem okay at first. As you try to increase your speed, however, you will soon find out what a problem it is. You won't be able to count the dots and dashes and then make the translation to a character fast enough!

People learning the Morse code with either of these methods often reach a learning plateau at about 10 words per minute. This is just below the 13-WPM speed required to upgrade to the General class license. That frustration (I just can't copy faster than 10 WPM!) is overcome by other methods, however.

LEARNING MORSE CODE

Many studies have been done, and various techniques tried, to teach Morse code. The method that has met with the most success is called the *Farnsworth method*. That is the method used on the ARRL code-teaching tapes included in the package, *Your Introduction to Morse Code*. With this technique, we send each character at a faster speed (we use an 18-WPM character speed at ARRL). At speeds in this range, the characters—and even some short words—begin to take on a distinctive rhythmic pattern.

With this faster character speed, we use longer spaces between characters and words to slow the overall code speed. The ARRL code-teaching tapes send code at an overall speed of 5 WPM. (You can still measure this timing by using the word PARIS, as described earlier.) Once you learn the character sounds, and can copy at 5 WPM, it will be easy to increase your speed. Just decrease the spaces between letters and words, and your code speed increases without changing the rhythmic pattern of the characters.

Learn to recognize that rhythmic pattern, and associate it directly with the character. You'll learn the code in the shortest possible time, and it will be much easier to increase your code speed. Decreasing the space between characters and words provides a natural progression to increase your code speed. The ARRL *Increasing Your Code Speed* series offers code-practice tapes from 5 to 10 WPM, 10 to 15 WPM and 15 to 22 WPM using this technique.

Morse code is a communications method that depends on sounds. To understand the communications you must hear the sounds and interpret their meaning. This is why most code-teaching methods repeat the sounds for you to listen to and associate the characters with the sounds. It is also why you will not find a copy of the Morse code dots and dashes printed in this book. Cassette tapes, computer programs and even classroom or individual practice with a **code-practice oscillator** and a **code key** all rely on sounds to teach you the Morse code characters. (A code key is sometimes called a **hand key**, **straight key** or a **telegraph key**.)

There are many computer programs around for teaching

Morse code, and some of them may be very helpful. The ARRL sells the GGTE programs *Morse Tutor* and *Advanced Morse Tutor* for the IBM PC and compatibles. Both are excellent teaching programs. Beware of programs written in BASIC, however. Because of the way BASIC operates, the timing is often not quite right. Also beware of programs written "to teach myself Morse code." While such writers may be excellent programmers, they are often unaware of the finer points of code training. Many people are using the Microsoft Windows operating environment on their IBM-compatible computers. Many Morse code programs do not run properly under multitasking software because of timing variations.

Several other techniques have also been successful for teaching Morse code. Some of these methods involve memorizing a printed copy of the code. There are even a few commercial packages that picture the dots and dashes of the code characters in various "creative" patterns to help you remember them. You can learn Morse code by following any of these methods. Most people will learn faster, and will be able to increase their code speed easier, however, by using a "listening" method. For those who seem unable to learn the Morse code using tapes or a computer program, one of these "printed" methods may prove helpful. Save that as a last resort, however. Practice faithfully with your cassette tapes or a computer program every day for at least three to four weeks. Then, if you have not learned many of the characters, you may want to try one of the visual, or printed, methods.

Morse Code: The Essential Language, by L. Peter Carron, Jr, W3DKV (published by ARRL) contains other suggestions for learning Morse code. That book also describes the history of Morse code. It includes several stories about lives saved because of emergency messages transmitted over the radio.

Sounding the Code

Some people find it helpful to say the sounds of Morse code characters, especially when they are first learning the code. Instead of saying the names of the Morse code elements, dot and dash, we use the sounds "dit" and "dah." If the dot is at the beginning or in the middle of the character, we sound it out as "di" instead of "dit."

Listen to the difference between the sounds you make saying the word "dit" and saying "dah." If you can tell the difference between those sounds, you have all the ability you need to learn the code. Being able to receive Morse code is really nothing more than being able to recognize a sound. Try it yourself. Say "didah." Now say "dididah." Can you hear the difference? Congratulations! You now know the sounds for the letters A and U, and are on your way to learning the Morse code.

Using this method, you hear the sound "didah" and associate that sound with the letter A. With practice, you'll learn all the sounds and associate them with the correct letters, numbers and punctuation.

The 26 letters of the alphabet and the numbers 0 through

MORSE CODE IS THE UNIVERSAL LANGUAGE

Clark J. Evans, Sr., WA4DLL and Lee Paulet, KB4FBX were operating the Amateur Radio display station at the Museum of Science and Industry in Tampa, Florida. A tour guide came into the station with two Japanese boys. When the woman introduced the boys to the operators, she said, "You won't be able to talk to these boys. They don't speak English at all." She also mentioned that one of the boys was a ham radio operator in Japan. That was the door opener.

Lee whistled CQ CQ CQ. That got the boy's attention, and the youngster was soon whistling back. Lee told the guide what the boy was saying. He was 13 years old from a town 40 miles north of Tokyo. He had been in the USA for two weeks.

The tour guide said, "You can't be communicating with him!" Lee's answer was, "Oh yes, we hams can. Morse code is a universal language!"

9 each have a different sound. There are also different sounds for the period, comma, question mark, double dash (=), fraction bar (/) and some procedural signals that hams use. The + sign, which hams call $\overline{AR}$, means "end of message." $\overline{SK}$ means "end of work," or "end of contact." Hams sometimes refer to the double dash as $\overline{BT}$ and the fraction bar as $\overline{DN}$. These two-letter combinations are written with a line over the letters to indicate that two letters are sent as one character to form these symbols. That's a total of 43 character sounds that you will have to learn for your 5-WPM Morse code exam. You'll learn the sounds of all these characters as you practice with your code cassettes or computer program.

Learning to Write

To learn the code, you train your hand to write a certain letter, number or punctuation mark whenever you hear a specific sound. You are forming a habit through your practice. After all, forming a habit is nothing more than doing something the same way time after time. Eventually, whenever you want to do that thing, you automatically do it the same way; it has become a habit.

You will need lots of practice writing the specific characters each time you hear a sound. Eventually you'll copy the code without thinking about it. In other words, you'll respond automatically to the sound by writing the corresponding character.

As you copy code sent at faster speeds, you may find that your ability to write the letters limits you. Practice writing the characters as quickly as possible. If you normally print, look for ways to avoid retracing lines. Don't allow yourself to be sloppy, though, because you may not be able to read your writing later.

Figure 11-3—A method of hand-printing letters and numbers with minimum effort and maximum speed.

Many people find that script, or cursive writing is faster than printing. Experiment with different writing methods, and find one that works for you. Then practice writing with that method so you don't have to think about forming the characters when you hear the sounds.

Figure 11-3 shows a systematic method of printing numbers and the letters of the alphabet. This system requires a minimum of pencil movement or retraced lines. You may find this technique helpful to increase your writing speed. Whatever method you choose, practice so it becomes second nature.

SOME STUDY SUGGESTIONS

The secret to easy and painless mastery of the Morse code is regular practice. Set aside two 15- to 30-minute periods every day to practice the code. If you try longer sessions, you may become over tired, and you will not learn as quickly. Likewise, if you only practice every other day or even less often, you will tend to forget more between practice sessions. It's a good idea to work your practice sessions into your daily routine. For instance, practice first thing in the morning and before dinner. Daily practice gives quick results.

Learn the sound of each letter. Morse code character elements sound like dits and dahs, so that's what we call them. Each character has its own pattern of dits and dahs, so learn to associate the sound of that pattern with the character. Don't try to remember how many dots and dashes make up each character. Practice until you automatically recognize each Morse code character.

Feel free to review. You're learning a new way of communicating, by using the Morse code. If you are having trouble with a particular character, spend some extra time with it. If you are using the cassette tapes from ARRL's *Your Introduction to Morse Code*, rewind the tape and play that

section again. If you are using one of the computer programs, spend some extra time drilling on the problem character. After you've listened to the practice on one character two or three times, however, go on to the next one. You will get more practice with the problem character later, because you will be constantly reviewing all of the characters learned so far. You can even come back to the problem character again in a later practice session. Be sure to listen to the practice for at least 3 characters during each practice session, however.

If you hear a Morse code character you don't immediately know, just draw a short line on your copy paper. Then get ready for the next letter. If you ignore your mistakes now, you'll make fewer of them. If you sit there worrying about the letter you missed, you'll miss a lot more! You shouldn't expect to copy perfectly while you are learning the code. You will get better with more practice.

When you think you've recognized a word after copying a few letters, concentrate all the more on the actual code sent. If you try to anticipate what comes next, your guess may turn out to be wrong. When that happens, you'll probably get confused and miss the next few letters as well. Write each letter just after it is sent. With more practice, you'll learn to "copy behind," hearing and writing whole words at one time. For now, when you are just learning the code, concentrate on writing each character as it is sent. This helps reinforce the association of a sound and a character in your mind.

Practice sending code. To communicate with Morse code you must be able to send it as well as receive it! You will need a **code-practice oscillator** and a **code key**. The sidebar, "Assembling A Code-Practice Oscillator" describes a simple oscillator that you can build. It even includes a simple key that you can use to practice sending.

One trick that some people use while learning the code is to whistle or hum the code while walking or driving. You can also say the sounds di, dit and dah to sound out the characters. Send the words on street signs, billboards and store windows. This extra practice may be just the help you need to master the code!

Morse code is a language. Eventually you'll begin to recognize common syllables and words. With practice you will know many complete words, and won't even listen to the individual letters. When you become this familiar with the code, it really starts to be fun!

Don't be discouraged if you don't seem to be breaking any speed records. Some people have an ear for code and can learn the entire code in a week or less. Others require a month or more to learn it. Be patient, continue to practice and you will reach your goal.

To pass the Novice class code examination, you must be able to understand a plain-language message sent at 5 WPM. As you know, in the English language some words are just one or two letters long, others 10 or more letters long. To standardize the code test, the FCC defines a word as a group of five letters. Numbers and punctuation marks normally count as two characters for this purpose. You will receive 25 characters in one minute for 5-WPM code.

ASSEMBLING A CODE-PRACTICE OSCILLATOR

It is not difficult to construct a code-practice oscillator. A complete oscillator that mounts on a small piece of wood is shown in Figure A. This figure also shows all the parts for this project laid out for assembly. The circuit board can be ordered from FAR Circuits, 18 N. 640 Field Ct., Dundee, IL 60118-9269. A complete parts kit, including circuit board, is available from Van Gorden Engineering, PO Box 21305, South Euclid, OH 44121.

Please read all instructions carefully before mounting any parts. Check the parts-placement diagram for the location of each part.

Figure A

Parts List

[ ] Check box as each part is installed and soldered

Quantity	Description	Radio Shack Part Number	Component Number	Used in Step Number
Capacitors				
[ ] 1	0.01-µF	272-131	C1	3
[ ] 1	0.01-µF	272-131	C2	6
[ ] 1	220-µF, 35-V electrolytic	272-1029	C3	5
Resistors				
[ ] 1	10-kilohm, ¼ W (brown-black-orange stripes)	271-1335	R2	4
[ ] 1	47-kilohm, ¼ W (yellow-violet-orange stripes)	271-1342	R3	8
[ ] 1	10-kilohm, ¼ W (brown-black-orange stripes)	271-1335	R1	9
[ ] 1	47-ohm, ¼ W (yellow-violet-black stripes)	271-009	R5	11
Miscellaneous				
[ ] 1	100-kilohm potentiometer	271 -284	R4	7
[ ] 1	7555 CMOS IC Timer (or 555 timer IC)	276-1718	U1	2
[ ] 1	Loudspeaker—3-inch, 8-ohm	40-245	LS1	11
[ ] 1	9-V Battery connector	270-325		10
[ ] 1	9-V Battery	23-553	BT1	10
[ ] 1	Brass rod 2 inches long, approximately 18 gauge (about the diameter of a wire coat hanger). (Available at hobby shops.)			
[ ] 1	8-pin IC socket	276-1995		2
[ ] 4	¼-inch spacers	64-3024		12
[ ] 1	U-shaped battery holder	270-326		12
[ ] 1	2 × 4 × ½-inch piece of wood for base			12
[ ] 5	No. 6 wood screws, ¾-inch long			12
[ ] 2	No. 6 wood screws, ⅜-inch long			12
[ ] 1	Five-lug tie point, used to mount speaker	274-688		11

What You Need

Besides the circuit board and parts, you will need a low-power (25-30 watt) soldering iron, some thin rosin-core solder, a straight-blade screwdriver and 000-gauge steel wool.

The circuit board has one blank side for components and one side with metal foil tracings. The foil tracings are the shiny metal paths on the board. Before you solder any part to the board, clean the part leads (long thin wires) with the steel wool. Rub lightly.

Figure B—Schematic diagram of a code-practice oscillator.

BRASS ROD SOLDERED IN POSITION

CIRCUIT–BOARD HANDLE

Figure C—Solder the brass rod in position on the foil side of the PC board. You can also use this figure as a circuit-board etching pattern if you want to make your own circuit-board, since the pattern is printed full size.

Brass Rod

FAR CIRCUITS DUNDEE ILL

Brass rod should not extend past this point.

Figure E

Figure F

Figure D—Parts-placement diagram

(Check box as each step is completed)

☐ Step 1: Attach the brass rod. Check parts-placement diagram (Figure D) for location.

Clean the brass rod with sandpaper or steel wool. Bend one end of the rod slightly less than 90 degrees. Lay the circuit board on the table with foil side up. Place the hooked end of the brass rod over the large hole near the handle (see Figure C). Make sure the rod extends out over the handle area. Solder the rod to the board on the foil side. The end of the brass rod should not extend past the marked oval on the handle. This is your contact point. If it does extend beyond this point, cut the rod off just before the end of the oval.

☐ Step 2: Solder the IC socket to the board.

The socket for the IC is placed on the component side of the board first. Do not plug the IC into the socket now. After all the other parts are soldered to the board you will be instructed to plug the IC into the socket (Step 13). Identify the notched end of the socket. Insert the socket into the circuit board on the component side. Make sure the notched end is facing the handle of the board. Turn the board over and gently spread the pins on the socket so they make contact with the foil side of the board. Solder the socket in place.

☐ Step 3: Place C1 (0.01-μF capacitor) on board.

Thread the wire leads on C1 through holes on board. (See Figure E.) Solder the wires onto the foil side of the board. Cut the extra wire off above the solder joint.

☐ Step 4: Place R2 (10-kilohm resistor) on board.

Prepare resistors for mounting by bending each lead (wire) of the resistor to approximately a 90° angle. (See Figure F.) Insert the leads into the board holes and bend them over to hold the resistor in place. Solder the leads to the foil and trim them close to the foil.

☐ Step 5: Place C3 (220-μF, 35-volt electrolytic capacitor) on board.

This capacitor has a plus (+) side and a negative (-) side. The (-) side is placed on the board facing away from the handle. Insert the capacitor leads into the circuit board holes, solder in place and trim off the extra wire.

☐ Step 6: Place C2 (0.01-μF capacitor) on board.

Thread C2 wire leads through holes on board. (See Step 3, and Figure E.) Solder the wires onto the board. Cut the extra wire off above the solder joint.

☐ Step 7: Place R4 (100-kilohm potentiometer) on board.

This component has three pins. All three pins must be plugged into the holes on the board. (It fits only one way.) Solder them in place.

☐ Step 8: Place R3 (47-kilohm resistor) on board.

Bend the wires on the resistor to plug it into the board. (See Step 4 and Figure F.) Plug the resistor into the board, spread the wires and solder in place.

☐ Step 9: Place R1 (10-kilohm resistor) on board.

Bend the wires on the resistor to plug it into the board. (See Step 4 and Figure F.) Plug the resistor into the board, spread the wires and solder in place.

☐ Step 10: Hook up the battery connector leads.

The battery connector consists of two wires, one red and one black, attached to a snap-on cap. Remove 1/4 inch of plastic insulation from the end of both wires. The black wire is negative and the red wire is positive. The positive and negative battery connections are marked on the solder side of the board. Be sure the red wire goes in the hole marked " + ", and the black wire goes in the hole marked " - ". Solder the wires in place.

☐ Step 11: Hook up the speaker.

Solder the speaker lugs to the tie-point lugs on either side of the center post. Cut the speaker wire into two equal lengths. Remove ¼ inch of plastic insulation from each end of both wires. Solder one end of each wire to one of the tie-point solder lugs below the speaker terminals. Solder one end of R5 (47-ohm resistor) to the circuit board as shown in Figure D. Solder one wire to the other end of this resistor, and the second speaker wire to the circuit board.

☐ Step 12: Attach the circuit board to the wood base.

Place the completed circuit board on the wood. Trace through the four holes with a pencil. Take the board off the wood and lay it aside. Place the spacers on the wood, standing upright. Carefully put the circuit board on top of the spacers. Put the ¾-inch screws through the holes in the circuit board and through the spacers and screw them into the board until snug. Be sure not to overtighten the screws, which could crack the circuit board. Attach the speaker to the front end of the board opposite the handle with a ⅜-inch screw through the tie-point mounting hole. Attach the U-shaped metal battery holder to the wooden base with a ⅜-inch screw.

☐ Step 13: Plug the integrated circuit (IC) into the socket.

CAUTION—The static electricity from your body could destroy the IC. Before touching the IC, be sure you have discharged any static that may be built up on your body. While sitting at your table or workbench, touch a metal pipe or other large metal object for a few seconds. Carefully remove the IC from its foam padding. Hold it by the black body and avoid touching the wires. Plug it into the socket, being sure that the notched end of the IC is facing towards the handle. The notch on the IC should line up with the notch on the socket.

Attach the battery to the snap-on battery connector, and place it in the U-shaped metal battery holder. This unit uses electricity only when the telegraph key handle is pushed down. No ON/OFF switch is necessary and you may leave the battery connected at all times.

You're done! The oscillator should produce a tone when you press the key. If your oscillator does not work, check all your connections carefully. Make sure the IC is positioned correctly in the socket, and that you have a fresh battery.

Once you have the oscillator working, you're ready to use it to practice Morse code. If you are studying with a friend, you can use the oscillator to send code to each other. If you are studying alone, tape record your sending and play it back later. Can you copy what you sent? How would it sound on the air? Good luck!

COMFORTABLE SENDING

There's more to the code than just learning to receive it; you'll also have to learn to send it. To accomplish this, you'll need a **telegraph key** and a code-practice oscillator. You can get these items at most electronics-parts stores, or you can build your own simple oscillator. The "Assembling A Code-Practice Oscillator" sidebar gives you step-by-step instructions for building one simple oscillator that even includes a key for you to practice with. The key with this project is not the best key you could use, but it will allow you to do some practice sending. You may want to consider adding a real telegraph key to the project—the circuit board has two holes to connect wires to a **straight key**.

Many experienced amateurs prefer to use an electronic keyer to send Morse code. An electronic keyer produces properly timed dots and dashes, because it uses one circuit to produce dots and another circuit to produce dashes. In general, it is probably better to learn to send Morse code with a hand key at first. Some students may have good success with a keyer. Commercial keyers range from simple, basic units to full-featured Morse code machines. While they are comparatively expensive, some of the full-featured machines offer features that are quite helpful to a beginner. The keyers shown in Figure 11-4 can send random-character code practice at any speed you desire. The AEA MM-3 keyer even has a program built in that will allow you to contact imaginary amateurs and carry on a typical amateur conversation with the keyer!

Figure 11-5 shows two different standard straight key models (one in the foreground and one to the left). There is a semiautomatic "bug" in the background and a popular Bencher paddle used with modern electronic keyers on the right. You'll want to obtain some type of code key to practice with before you're ready for your Novice code test. You'll need the key for some of your on-the-air operating after the license arrives from the FCC as well! Most new hams start with an inexpensive straight key.

Just as with receiving, it is important that you be comfortable when sending. It helps to rest your arm on the table, letting your wrist and hand do all the work. Grasp the key lightly with your fingertips. Don't grip it tightly. If you do, you'll soon discover a few muscles you didn't know you had, and each one will ache. With a light grasp, you'll be able to send for long periods without fatigue. See Figure 11-6.

Another important part of sending code is the proper adjustment of your telegraph key. There are only two adjustments to make on a straight key, but you'll find they are very important. Figure 11-7 illustrates these adjustments.

The first adjustment is the spacing between the contacts, which determines the distance the key knob must move to send a letter. Adjust the contacts so the knob moves about the thickness of a dime (1/16 inch). Try it. If you're not satisfied with this setting, try a wider space. If that doesn't do it, try reducing the space. Eventually, you'll find a spacing that works best for you. Don't be surprised if your feelings change

Figure 11-4—Electronic keyers shown here can store messages in memory to be sent at the touch of a button and also offer various training features to help you learn Morse code and increase your speed. The MFJ-486 keyer sends 5-character random-letter groups, random 1- to 8-character groups and plain text in the format of an on-the-air Amateur Radio contact (QSO). In addition to sending random characters, the AEA MM-3 can connect to a computer to send text from a keyboard or text file. The MM-3 also contains a program that allows you to "contact" other stations in the keyer and exchange QSO information.

from time to time, however, especially when your sending speed increases.

The second adjustment you must make is the spring tension that keeps the contacts apart. Just as there is no "correct" contact spacing, there is no correct tension adjustment for everyone. You will find, however, that adjusting the spacing may also require a tension adjustment. Adjust the tension to provide what you think is the best "feel" when sending.

Some straight keys have ball bearing pivot points on either side of the crossarms. These normally need no adjustment, but you should be sure the crossarms move freely in these pivots. If the bearings are too tight the key will bind or stick. If they are too loose there will be excessive play in the bearings. The side screws also adjust the crossarms from side to side so the contact points line up.

All of these adjustment screws have lock nuts, so be sure you loosen them before you make any adjustments. Tighten the lock nuts securely after making the adjustment.

Some better-quality keys have a shorting bar like the one shown in Figure 11-7. This shorting bar can be used to close the key contacts for transmitter tuning or adjustments.

You'll probably want to fasten the key to a piece of wood or other heavy weight to prevent it from sliding around as you send. You might even want to fasten the key directly to the table. It's best to experiment with different positions before permanently attaching the key to any surface, however.

Some operators use a board that extends under their

Figure 11-5—Older than radio itself, code still reigns as the most efficient and effective communications mode; many hams use it almost exclusively. A modern straight key, the device most beginners use, is in the foreground.

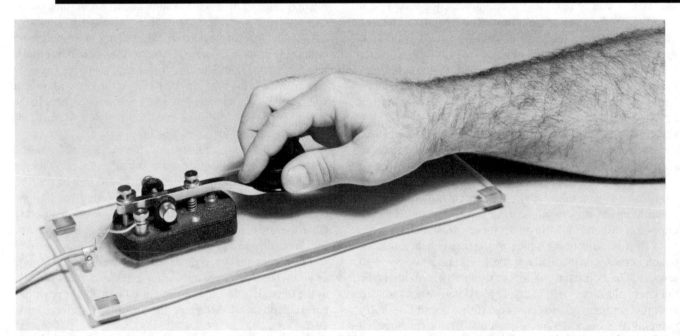

Figure 11-6—Proper forearm support, with the wrist off the table, a gentle grip and a smooth up-and-down motion make for clean, effortless sending.

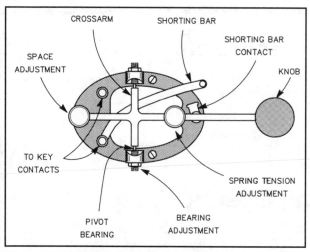

Figure 11-7—A few simple adjustments to suit your style allow you to send for hours without fatigue. The contact spacing and spring tension should be set to provide the most comfortable feel.

forearm and allows their arm to hold the key in position. This way the key can be moved aside for storage, or to clean off the table for other activities. A piece of ¼-inch Plexiglas about 12 inches long works well for this type of key base. This technique also allows you to reposition the key to find the most comfortable sending position.

Learning to send good code, like learning to receive, requires practice. A good way to start is to send along with your code tapes or computer program. Try to duplicate the sounds as much as possible.

Always remember that you're trying to send a complete sound, not a series of dots and dashes. With that in mind, take a moment to think about the sound you're trying to send. It consists of dits, dahs and pauses (or spaces). The key to good sending, then, is timing. If you're not convinced, listen to some code on the Novice bands. What transmissions would

you prefer to receive? Why? That's right—because they have good timing.

Each person develops a unique rhythm in his or her sending. It's almost impossible to describe what makes one person's rhythm different from another person's. No two people send code exactly the same way, though. This unique rhythm, or **fist** is like the signature of the operator sending the code. Some experienced CW operators can even identify their friends without hearing call signs.

Some people send code that is very easy and enjoyable to copy. Others are not so good, and require a lot of concentration to understand. Since your fist is your on-the-air signature, try to make it as easy to read as possible. Learn to send code that is easy to copy. Then you'll have a fist you can be proud of. It can make a big difference in your success as an amateur operator.

One of the best ways to learn to send and receive code is to work with another person. That person may be another member of your own family, a friend who is also studying the code or a licensed ham. If you're attending an organized class, you may be able to get together with another student several times a week.

If you must work alone, make good use of your tape recorder. Try recording your sending. After waiting a day or two, try receiving what you sent. The wait between sending and receiving will help prevent you from writing down the message from memory. Not only will this procedure provide practice in receiving, but it will also let you hear exactly how your sending sounds. If your timing is off, you'll hear it. If you're having trouble with a specific letter or number or punctuation mark, you'll soon know about it. If you can't understand your own sending, neither will anyone else!

Regardless of whether you're sending or receiving, the key to your success with Morse code is regular practice. After you've learned all the characters you will want to continue your regular practice to gain confidence and increase your speed.

If you can, listen to actual contacts between hams. Make use of the code-practice material transmitted daily by **W1AW**, the ARRL station heard nationwide from Newington, Connecticut. Table 11-1 lists the code-practice schedule for W1AW. When we change from Standard Time to Daylight Saving Time, the code-practice is sent at the same local times.

THE CODE TEST

To pass the Novice code test, you must demonstrate your ability to send and receive Morse code at 5 words per minute. Your Novice test examiners will give you the code test. The FCC specifies that the exam message should be at least 5 minutes long. The message must contain all letters of the alphabet, all numbers, period, comma, question mark and fraction bar (/). The exam message must also include some common procedural signals. These are the + sign ($\overline{AR}$), the = sign ($\overline{BT}$) and $\overline{SK}$.

The Volunteer Examiners giving you the test have sev-

Table 11-1

W1AW schedule

Pacific	Mtn	Cent	East	Sun	Mon	Tue	Wed	Thu	Fri	Sat
6 am	7 am	8 am	9 am			Fast Code	Slow Code	Fast Code	Slow Code	
7 am	8 am	9 am	10 am			Code Bulletin				
8 am	9 am	10 am	11 am			Teleprinter Bulletin				
9 am	10 am	11 am	noon							
10 am	11 am	noon	1 pm			Visiting Operator Time				
11 am	noon	1 pm	2 pm							
noon	1 pm	2 pm	3 pm							
1 pm	2 pm	3 pm	4 pm	Slow Code	Fast Code	Slow Code	Fast Code	Slow Code	Fast Code	Slow Code
2 pm	3 pm	4 pm	5 pm	Code Bulletin						
3 pm	4 pm	5 pm	6 pm	Teleprinter Bulletin						
4 pm	5 pm	6 pm	7 pm	Fast Code	Slow Code	Fast Code	Slow Code	Fast Code	Slow Code	Fast Code
5 pm	6 pm	7 pm	8 pm	Code Bulletin						
6 pm	7 pm	8 pm	9 pm	Teleprinter Bulletin						
6:45 pm	7:45 pm	8:45 pm	9:45 pm	Voice Bulletin						
7 pm	8 pm	9 pm	10 pm	Slow Code	Fast Code	Slow Code	Fast Code	Slow Code	Fast Code	Slow Code
8 pm	9 pm	10 pm	11 pm	Code Bulletin						
9 pm	10 pm	11 pm	Mdnte	Teleprinter Bulletin						
9:45 pm	10:45 pm	11:45 pm	12:45 am	Voice Bulletin						

❏ *Morse Code transmissions:*

Frequencies are 1.818, 3.5815, 7.0475, 14.0475, 18.0975, 21.0675, 28.0675 and 147.555 MHz.

Slow Code = practice sent at 5, 7-1/2, 10, 13 and 15 WPM.

Fast Code = practice sent at 35, 30, 25, 20, 15, 13 and 10 WPM.

Code practice text is from the pages of *QST*. The source is given at the beginning of each practice session and alternate speeds within each session. For example, "Text is from July 1992 *QST*, pages 9 and 81" indicates that the plain text is from the article on page 9 and mixed number/letter groups are from page 81.

Code bulletins are sent at 18 WPM.

❏ *Teleprinter transmissions:*

Frequencies are 3.625, 7.095, 14.095, 18.1025, 21.095, 28.095 and 147.555 MHz.

Bulletins are sent at 45.45-baud Baudot and 100 baud AMTOR, FEC Mode B. 110-baud ASCII will be sent only as time allows.

On Tuesdays and Saturdays at 6:30 P.M. Eastern time, Keplerian Elements for many amateur satellites are sent on the regular teleprinter frequencies.

❏ *Voice transmissions:*

Frequencies are 3.99, 7.29, 14.29, 18.16, 21.39, 28.59 and 147.555 MHz.

❏ *Miscellanea:*

On Fridays, UTC, a DX bulletin replaces the regular bulletins.

W1AW is open to visitors during normal operating hours: from 1:00 P.M. until 1 A.M. on Mondays, 9 A.M. until 1 A.M. Tuesday through Friday, and from 3:30 P.M. to 1 A.M. on Saturdays and Sundays. FCC licensed amateurs may operate the station from 1 - 4 P.M. Monday through Friday. Be sure to bring your current FCC amateur license or a photocopy. Special arrangements must be made for weekend operation; please call or write at least a week in advance.

In a communications emergency, monitor W1AW for special bulletins as follows: voice on the hour, teleprinter at 15 minutes past the hour, and CW on the half hour.

Headquarters and W1AW are closed on New Year's Day, President's Day, Good Friday, Memorial Day, Independence Day, Labor Day, Thanksgiving and the following Friday, and Christmas Day. On the first Thursday of September, Headquarters and W1AW will be closed during the afternoon.

AMATEUR RADIO EXCITEMENT

Seven-year-old Luke Ward of Alexandria, Virginia and his dad Keith saw an article about how to convert a small AM radio to receive shortwave broadcasts. After completing the project over a weekend, they were able to hear shortwave stations from all around the world! When Luke asked about the strange sounds they often heard, his dad explained that they were Morse code messages being sent by Amateur Radio operators. (Keith had been a ham when he was a boy, but was no longer licensed.) When Luke learned that there was no minimum age to become a ham, he and his dad began studying for their Novice exams. They also bought a used shortwave receiver and became avid shortwave listeners (SWLs). Luke and Keith also joined the Mt Vernon (VA) Amateur Radio Club.

About six weeks later, in August 1990, Luke and Keith passed their Novice exams. The next morning they left for a vacation trip to Cape Cod, and stopped at ARRL Headquarters in Newington, Connecticut on the way. Luke and his parents were thrilled to see the W1AW facilities, many of which they had read about in *Tune in the World with Ham Radio* (the ARRL's previous book for the Novice license). Luke was in third grade that fall when the tickets arrived from the FCC, and Luke (now KO4IQ) and Keith (now KO4IO) were soon on the air. Luke's first contact was with VE1VAZ in Nova Scotia, Canada! Within a month Luke had 60 CW contacts on 80, 40 and 15 meters. Luke and his dad have been competing for most states and countries worked.

Luke wasted no time in studying for his Technician license, and he and Keith passed their Technician exams in October 1990, about six weeks before Luke's eighth birthday.

"Learning Morse code took a lot of practice and so was harder than learning the other stuff necessary to pass the written exams," says Luke, who was already technically inclined before he got involved with ham radio. "Besides ham radio, I enjoy BASIC computer programming. Math and PE are my favorite school subjects," he adds. Luke also enjoys soccer and Cub Scouts.

After their 10-meter antenna was tuned, Luke tried SSB phone, and came to enjoy that as much as CW. His first phone contact was with CU1AC on Santa Maria Island in the Azores (a chain of islands in the North Atlantic that belong to Portugal)! "Eighty meters used to be my favorite band, but now I like 10 meters best because of the DX. I really like to get QSL cards, especially from foreign countries. My CW QSOs helped improve my code speed. I think I am about ready to take my General exam, once I finish studying the radio theory part. I want a General ticket so that I can use the extra band space, and I hope I can do moonbounce QSOs with it, too." (Luke did pass his General class exam in February 1991, at the age of eight. He and Keith have since upgraded to Advanced.)

---

eral options in deciding if you pass the exam. If you have one minute of solid (perfect) copy out of a five-minute test, you pass. The examiners may also give you a 10-question fill-in-the-blank or multiple-choice test about the exam content. A passing grade on such a written exam is 70%. Your examiners may devise other fair and reasonable standards for taking the code test. Again, the examiners are simply certifying that you have proven your ability to send and receive code at 5 words per minute.

EXAMINATION NERVES

If you have prepared properly, the code exam should be easy—a pleasant experience, not an agonizing one. The key to a successful exam is to relax. One of the biggest reasons people fail the code examination is because they become nervous. You can be confident of passing, however, because you have studied hard and are properly prepared.

The exam isn't *that* important. You can always take the test again soon. It won't kill you even if you don't pass. But you are going to pass! You can copy 5 WPM at home with only a couple of errors every few minutes. The code test will be even easier.

Think positively: "I will pass!" Relax. Before you know it, you'll be grinning from ear to ear as you hear the words, "You've passed!"

HEADING ON UP

Once you obtain your Novice ticket, you'll soon set your sights on a Technician or General class license. The ARRL's *Increasing Your Code Speed* series of code-practice tapes will help you increase your code speed. There are tapes to go from 5 to 10 WPM, from 10 to 15 WPM and from 15 to 22 WPM. Each set of tapes includes two 90-minute cassettes, which provide practice with words, sentences, random code groups, related text and QSO practice. The space between characters decreases gradually to increase the overall code speed. You'll be copying faster code before you know it! W1AW also provides practice at a variety of speeds. Contact the ARRL, 225 Main Street, Newington, CT 06111 for the tapes you'll need to reach your desired code speed.

Chapter 12

Novice (Element 2) Question Pool—With Answers

DON'T START HERE

This chapter contains the complete question pool for the Element 2 exam. Element 2 is the written part of the Novice exam. To obtain a Novice license you must also pass a 5-word-per-minute Morse code test.

The Element 2 exam is also part of the Technician license exam. The Technician exam also includes Element 3A. Chapter 13 contains the entire Element 3A question pool, with answers. The Technician license does not require a code test.

Before you read the questions and answers printed in this chapter, be sure to read the text in the previous chapters. Use these questions as review exercises, when the text tells you to study them. Don't try to memorize all the questions and answers. This book was carefully written and prepared to guide you step-by-step as you learn about Amateur Radio. By understanding the electronics principles and Amateur Radio concepts in this book, you will enjoy our hobby more. You will also better appreciate the privileges granted by an Amateur Radio license.

This pool, released by the Volunteer Examiner Coordinators' Question Pool Committee, went into effect July 1, 1993. The pool is scheduled to be used until June 30, 1997. Changes to FCC Rules and other factors may result in earlier revisions. Such changes will be announced in *QST* and other Amateur Radio publications.

HOW MANY QUESTIONS?

Your Element 2 exam will consist of 30 of these questions, selected by the Volunteer Examiners giving you the test, or by the Volunteer Examiner Coordinator that coordinates the test session you attend. All VECs and examiners must use the questions, answers and distracters (incorrect answers) printed here.

This question pool contains more than ten times the number of questions necessary to make up an exam. This ensures that the examiners have sufficient questions to choose from when they make up an exam.

WHO GIVES THE TEST?

You will take your exam at a session coordinated by a Volunteer Examiner Coordinator. The examiners may select your questions or they may use a question set provided by the VEC or another provider. In any case the questions will be selected from the question pool printed in this chapter.

QUESTION POOL SECTIONS

The question pool is divided into nine sections, called subelements. (A subelement is a portion of the exam element, in this case Element 2.) The FCC specifies how many questions from each section must be on your test. For example, there must be ten questions from the Commission's Rules section, Subelement N1. There must also be two questions from the Operating Procedures section, Subelement N2, and so on. Table 12-1 summarizes the number of questions that make up an exam. The number of questions to be used from each subelement appears at the beginning of that subelement in the question pool, too.

The subelements are broken into smaller groups, with one question to come from each smaller group. These groups are called out with a list of topics given in bold type. They are labeled alphabetically within each subelement. For example, since there are ten exam questions from Subelement N1, these sections are labeled N1A through N1J. Two exam questions come from the Circuit Components subelement, N6, so that subelement has sections N6A and N6B.

The question numbers used in the question pool relate to the syllabus or study guide printed at the end of Chapter 1. The syllabus is an outline of topics covered by the exam. Each question number begins with an N. This indicates the

Table 12-1

Novice Exam Content

Subelement	Topic	Number of Questions
N1	Commission's Rules	10
N2	Operating Procedures	2
N3	Radio-Wave Propagation	1
N4	Amateur Radio Practices	4
N5	Electrical Principles	4
N6	Circuit Components	2
N7	Practical Circuits	2
N8	Signals and Emissions	2
N9	Antennas and Feed Lines	3

question is from the Novice question pool. Next is a number to indicate which subelement the question is from. These numbers will range from 1 to 9. Following this number is a letter to indicate which subelement group the question is from. Each question number ends with a two-digit number to specify its position in the set. So question N1A01 is the first question in the A group of the first subelement. Question N5D09 is the ninth question in the D group of the fifth subelement.

QUESTION POOL FORMAT

The rest of this chapter contains the entire Element 2 question pool. We have printed the answer key to these questions along the edge of the page. There is a line to indicate where you should fold the page to hide the answer key while you study. After making your best effort to answer the questions, you can look at the answers to check your understanding. We also have included page references along with the answers. These page numbers indicate where you will find the text discussion related to each question. If you have any problems with a question, refer to the page listed for that question. You may have to study beyond the listed page number to review all the related material. We also have included references to sections of Part 97 for the Commission's Rules Subelement, T1. This is to help you identify the particular rule citations for these questions. The complete text of the FCC rules in Part 97 is included in the ARRL publication, *The FCC Rule Book*.

The Question Pool Committee included all the drawings for the question pool on a single page. We placed a copy of this page at the back of the question pool, for your informa-

tion. You should expect to receive a page like this as part of your exam papers. Notice there are several equations included at the bottom of the page. The first one

$$\frac{E}{I\,R}$$

represents the Ohm's Law Circle you learned about in Chapter 3. This will help you remember the Ohm's Law equations for your exam. The next three equations relate to antenna length calculations.

$$\frac{984}{F_{MHz}} = Ft$$

$$\frac{468}{F_{MHz}} = Ft$$

$$\frac{234}{F_{MHz}} = Ft$$

These give the length in feet of a full-wavelength, half-wavelength and quarter-wavelength antenna. The last equation

$$\frac{300}{F_{MHz}} = M$$

gives the wavelength, measured in meters, of a radio signal in space. You may not need all these equations for your exam, but they will be available for you on the Figure sheet. For your convenience, we also placed the individual figures near the questions that refer to them in the question pool.

Good luck with your studies.

Element 2 (Novice Class) Question Pool

SUBELEMENT N1—COMMISSION'S RULES
[10 exam questions—10 groups]

N1A Basis and purpose of amateur service and definitions.

ANSWER KEY
Page numbers tell you where to look in this book for more information.

Subelement N1
Numbers in [square brackets] indicate sections in FCC Part 97, the Amateur Radio rules.

N1A01

What document contains the rules and regulations for the amateur service in the US?

A. Part 97 of Title 47 CFR (Code of Federal Regulations)
B. The Communications Act of 1934 (as amended)
C. The Radio Amateur's Handbook
D. The minutes of the International Telecommunication Union meetings

N1A01
(A)
[97]
Page 2-3

N1A02

Who makes and enforces the rules and regulations of the amateur service in the US?

A. The Congress of the United States
B. The Federal Communications Commission (FCC)
C. The Volunteer Examiner Coordinators (VECs)
D. The Federal Bureau of Investigation (FBI)

N1A02
(B)
[97]
Page 2-3

N1A03

Which three topics are part of the rules and regulations of the amateur service?

A. Station operation standards, technical standards, emergency communications
B. Notice of Violation, common operating procedures, antenna lengths
C. Frequency band plans, repeater locations, Ohm's Law
D. Station construction standards, FCC approved radios, FCC approved antennas

N1A03
(A)
[97]
Page 2-3

N1A04

Which of these topics is NOT part of the rules and regulations of the amateur service?

A. Qualifying examination systems
B. Technical standards
C. Providing emergency communications
D. Station construction standards

N1A04
(D)
[97]
Page 2-4

N1A05
(A)
[97.1]
Page 2-3

N1A05
What are three reasons that the amateur service exists?
A. To recognize the value of emergency communications, advance the radio art, and improve communication and technical skills
B. To learn about business communications, increase testing by trained technicians, and improve amateur communications
C. To preserve old radio techniques, maintain a pool of people familiar with early tube-type equipment, and improve tube radios
D. To improve patriotism, preserve nationalism, and promote world peace

N1A06
(D)
[97.1]
Page 2-3

N1A06
What are two of the five purposes for the amateur service?
A. To protect historical radio data, and help the public understand radio history
B. To help foreign countries improve communication and technical skills, and encourage visits from foreign hams
C. To modernize radio schematic drawings, and increase the pool of electrical drafting people
D. To increase the number of trained radio operators and electronics experts, and improve international goodwill

N1A07
(B)
[97.3a1]
Page 2-5

N1A07
What is the definition of an amateur operator?
A. A person who has not received any training in radio operations
B. A person who has a written authorization to be the control operator of an amateur station
C. A person who has very little practice operating a radio station
D. A person who is in training to become the control operator of a radio station

N1A08
(C)
[97.3a4]
Page 2-5

N1A08
What is the definition of the amateur service?
A. A private radio service used for profit and public benefit
B. A public radio service for US citizens which requires no exam
C. A personal radio service used for self-training, communication, and technical studies
D. A private radio service used for self-training of radio announcers and technicians

N1A09
(D)
[97.3a5]
Page 2-5

N1A09
What is the definition of an amateur station?
A. A station in a public radio service used for radiocommunications
B. A station using radiocommunications for a commercial purpose
C. A station using equipment for training new radiocommunications operators
D. A station in an Amateur Radio service used for radiocommunications

N1A10
(C)
[97.3a11]
Page 2-6

N1A10
What is the definition of a control operator of an amateur station?
A. Anyone who operates the controls of the station
B. Anyone who is responsible for the station's equipment
C. Any licensed amateur operator who is responsible for the station's transmissions
D. The amateur operator with the highest class of license who is near the controls of the station

N1A11

What is a Volunteer Examiner (VE)?

A. An amateur who volunteers to check amateur teaching manuals
B. An amateur who volunteers to teach amateur classes
C. An amateur who volunteers to test others for amateur licenses
D. An amateur who volunteers to examine amateur station equipment

N1B Station/Operator license.

N1B01

Which one of these must you have an amateur license to do?

A. Transmit on public-service frequencies
B. Retransmit shortwave broadcasts
C. Repair broadcast station equipment
D. Transmit on amateur service frequencies

N1B02

What does an amateur license allow you to control?

A. A shortwave-broadcast station's transmissions
B. An amateur station's transmissions
C. Non-commercial FM broadcast transmissions
D. Any type of transmitter, as long as it is used for non-commercial transmissions

N1B03

What allows someone to operate an amateur station in the US?

A. An FCC operator's training permit for a licensed radio station
B. An FCC Form 610 together with a license examination fee
C. An FCC amateur operator/primary station license
D. An FCC Certificate of Successful Completion of Amateur Training

N1B04

Where does a US amateur license allow you to operate?

A. Anywhere in the world
B. Wherever the amateur service is regulated by the FCC
C. Within 50 km of your primary station location
D. Only at your primary station location

N1B05

If you have a Novice license, how many transmitters may you control in your station at the same time?

A. Only one at a time
B. Only one at a time, except for emergency communications
C. Any number
D. Any number, as long as they are transmitting on different bands

N1B06

What document must you keep at your amateur station?

A. A copy of your written authorization for an amateur station
B. A copy of the Rules and Regulations of the Amateur Service (Part 97)
C. A copy of the Amateur Radio Handbook for instant reference
D. A chart of the frequencies allowed for your class of license

N1A11
(C)
[97.513a]
Page 2-10

N1B01
(D)
[97.5a]
Page 2-5

N1B02
(B)
[97.5]
Page 2-5

N1B03
(C)
[97.5a]
Page 2-5

N1B04
(B)
[97.5d]
Page 2-5

N1B05
(C)
[97.5e]
Page 2-6

N1B06
(A)
[97.5e]
Page 2-5

N1B07
(C)
[97.7]
Page 2-6

N1B07
Which one of the following does not allow a person to control a US amateur station?
A. An operator/primary station license from the FCC
B. A reciprocal permit for alien amateur licensee from the FCC
C. An amateur service license from any government which is a member of the European Community (EC)
D. An amateur service license from the Government of Canada, if it is held by a Canadian citizen

N1B08
(D)
[97.9a]
Page 2-6

N1B08
What are the five US amateur operator license classes?
A. Novice, Communicator, General, Advanced, Amateur Extra
B. Novice, Technician, General, Advanced, Expert
C. Novice, Communicator, General, Amateur, Extra
D. Novice, Technician, General, Advanced, Amateur Extra

N1B09
(A)
[97.9]
Page 2-6

N1B09
What does the FCC consider to be the first two classes of US amateur operator licenses (one of which most new amateurs initially hold)?
A. Novice and Technician
B. CB and Communicator
C. Novice and General
D. CB and Novice

N1B10
(B)
[97.9]
Page 2-5

N1B10
What must you have with you when you are the control operator of an amateur station?
A. A copy of the Rules and Regulations of the Amateur Service (Part 97)
B. The original or a photocopy of your amateur license
C. A list of countries which allow third-party communications from the US
D. A chart of the frequencies allowed for your class of license

N1B11
(D)
[97.501d]
Page 2-6

N1B11
Which US amateur license has no Morse code requirements?
A. Amateur Extra
B. Advanced
C. General
D. Technician

N1C Novice control operator frequency privileges.

N1C01
(B)
[97.301e]
Page 2-8

N1C01
What are the frequency limits of the 80-meter Novice band?
A. 3500 - 4000 kHz
B. 3675 - 3725 kHz
C. 7100 - 7150 kHz
D. 7000 - 7300 kHz

N1C02

What are the frequency limits of the 40-meter Novice band (ITU Region 2)?
A. 3500 - 4000 kHz
B. 3700 - 3750 kHz
C. 7100 - 7150 kHz
D. 7000 - 7300 kHz

N1C03

What are the frequency limits of the 15-meter Novice band?
A. 21.100 - 21.200 MHz
B. 21.000 - 21.450 MHz
C. 28.000 - 29.700 MHz
D. 28.100 - 28.200 MHz

N1C04

What are the frequency limits of the 10-meter Novice band?
A. 28.000 - 28.500 MHz
B. 28.100 - 29.500 MHz
C. 28.100 - 28.500 MHz
D. 29.100 - 29.500 MHz

N1C05

This question has been withdrawn.

N1C06

What are the frequency limits of the 23-centimeter Novice band?
A. 1260 - 1270 MHz
B. 1240 - 1300 MHz
C. 1270 - 1295 MHz
D. 1240 - 1246 MHz

N1C07

If you are operating on 3700 kHz, in what amateur band are you operating?
A. 80 meters
B. 40 meters
C. 15 meters
D. 10 meters

N1C08

If you are operating on 7125 kHz, in what amateur band are you operating?
A. 80 meters
B. 40 meters
C. 15 meters
D. 10 meters

N1C02
(C)
[97.301e]
Page 2-8

N1C03
(A)
[97.301e]
Page 2-8

N1C04
(C)
[97.301e]
Page 2-8

N1C05

N1C06
(C)
[97.301f]
Page 2-8

N1C07
(A)
[97.301e]
Page 2-8

N1C08
(B)
[97.301e]
Page 2-8

N1C09
(C)
[97.301e]
Page 2-8

N1C09
If you are operating on 21.150 MHz, in what amateur band are you operating?
A. 80 meters
B. 40 meters
C. 15 meters
D. 10 meters

N1C10
(D)
[97.301e]
Page 2-8

N1C10
If you are operating on 28.150 MHz, in what amateur band are you operating?
A. 80 meters
B. 40 meters
C. 15 meters
D. 10 meters

N1C11
(D)
[97.301f]
Page 2-8

N1C11
If you are operating on 223 MHz, in what amateur band are you operating?
A. 15 meters
B. 10 meters
C. 2 meters
D. 1.25 meters

N1D Novice eligibility, exam elements, mailing addresses, US call sign assignment and life of license.

N1D01
(A)
[97.5d1]
Page 2-6

N1D01
Who can become an amateur licensee in the US?
A. Anyone except a representative of a foreign government
B. Only a citizen of the United States
C. Anyone except an employee of the US government
D. Anyone

N1D02
(D)
[No Part 97 Ref]
Page 2-6

N1D02
What age must you be to hold an amateur license?
A. 14 years or older
B. 18 years or older
C. 70 years or younger
D. There are no age limits

N1D03
(C)
[97.501e]
Page 2-10

N1D03
What minimum examinations must you pass for a Novice amateur license?
A. A written exam, Element 1(A); and a 5 WPM code exam, Element 2(A)
B. A 5 WPM code exam, Element 1(A); and a written exam, Element 3(A)
C. A 5 WPM code exam, Element 1(A); and a written exam, Element 2
D. A written exam, Element 2; and a 5 WPM code exam, Element 4

N1D04

Why must an amateur operator have a current US Postal mailing address?

A. So the FCC has a record of the location of each amateur station
B. To follow the FCC rules and so the licensee can receive mail from the FCC
C. So the FCC can send license-renewal notices
D. So the FCC can publish a call-sign directory

N1D05

What must you do to replace your license if it is lost, mutilated or destroyed?

A. Nothing; no replacement is needed
B. Send a change of address to the FCC using a current FCC Form 610
C. Retake all examination elements for your license
D. Request a new one from the FCC, explaining what happened to the original

N1D06

What must you do to notify the FCC if your mailing address changes?

A. Fill out an FCC Form 610 using your new address, attach a copy of your license, and mail it to your local FCC Field Office
B. Fill out an FCC Form 610 using your new address, attach a copy of your license, and mail it to the FCC office in Gettysburg, PA
C. Call your local FCC Field Office and give them your new address over the phone
D. Call the FCC office in Gettysburg, PA, and give them your new address over the phone

N1D07

Which of the following call signs is a valid US amateur call?

A. UA4HAK
B. KBL7766
C. KA9OLS
D. BY7HY

N1D08

What letters must be used for the first letter in US amateur call signs?

A. K, N, U and W
B. A, K, N and W
C. A, B, C and D
D. A, N, V and W

N1D09

What numbers are normally used in US amateur call signs?

A. Any two-digit number, 10 through 99
B. Any two-digit number, 22 through 45
C. A single digit, 1 though 9
D. A single digit, 0 through 9

N1D10

For how many years is an amateur license normally issued?

A. 2
B. 5
C. 10
D. 15

N1D04	(B) [97.21] Page 2-11
N1D05	(D) [97.27] Page 2-6
N1D06	(B) [97.19] Page 2-6
N1D07	(C) [No Part 97 Ref] Page 2-15
N1D08	(B) [No Part 97 Ref] Page 2-15
N1D09	(D) [No Part 97 Ref] Page 2-15
N1D10	(C) [97.23] Page 2-6

N1D11
(A)
[97.19c]
Page 2-6

N1D11
How soon before your license expires should you send the FCC a completed 610 for a renewal?
A. 60 to 90 days
B. Within 21 days of the expiration date
C. 6 to 9 months
D. 6 months to a year

N1E Novice control operator emission privileges.

N1E01
(A)
[97.305/
307f9]
Page 2-8

N1E01
What emission types are Novice control operators allowed to use in the 80-meter band?
A. CW only
B. Data only
C. RTTY only
D. Phone only

N1E02
(A)
[97.305/
307f9]
Page 2-8

N1E02
What emission types are Novice control operators allowed to use in the 40-meter band?
A. CW only
B. Data only
C. RTTY only
D. Phone only

N1E03
(A)
[97.305/
307f9]
Page 2-8

N1E03
What emission types are Novice control operators allowed to use in the 15-meter band?
A. CW only
B. Data only
C. RTTY only
D. Phone only

N1E04
(D)
[97.305/
307f9]
Page 2-8

N1E04
What emission types are Novice control operators allowed to use from 3675 to 3725 kHz?
A. Phone only
B. Image only
C. Data only
D. CW only

N1E05
(D)
[97.305/
307f9]
Page 2-8

N1E05
What emission types are Novice control operators allowed to use from 7100 to 7150 kHz in ITU Region 2?
A. CW and data
B. Phone
C. Data only
D. CW only

N1E06
(D)
[97.305/
307f9]
Page 2-8

N1E06
What emission types are Novice control operators allowed to use on frequencies from 21.1 to 21.2 MHz?
A. CW and data
B. CW and phone
C. Data only
D. CW only

N1E07
What emission types are Novice control operators allowed to use on
frequencies from 28.1 to 28.3 MHz?
A. All authorized amateur emission privileges
B. Data or phone
C. CW, RTTY and data
D. CW and phone

N1E08
What emission types are Novice control operators allowed to use on
frequencies from 28.3 to 28.5 MHz?
A. All authorized amateur emission privileges
B. CW and data
C. CW and single-sideband phone
D. Data and phone

N1E09
What emission types are Novice control operators allowed to use on the
amateur 1.25-meter band in ITU Region 2?
A. CW and phone
B. CW and data
C. Data and phone
D. All amateur emission privileges authorized for use on the band

N1E10
What emission types are Novice control operators allowed to use on the
amateur 23-centimeter band?
A. Data and phone
B. CW and data
C. CW and phone
D. All amateur emission privileges authorized for use on the band

N1E11
On what HF frequencies may Novice control operators use single-
sideband (SSB) phone?
A. 3700 - 3750 kHz
B. 7100 - 7150 kHz
C. 21100 - 21200 kHz
D. 28300 - 28500 kHz

N1E12
This question has been withdrawn.

N1E13
On what frequencies in the 10-meter band may Novice control operators
use RTTY?
A. 28.0 - 28.3 MHz
B. 28.1 - 28.3 MHz
C. 28.0 - 29.3 MHz
D. 29.1 - 29.3 MHz

N1E07
(C)
[97.305]
Page 2-8

N1E08
(C)
[97.305/
307f10]
Page 2-8

N1E09
(D)
[97.305]
Page 2-8

N1E10
(D)
[97.305]
Page 2-8

N1E11
(D)
[97.305/
307f10]
Page 2-8

N1E12

N1E13
(B)
[97.301e/
305]
Page 2-8

N1E14
(B)
[97.301e/
305]
Page 2-8

N1E14
On what frequencies in the 10-meter band may Novice control operators use data emissions?
A. 28.0 - 28.3 MHz
B. 28.1 - 28.3 MHz
C. 28.0 - 29.3 MHz
D. 29.1 - 29.3 MHz

N1F Transmitter power on Novice sub-bands and digital communications (limited to concepts only).

N1F01
(D)
[97.313a]
Page 2-10

N1F01
What amount of transmitter power must amateur stations use at all times?
A. 25 watts PEP output
B. 250 watts PEP output
C. 1500 watts PEP output
D. The minimum legal power necessary to communicate

N1F02
(C)
[97.313c1]
Page 2-8

N1F02
What is the most transmitter power an amateur station may use on 3700 kHz?
A. 5 watts PEP output
B. 25 watts PEP output
C. 200 watts PEP output
D. 1500 watts PEP output

N1F03
(C)
[97.313c1]
Page 2-8

N1F03
What is the most transmitter power an amateur station may use on 7125 kHz?
A. 5 watts PEP output
B. 25 watts PEP output
C. 200 watts PEP output
D. 1500 watts PEP output

N1F04
(C)
[97.313c1]
Page 2-8

N1F04
What is the most transmitter power an amateur station may use on 21.125 MHz?
A. 5 watts PEP output
B. 25 watts PEP output
C. 200 watts PEP output
D. 1500 watts PEP output

N1F05
(C)
[97.313c2]
Page 2-10

N1F05
What is the most transmitter power a Novice station may use on 28.125 MHz?
A. 5 watts PEP output
B. 25 watts PEP output
C. 200 watts PEP output
D. 1500 watts PEP output

N1F06
(C)
[97.313c2]
Page 2-10

N1F06
What is the most transmitter power a Novice station may use on the 10-meter band?
A. 5 watts PEP output
B. 25 watts PEP output
C. 200 watts PEP output
D. 1500 watts PEP output

N1F07

What is the most transmitter power a Novice station may use on the
1.25-meter band?
A. 5 watts PEP output
B. 25 watts PEP output
C. 200 watts PEP output
D. 1500 watts PEP output

N1F07
(B)
[97.313d]
Page 2-8

N1F08

What is the most transmitter power a Novice station may use on the
23-centimeter band?
A. 5 watts PEP output
B. 25 watts PEP output
C. 200 watts PEP output
D. 1500 watts PEP output

N1F08
(A)
[97.313e]
Page 2-8

N1F09

On which band(s) may a Novice station use up to 200 watts PEP output
power?
A. 80, 40, 15, and 10 meters
B. 80, 40, 20, and 10 meters
C. 1.25 meters
D. 23 centimeters

N1F09
(A)
[97.313c]
Page 2-8

N1F10

On which band(s) must a Novice station use no more than 25 watts PEP
output power?
A. 80, 40, 15, and 10 meters
B. 80, 40, 20, and 10 meters
C. 1.25 meters
D. 23 centimeters

N1F10
(C)
[97.313d]
Page 2-8

N1F11

On which band(s) must a Novice station use no more than 5 watts PEP
output power?
A. 80, 40, 15, and 10 meters
B. 80, 40, 20, and 10 meters
C. 1.25 meters
D. 23 centimeters

N1F11
(D)
[97.313e]
Page 2-8

N1G Responsibility of licensee, control operator requirements.

N1G01

If you allow another amateur to be responsible for the transmissions
from your station, what is the other operator called?
A. An auxiliary operator
B. The operations coordinator
C. A third-party operator
D. A control operator

N1G01
(D)
[97.3a11]
Page 2-6

N1G02

Who is responsible for the proper operation of an amateur station?
A. Only the control operator
B. Only the station licensee
C. Both the control operator and the station licensee
D. The person who owns the station equipment

N1G02
(C)
[97.103a]
Page 2-6

N1G03
(A)
[97.103a]
Page 2-6

N1G03

If you transmit from another amateur's station, who is responsible for its proper operation?

A. Both of you
B. The other amateur (the station licensee)
C. You, the control operator
D. The station licensee, unless the station records show that you were the control operator at the time

N1G04
(D)
[97.103a]
Page 2-6

N1G04

What is your responsibility as a station licensee?

A. You must allow another amateur to operate your station upon request
B. You must be present whenever the station is operated
C. You must notify the FCC if another amateur acts as the control operator
D. You are responsible for the proper operation of the station in accordance with the FCC rules

N1G05
(C)
[97.103b]
Page 2-6

N1G05

Who may be the control operator of an amateur station?

A. Any person over 21 years of age
B. Any person over 21 years of age with a General class license or higher
C. Any licensed amateur chosen by the station licensee
D. Any licensed amateur with a Technician class license or higher

N1G06
(B)
[97.103]
Page 2-6

N1G06

If another amateur transmits from your station, which of these is NOT true?

A. You must first give permission for the other amateur to use your station
B. You must keep the call sign of the other amateur, together with the time and date of transmissions, in your station records
C. The FCC will think that you are the station's control operator unless your station records show that you were not
D. Both of you are equally responsible for the proper operation of the station

N1G07
(A)
[97.105b]
Page 2-6

N1G07

If you let another amateur with a higher class license than yours control your station, what operating privileges are allowed?

A. Any privileges allowed by the higher license
B. Only the privileges allowed by your license
C. All the emission privileges of the higher license, but only the frequency privileges of your license
D. All the frequency privileges of the higher license, but only the emission privileges of your license

N1G08
If you are the control operator at the station of another amateur who has a higher class license than yours, what operating privileges are you allowed?
A. Any privileges allowed by the higher license
B. Only the privileges allowed by your license
C. All the emission privileges of the higher license, but only the frequency privileges of your license
D. All the frequency privileges of the higher license, but only the emission privileges of your license

N1G08
(B)
[97.105b]
Page 2-6

N1G09
When must an amateur station have a control operator?
A. Only when training another amateur
B. Whenever the station receiver is operated
C. Whenever the station is transmitting
D. A control operator is not needed

N1G09
(C)
[97.7]
Page 2-6

N1G10
When a Novice station is transmitting, where must its control operator be?
A. At the station's control point
B. Anywhere in the same building as the transmitter
C. At the station's entrance, to control entry to the room
D. Anywhere within 50 km of the station location

N1G10
(A)
[97.109b]
Page 2-6

N1G11
Why can't unlicensed persons in your family transmit using your amateur station if they are alone with your equipment?
A. They must not use your equipment without your permission
B. They must be licensed before they are allowed to be control operators
C. They must first know how to use the right abbreviations and Q signals
D. They must first know the right frequencies and emissions for transmitting

N1G11
(B)
[97.109b]
Page 2-6

N1H Station identification, points of communication and operation, and business communications.

N1H01
When may you operate your amateur station somewhere in the US besides the location listed on your license?
A. Only during times of emergency
B. Only after giving proper notice to the FCC
C. During an emergency or an FCC-approved emergency practice
D. Whenever you want to

N1H01
(D)
[97.5a]
Page 2-20

N1H02
With which non-amateur stations is a US amateur station allowed to communicate?
A. No non-amateur stations
B. All non-amateur stations
C. Only those authorized by the FCC
D. Only those who use international Morse code

N1H02
(C)
[97.111]
Page 2-17

N1H03

N1H03
This question has been withdrawn.

N1H04

N1H04
This question has been withdrawn.

N1H05

N1H05
This question has been withdrawn.

N1H06
(C)
[97.119a]
Page 2-16

N1H06
How often must an amateur station be identified?
A. At the beginning of a contact and at least every ten minutes after that
B. At least once during each transmission
C. At least every ten minutes during and at the end of a contact
D. At the beginning and end of each transmission

N1H07
(B)
[97.119a]
Page 2-16

N1H07
What do you transmit to identify your amateur station?
A. Your "handle"
B. Your call sign
C. Your first name and your location
D. Your full name

N1H08
(A)
[97.119a]
Page 2-16

N1H08
What identification, if any, is required when two amateur stations begin communications?
A. No identification is required
B. One of the stations must give both stations' call signs
C. Each station must transmit its own call sign
D. Both stations must transmit both call signs

N1H09
(C)
[97.119a]
Page 2-16

N1H09
What identification, if any, is required when two amateur stations end communications?
A. No identification is required
B. One of the stations must transmit both stations' call signs
C. Each station must transmit its own call sign
D. Both stations must transmit both call signs

N1H10

Besides normal identification, what else must a US station do when sending third-party communications internationally?

A. The US station must transmit its own call sign at the beginning of each communication, and at least every ten minutes after that

B. The US station must transmit both call signs at the end of each communication

C. The US station must transmit its own call sign at the beginning of each communication, and at least every five minutes after that

D. Each station must transmit its own call sign at the end of each communication, and at least every five minutes after that

N1H10
(B)
[97.115c]
Page 2-17

N1H11

What is the longest period of time an amateur station can operate without transmitting its call sign?

A. 5 minutes

B. 10 minutes

C. 15 minutes

D. 20 minutes

N1H11
(B)
[97.119a]
Page 2-16

N1I International and space communications, authorized and prohibited transmissions.

N1I01

What is the definition of third-party communications?

A. A message sent between two amateur stations for someone else

B. Public service communications for a political party

C. Any messages sent by amateur stations

D. A three-minute transmission to another amateur

N1I01
(A)
[97.3a39]
Page 2-18

N1I02

When are you allowed to communicate with an amateur in a foreign country?

A. Only when the foreign amateur uses English

B. Only when you have permission from the FCC

C. Only when a third-party agreement exists between the US and the foreign country

D. At any time, unless it is not allowed by either government

N1I02
(D)
[97.111a1]
Page 2-17

N1I03

What is an amateur space station?

A. An amateur station operated on an unused frequency

B. An amateur station awaiting its new call letters from the FCC

C. An amateur station located more than 50 kilometers above the Earth's surface

D. An amateur station that communicates with Space Shuttles

N1I03
(C)
[97.3a36]
Page 2-20

N1I04

Who may be the licensee of an amateur space station?

A. An amateur holding an Amateur Extra class operator license

B. Any licensed amateur operator

C. Anyone designated by the commander of the spacecraft

D. No one unless specifically authorized by the government

N1I04
(B)
[New
97.207a per
FCC 92-310]
Page 2-20

N1I05
(D)
[97.113b]
Page 2-19

N1I05
When may someone be paid to transmit messages from an amateur station?
A. Only if he or she works for a public service agency such as the Red Cross
B. Under no circumstances
C. Only if he or she reports all such payments to the IRS
D. Only if he or she works for a club station and special requirements are met

N1I06
(A)
[97.113c]
Page 2-17

N1I06
When is an amateur allowed to broadcast information to the general public?
A. Never
B. Only when the operator is being paid
C. Only when broadcasts last less than 1 hour
D. Only when broadcasts last longer than 15 minutes

N1I07
(A)
[97.113d]
Page 2-19

N1I07
When is an amateur station permitted to transmit music?
A. Never
B. Only if the music played produces no spurious emissions
C. Only if it is used to jam an illegal transmission
D. Only if it is above 1280 MHz

N1I08
(C)
[97.113d]
Page 2-19

N1I08
When is the use of codes or ciphers allowed to hide the meaning of an amateur message?
A. Only during contests
B. Only during nationally declared emergencies
C. Never, except when special requirements are met
D. Only on frequencies above 1280 MHz

N1I09
(B)
[97.3a42]
Page 2-18

N1I09
What is a "third-party" in amateur communications?
A. An amateur station that breaks in to talk
B. A person who is sent a message by amateur communications other than a control operator who handles the message
C. A shortwave listener who monitors amateur communications
D. An unlicensed control operator

N1I10
(A)
[97.115a2]
Page 2-18

N1I10
If you are allowing a non-amateur friend to use your station to talk to someone in the US, and a foreign station breaks in to talk to your friend, what should you do?
A. Have your friend wait until you find out if the US has a third-party agreement with the foreign station's government
B. Stop all discussions and quickly sign off
C. Since you can talk to any foreign amateurs, your friend may keep talking as long as you are the control operator
D. Report the incident to the foreign amateur's government

N1I11

When are you allowed to transmit a message to a station in a foreign country for a third party?

A. Anytime

B. Never

C. Anytime, unless there is a third-party agreement between the US and the foreign government

D. If there is a third-party agreement with the US government, or if the third party could be the control operator

N1J False signals or unidentified communications and malicious interference.

N1J01

What is a transmission called that disturbs other communications?

A. Interrupted CW

B. Harmful interference

C. Transponder signals

D. Unidentified transmissions

N1J02

Why is transmitting on a police frequency as a "joke" called harmful interference that deserves a large penalty?

A. It annoys everyone who listens

B. It blocks police calls which might be an emergency and interrupts police communications

C. It is in bad taste to communicate with non-amateurs, even as a joke

D. It is poor amateur practice to transmit outside the amateur bands

N1J03

When may you deliberately interfere with another station's communications?

A. Only if the station is operating illegally

B. Only if the station begins transmitting on a frequency you are using

C. Never

D. You may expect, and cause, deliberate interference because it can't be helped during crowded band conditions

N1J04

When may false or deceptive amateur signals or communications be transmitted?

A. Never

B. When operating a beacon transmitter in a "fox hunt" exercise

C. When playing a harmless "practical joke"

D. When you need to hide the meaning of a message for secrecy

N1J05

If an amateur pretends there is an emergency and transmits the word "MAYDAY," what is this called?

A. A traditional greeting in May

B. An emergency test transmission

C. False or deceptive signals

D. Nothing special; "MAYDAY" has no meaning in an emergency

N1I11
(D)
[97.115a2]
Page 2-18

N1J01
(B)
[97.3a21]
Page 2-19

N1J02
(B)
[97.3a21]
Page 2-19

N1J03
(C)
[97.101d]
Page 2-19

N1J04
(A)
[97.113d]
Page 2-19

N1J05
(C)
[97.113d]
Page 2-19

N1J06
(C)
[97.119a]
Page 2-16

N1J06
When may an amateur transmit unidentified communications?
A. Only for brief tests not meant as messages
B. Only if it does not interfere with others
C. Never, except to control a model craft
D. Only for two-way or third-party communications

N1J07
(A)
[97.119a]
Page 2-16

N1J07
What is an amateur communication called that does not have the required station identification?
A. Unidentified communications or signals
B. Reluctance modulation
C. Test emission
D. Tactical communication

N1J08
(D)
[97.405a]
Page 2-20

N1J08
If you hear a voice distress signal on a frequency outside of your license privileges, what are you allowed to do to help the station in distress?
A. You are NOT allowed to help because the frequency of the signal is outside your privileges
B. You are allowed to help only if you keep your signals within the nearest frequency band of your privileges
C. You are allowed to help on a frequency outside your privileges only if you use international Morse code
D. You are allowed to help on a frequency outside your privileges in any way possible

N1J09
(D)
[97.119a]
Page 2-16

N1J09
If you answer someone on the air without giving your call sign, what type of communication have you just conducted?
A. Test transmission
B. Tactical signal
C. Packet communication
D. Unidentified communication

N1J10
(C)
[97.403]
Page 2-19

N1J10
When may you use your amateur station to transmit an "SOS" or "MAYDAY"?
A. Never
B. Only at specific times (at 15 and 30 minutes after the hour)
C. In a life- or property-threatening emergency
D. When the National Weather Service has announced a severe weather watch

N1J11
(B)
[97.405a]
Page 2-20

N1J11
When may you send a distress signal on any frequency?
A. Never
B. In a life- or property-threatening emergency
C. Only at specific times (at 15 and 30 minutes after the hour)
D. When the National Weather Service has announced a severe weather watch

SUBELEMENT N2—OPERATING PROCEDURES
[2 exam questions—2 groups]

N2A Choosing a frequency for tune-up, operating or emergencies; understanding the Morse code; RST signal reports; Q signals; voice communications and phonetics.

N2A01
What should you do before you transmit on any frequency?
A. Listen to make sure others are not using the frequency
B. Listen to make sure that someone will be able to hear you
C. Check your antenna for resonance at the selected frequency
D. Make sure the SWR on your antenna feed line is high enough

N2A02
If you make contact with another station and your signal is extremely strong and perfectly readable, what adjustment might you make to your transmitter?
A. Turn on your speech processor
B. Reduce your SWR
C. Continue with your contact, making no changes
D. Turn down your power output to the minimum necessary

N2A03
What is one way to shorten transmitter tune-up time on the air to cut down on interference?
A. Use a random wire antenna
B. Tune up on 40 meters first, then switch to the desired band
C. Tune the transmitter into a dummy load
D. Use twin lead instead of coaxial-cable feed lines

N2A04
If you are in contact with another station and you hear an emergency call for help on your frequency, what should you do?
A. Tell the calling station that the frequency is in use
B. Direct the calling station to the nearest emergency net frequency
C. Call your local Civil Preparedness Office and inform them of the emergency
D. Stop your QSO immediately and take the emergency call

N2A05
What is the correct way to call CQ when using Morse code?
A. Send the letters "CQ" three times, followed by "DE," followed by your call sign sent once
B. Send the letters "CQ" three times, followed by "DE," followed by your call sign sent three times
C. Send the letters "CQ" ten times, followed by "DE," followed by your call sign sent once
D. Send the letters "CQ" over and over

N2A06
(B)
Page 9-4

N2A06
How should you answer a Morse code CQ call?
A. Send your call sign four times
B. Send the other station's call sign twice, followed by "DE," followed by your call sign twice
C. Send the other station's call sign once, followed by "DE," followed by your call sign four times
D. Send your call sign followed by your name, station location and a signal report

N2A07
(C)
Page 9-3

N2A07
At what speed should a Morse code CQ call be transmitted?
A. Only speeds below five WPM
B. The highest speed your keyer will operate
C. Any speed at which you can reliably receive
D. The highest speed at which you can control the keyer

N2A08
(D)
Page 9-3

N2A08
What is the meaning of the procedural signal "CQ"?
A. "Call on the quarter hour"
B. "New antenna is being tested" (no station should answer)
C. "Only the called station should transmit"
D. "Calling any station"

N2A09
(A)
Page 9-3

N2A09
What is the meaning of the procedural signal "DE"?
A. "From" or "this is," as in "W9NGT DE N9BTT"
B. "Directional Emissions" from your antenna
C. "Received all correctly"
D. "Calling any station"

N2A10
(A)
Page 9-5

N2A10
What is the meaning of the procedural signal "K"?
A. "Any station transmit"
B. "All received correctly"
C. "End of message"
D. "Called station only transmit"

N2A11
(B)
Page 9-2

N2A11
What is meant by the term "DX"?
A. Best regards
B. Distant station
C. Calling any station
D. Go ahead

N2A12
(B)
Page 9-5

N2A12
What is the meaning of the term "73"?
A. Long distance
B. Best regards
C. Love and kisses
D. Go ahead

N2A13

What are RST signal reports?

A. A short way to describe ionospheric conditions
B. A short way to describe transmitter power
C. A short way to describe signal reception
D. A short way to describe sunspot activity

N2A14

What does RST mean in a signal report?

A. Recovery, signal strength, tempo
B. Recovery, signal speed, tone
C. Readability, signal speed, tempo
D. Readability, signal strength, tone

N2A15

What is one meaning of the Q signal "QRS"?

A. Interference from static
B. Send more slowly
C. Send RST report
D. Radio station location is

N2A16

What is one meaning of the Q signal "QTH"?

A. Time here is
B. My name is
C. Stop sending
D. My location is

N2A17

What is a QSL card?

A. A letter or postcard from an amateur pen pal
B. A Notice of Violation from the FCC
C. A written proof of communication between two amateurs
D. A postcard reminding you when your license will expire

N2A18

What is the correct way to call CQ when using voice?

A. Say "CQ" once, followed by "this is," followed by your call sign spoken three times
B. Say "CQ" at least five times, followed by "this is," followed by your call sign spoken once
C. Say "CQ" three times, followed by "this is," followed by your call sign spoken three times
D. Say "CQ" at least ten times, followed by "this is," followed by your call sign spoken once

N2A19

How should you answer a voice CQ call?

A. Say the other station's call sign at least ten times, followed by "this is," then your call sign at least twice
B. Say the other station's call sign at least five times phonetically, followed by "this is," then your call sign at least once
C. Say the other station's call sign at least three times, followed by "this is," then your call sign at least five times phonetically
D. Say the other station's call sign once, followed by "this is," then your call sign given phonetically

N2A13
(C)
Page 9-5

N2A14
(D)
Page 9-5

N2A15
(B)
Page 9-3

N2A16
(D)
Page 9-3

N2A17
(C)
Page 9-3

N2A18
(C)
Page 9-8

N2A19
(D)
Page 9-8

N2A20

To make your call sign better understood when using voice transmissions, what should you do?

A. Use Standard International Phonetics for each letter of your call
B. Use any words which start with the same letters as your call sign for each letter of your call
C. Talk louder
D. Turn up your microphone gain

N2B Radio teleprinting; packet; repeater operating procedures; special operations.

N2B01

What is the correct way to call CQ when using RTTY?

A. Send the letters "CQ" three times, followed by "DE," followed by your call sign sent once
B. Send the letters "CQ" three to six times, followed by "DE," followed by your call sign sent three times
C. Send the letters "CQ" ten times, followed by the procedural signal "DE," followed by your call sent one time
D. Send the letters "CQ" over and over

N2B02

What speed should you use when answering a CQ call using RTTY?

A. Half the speed of the received signal
B. The same speed as the received signal
C. Twice the speed of the received signal
D. Any speed, since RTTY systems adjust to any signal speed

N2B03

What does "connected" mean in a packet-radio link?

A. A telephone link is working between two stations
B. A message has reached an amateur station for local delivery
C. A transmitting station is sending data to only one receiving station; it replies that the data is being received correctly
D. A transmitting and receiving station are using a digipeater, so no other contacts can take place until they are finished

N2B04

What does "monitoring" mean on a packet-radio frequency?

A. The FCC is copying all messages
B. A member of the Amateur Auxiliary to the FCC's Field Operations Bureau is copying all messages
C. A receiving station is displaying all messages sent to it, and replying that the messages are being received correctly
D. A receiving station is displaying messages that may not be sent to it, and is not replying to any message

N2B05

What is a digipeater?

A. A packet-radio station that retransmits only data that is marked to be retransmitted
B. A packet-radio station that retransmits any data that it receives
C. A repeater that changes audio signals to digital data
D. A repeater built using only digital electronics parts

N2B06

What does "network" mean in packet radio?

A. A way of connecting terminal-node controllers by telephone so data can be sent over long distances

B. A way of connecting packet-radio stations so data can be sent over long distances

C. The wiring connections on a terminal-node controller board

D. The programming in a terminal-node controller that rejects other callers if a station is already connected

N2B07

What is simplex operation?

A. Transmitting and receiving on the same frequency

B. Transmitting and receiving over a wide area

C. Transmitting on one frequency and receiving on another

D. Transmitting one-way communications

N2B08

When should you use simplex operation instead of a repeater?

A. When the most reliable communications are needed

B. When a contact is possible without using a repeater

C. When an emergency telephone call is needed

D. When you are traveling and need some local information

N2B09

What is a good way to make contact on a repeater?

A. Say the call sign of the station you want to contact three times

B. Say the other operator's name, then your call sign three times

C. Say the call sign of the station you want to contact, then your call sign

D. Say, "Breaker, breaker," then your call sign

N2B10

When using a repeater to communicate, what do you need to know about the repeater besides its output frequency?

A. Its input frequency

B. Its call sign

C. Its power level

D. Whether or not it has a phone patch

N2B11

What is the main purpose of a repeater?

A. To make local information available 24 hours a day

B. To link amateur stations with the telephone system

C. To retransmit NOAA weather information during severe storm warnings

D. To increase the range of portable and mobile stations

N2B06
(B)
Page 9-23

N2B07
(A)
Page 9-12

N2B08
(B)
Page 9-14

N2B09
(C)
Page 9-13

N2B10
(A)
Page 9-12

N2B11
(D)
Page 9-12

N2B12
(A)
Page 9-12

N2B12
What does it mean to say that a repeater has an input and an output frequency?
A. The repeater receives on one frequency and transmits on another
B. The repeater offers a choice of operating frequency, in case one is busy
C. One frequency is used to control the repeater and another is used to retransmit received signals
D. The repeater must receive an access code on one frequency before retransmitting received signals

N2B13
(C)
Page 9-15

N2B13
What is an autopatch?
A. Something that automatically selects the strongest signal to be repeated
B. A device which connects a mobile station to the next repeater if it moves out of range of the first
C. A device that allows repeater users to make telephone calls from their stations
D. A device which locks other stations out of a repeater when there is an important conversation in progress

N2B14
(D)
Page 9-14

N2B14
What is the purpose of a repeater time-out timer?
A. It lets a repeater have a rest period after heavy use
B. It logs repeater transmit time to predict when a repeater will fail
C. It tells how long someone has been using a repeater
D. It limits the amount of time someone can transmit on a repeater

N2B15
(B)
Page 9-13

N2B15
What is a CTCSS (or PL) tone?
A. A special signal used for telecommand control of model craft
B. A sub-audible tone added to a carrier which may cause a receiver to accept a signal
C. A tone used by repeaters to mark the end of a transmission
D. A special signal used for telemetry between amateur space stations and Earth stations

Subelement N3

SUBELEMENT N3—RADIO WAVE PROPAGATION
[1 exam question—1 group]

N3A Radio wave propagation, line of sight, ground wave, sky wave, sunspots and the sunspot cycle, reflection of VHF/UHF signals.

N3A01
(A)
Page 9-29

N3A01
When a signal travels in a straight line from one antenna to another, what is this called?
A. Line-of-sight propagation
B. Straight-line propagation
C. Knife-edge diffraction
D. Tunnel propagation

N3A02

What type of propagation usually occurs from one hand-held VHF transceiver to another nearby?

A. Tunnel propagation
B. Sky-wave propagation
C. Line-of-sight propagation
D. Auroral propagation

N3A02
(C)
Page 9-29

N3A03

How do VHF and UHF radio waves usually travel from a transmitting antenna to a receiving antenna?

A. They bend through the ionosphere
B. They go in a straight line
C. They wander in any direction
D. They move in a circle going either east or west from the transmitter

N3A03
(B)
Page 9-29

N3A04

What can happen to VHF or UHF signals going towards a metal-framed building?

A. They will go around the building
B. They can be bent by the ionosphere
C. They can be easily reflected by the building
D. They are sometimes scattered in the ectosphere

N3A04
(C)
Page 9-34

N3A05

When a signal travels along the surface of the Earth, what is this called?

A. Sky-wave propagation
B. Knife-edge diffraction
C. E-region propagation
D. Ground-wave propagation

N3A05
(D)
Page 9-29

N3A06

How does the range of sky-wave propagation compare to ground-wave propagation?

A. It is much shorter
B. It is much longer
C. It is about the same
D. It depends on the weather

N3A06
(B)
Page 9-29

N3A07

When a signal is returned to earth by the ionosphere, what is this called?

A. Sky-wave propagation
B. Earth-moon-earth propagation
C. Ground-wave propagation
D. Tropospheric propagation

N3A07
(A)
Page 9-29

N3A08

What is the usual cause of sky-wave propagation?

A. Signals are reflected by a mountain
B. Signals are reflected by the moon
C. Signals are bent back to earth by the ionosphere
D. Signals are repeated by a repeater

N3A08
(C)
Page 9-29

N3A09
(C)
Page 9-30

N3A09
What is a skip zone?
A. An area covered by ground-wave propagation
B. An area covered by sky-wave propagation
C. An area which is too far away for ground-wave propagation, but too close for sky-wave propagation
D. An area which is too far away for ground-wave or sky-wave propagation

N3A10
(A)
Page 9-29

N3A10
What are the regions of ionized gases high above the earth called?
A. The ionosphere
B. The troposphere
C. The gas region
D. The ion zone

N3A11
(A)
Page 9-30

N3A11
How do sunspots change the ionization of the atmosphere?
A. The more sunspots there are, the greater the ionization
B. The more sunspots there are, the less the ionization
C. Unless there are sunspots, the ionization is zero
D. They have no effect

N3A12
(C)
Page 9-30

N3A12
How long is an average sunspot cycle?
A. 2 years
B. 5 years
C. 11 years
D. 17 years

**Subelement
N4**

SUBELEMENT N4—AMATEUR RADIO PRACTICES
[4 exam questions—4 groups]

N4A Unauthorized use prevention, lightning protection, and station grounding.

N4A01
(B)
Page 8-3

N4A01
How could you best keep unauthorized persons from using your amateur station at home?
A. Use a carrier-operated relay in the main power line
B. Use a key-operated on/off switch in the main power line
C. Put a "Danger - High Voltage" sign in the station
D. Put fuses in the main power line

N4A02
(A)
Page 8-5

N4A02
How could you best keep unauthorized persons from using a mobile amateur station in your car?
A. Disconnect the microphone when you are not using it
B. Put a "do not touch" sign on the radio
C. Turn the radio off when you are not using it
D. Tune the radio to an unused frequency when you are done using it

N4A03
Why would you use a key-operated on/off switch in the main power line of your station?
A. To keep unauthorized persons from using your station
B. For safety, in case the main fuses fail
C. To keep the power company from turning off your electricity during an emergency
D. For safety, to turn off the station in the event of an emergency

N4A03
(A)
Page 8-3

N4A04
Why should you ground all antenna and rotator cables when your amateur station is not in use?
A. To lock the antenna system in one position
B. To avoid radio frequency interference
C. To save electricity
D. To protect the station and building from lightning damage

N4A04
(D)
Page 8-4

N4A05
How can an antenna system best be protected from lightning damage?
A. Install a balun at the antenna feed point
B. Install an RF choke in the antenna feed line
C. Ground all antennas when they are not in use
D. Install a fuse in the antenna feed line

N4A05
(C)
Page 8-4

N4A06
How can amateur station equipment best be protected from lightning damage?
A. Use heavy insulation on the wiring
B. Never turn off the equipment
C. Disconnect the ground system from all radios
D. Disconnect all equipment from the power lines and antenna cables

N4A06
(D)
Page 8-6

N4A07
For best protection from electrical shock, what should be grounded in an amateur station?
A. The power supply primary
B. All station equipment
C. The antenna feed line
D. The AC power mains

N4A07
(B)
Page 8-6

N4A08
What is usually a good indoor grounding point for an amateur station?
A. A metallic cold water pipe
B. A plastic cold water pipe
C. A window screen
D. A metallic natural gas pipe

N4A08
(A)
Page 8-4

N4A09
Where should you connect the chassis of each piece of your station equipment to best protect against electrical shock?
A. To insulated shock mounts
B. To the antenna
C. To a good ground connection
D. To a circuit breaker

N4A09
(C)
Page 8-4

N4A10
(B)
Page 8-4

N4A10
Which of these materials is best for a ground rod driven into the earth?
A. Hard plastic
B. Copper or copper-clad steel
C. Iron or steel
D. Fiberglass

N4A11
(C)
Page 8-4

N4A11
If you ground your station equipment to a ground rod driven into the earth, what is the shortest length the rod should be?
A. 4 feet
B. 6 feet
C. 8 feet
D. 10 feet

N4B Radio frequency safety precautions, safety interlocks, antenna installation safety procedures.

N4B01
(B)
Page 8-7

N4B01
What should you do for safety when operating at 1270 MHz?
A. Make sure that an RF leakage filter is installed at the antenna feed point
B. Keep antenna away from your eyes when RF is applied
C. Make sure the standing wave ratio is low before you conduct a test
D. Never use a shielded horizontally polarized antenna

N4B02
(A)
Page 8-8

N4B02
What should you do for safety if you put up a UHF transmitting antenna?
A. Make sure the antenna will be in a place where no one can get near it when you are transmitting
B. Make sure that RF field screens are in place
C. Make sure the antenna is near the ground to keep its RF energy pointing in the correct direction
D. Make sure you connect an RF leakage filter at the antenna feed point

N4B03
(C)
Page 8-7

N4B03
What should you do for safety before removing the shielding on a UHF power amplifier?
A. Make sure all RF screens are in place at the antenna feed line
B. Make sure the antenna feed line is properly grounded
C. Make sure the amplifier cannot accidentally be turned on
D. Make sure that RF leakage filters are connected

N4B04
(A)
Page 8-7

N4B04
Why should you use only good quality coaxial cable and connectors for a UHF antenna system?
A. To keep RF loss low
B. To keep television interference high
C. To keep the power going to your antenna system from getting too high
D. To keep the standing wave ratio of your antenna system high

N4B05

Why should you make sure the antenna of a hand-held transceiver is not close to your head when transmitting?

A. To help the antenna radiate energy equally in all directions
B. To reduce your exposure to the radio-frequency energy
C. To use your body to reflect the signal in one direction
D. To keep static charges from building up

N4B05
(B)
Page 8-8

N4B06

Microwave oven radiation is similar to what type of amateur station RF radiation?

A. Signals in the 3.5 MHz range
B. Signals in the 21 MHz range
C. Signals in the 50 MHz range
D. Signals in the 1270 MHz range

N4B06
(D)
Page 8-7

N4B07

Why would there be a switch in a high-voltage power supply to turn off the power if its cabinet is opened?

A. To keep dangerous RF radiation from leaking out through an open cabinet
B. To keep dangerous RF radiation from coming in through an open cabinet
C. To turn the power supply off when it is not being used
D. To keep anyone opening the cabinet from getting shocked by dangerous high voltages

N4B07
(D)
Page 8-6

N4B08

What kind of safety equipment should you wear if you are working on an antenna tower?

A. A grounding chain
B. A reflective vest of approved color
C. A flashing red, yellow or white light
D. A carefully inspected safety belt, hard hat and safety glasses

N4B08
(D)
Page 7-25

N4B09

Why should you wear a safety belt if you are working on an antenna tower?

A. To safely hold your tools so they don't fall and injure someone on the ground
B. To keep the tower from becoming unbalanced while you are working
C. To safely bring any tools you might use up and down the tower
D. To prevent you from accidentally falling

N4B09
(D)
Page 7-25

N4B10

For safety, how high should you place a horizontal wire antenna?

A. High enough so that no one can touch any part of it from the ground
B. As close to the ground as possible
C. Just high enough so you can easily reach it for adjustments or repairs
D. Above high-voltage electrical lines

N4B10
(A)
Page 7-25

N4B11
(C)
Page 7-25

N4B11
Why should you wear a hard hat if you are on the ground helping someone work on an antenna tower?
A. So you won't be hurt if the tower should accidentally fall
B. To keep RF energy away from your head during antenna testing
C. To protect your head from something dropped from the tower
D. So someone passing by will know that work is being done on the tower and will stay away

N4C SWR meaning and measurements.

N4C01
(C)
Page 7-6

N4C01
What instrument is used to measure standing wave ratio?
A. An ohmmeter
B. An ammeter
C. An SWR meter
D. A current bridge

N4C02
(D)
Page 7-6

N4C02
What instrument is used to measure the relative impedance match between an antenna and its feed line?
A. An ammeter
B. An ohmmeter
C. A voltmeter
D. An SWR meter

N4C03
(A)
Page 7-6

N4C03
Where would you connect an SWR meter to measure standing wave ratio?
A. Between the feed line and the antenna
B. Between the transmitter and the power supply
C. Between the transmitter and the receiver
D. Between the transmitter and the ground

N4C04
(B)
Page 7-6

N4C04
What does an SWR reading of 1:1 mean?
A. An antenna for another frequency band is probably connected
B. The best impedance match has been attained
C. No power is going to the antenna
D. The SWR meter is broken

N4C05
(C)
Page 7-6

N4C05
What does an SWR reading of less than 1.5:1 mean?
A. An impedance match which is too low
B. An impedance mismatch; something may be wrong with the antenna system
C. A fairly good impedance match
D. An antenna gain of 1.5

N4C06
(D)
Page 7-7

N4C06
What does an SWR reading of 4:1 mean?
A. An impedance match which is too low
B. An impedance match which is good, but not the best
C. An antenna gain of 4
D. An impedance mismatch; something may be wrong with the antenna system

N4C07

What kind of SWR reading may mean poor electrical contact between parts of an antenna system?
A. A jumpy reading
B. A very low reading
C. No reading at all
D. A negative reading

N4C07
(A)
Page 7-8

N4C08

What does a very high SWR reading mean?
A. The antenna is the wrong length, or there may be an open or shorted connection somewhere in the feed line
B. The signals coming from the antenna are unusually strong, which means very good radio conditions
C. The transmitter is putting out more power than normal, showing that it is about to go bad
D. There is a large amount of solar radiation, which means very poor radio conditions

N4C08
(A)
Page 7-8

N4C09

If an SWR reading at the low frequency end of an amateur band is 2.5:1, and is 5:1 at the high frequency end of the same band, what does this tell you about your 1/2-wavelength dipole antenna?
A. The antenna is broadbanded
B. The antenna is too long for operation on the band
C. The antenna is too short for operation on the band
D. The antenna is just right for operation on the band

N4C09
(B)
Page 7-8

N4C10

If an SWR reading at the low frequency end of an amateur band is 5:1, and 2.5:1 at the high frequency end of the same band, what does this tell you about your 1/2-wavelength dipole antenna?
A. The antenna is broadbanded
B. The antenna is too long for operation on the band
C. The antenna is too short for operation on the band
D. The antenna is just right for operation on the band

N4C10
(C)
Page 7-8

N4C11

If you use a 3-30 MHz RF-power meter at UHF frequencies, how accurate will its readings be?
A. They may not be accurate at all
B. They will be accurate enough to get by
C. They will be accurate but the readings must be divided by two
D. They will be accurate but the readings must be multiplied by two

N4C11
(A)
Page 7-8

N4D RFI and its complications.

N4D01

What is meant by receiver overload?
A. Too much voltage from the power supply
B. Too much current from the power supply
C. Interference caused by strong signals from a nearby transmitter
D. Interference caused by turning the volume up too high

N4D01
(C)
Page 10-4

N4D02
(B)
Page 10-4

N4D02
What is one way to tell if radio-frequency interference to a receiver is caused by front-end overload?
A. If connecting a low-pass filter to the transmitter greatly cuts down the interference
B. If the interference is about the same no matter what frequency is used for the transmitter
C. If connecting a low-pass filter to the receiver greatly cuts down the interference
D. If grounding the receiver makes the problem worse

N4D03
(C)
Page 10-4

N4D03
If your neighbor reports television interference whenever you are transmitting from your amateur station, no matter what frequency band you use, what is probably the cause of the interference?
A. Too little transmitter harmonic suppression
B. Receiver VR tube discharge
C. Receiver overload
D. Incorrect antenna length

N4D04
(D)
Page 10-5

N4D04
If your neighbor reports television interference on one or two channels only when you are transmitting on the 15-meter band, what is probably the cause of the interference?
A. Too much low-pass filtering on the transmitter
B. De-ionization of the ionosphere near your neighbor's TV antenna
C. TV receiver front-end overload
D. Harmonic radiation from your transmitter

N4D05
(B)
Page 10-4

N4D05
What type of filter should be connected to a TV receiver as the first step in trying to prevent RF overload from an amateur HF station transmission?
A. Low-pass
B. High-pass
C. Band pass
D. Notch

N4D06
(B)
Page 10-5

N4D06
What type of filter might be connected to an amateur HF transmitter to cut down on harmonic radiation?
A. A key-click filter
B. A low-pass filter
C. A high-pass filter
D. A CW filter

N4D07
(A)
Page 10-5

N4D07
What is meant by harmonic radiation?
A. Unwanted signals at frequencies which are multiples of the fundamental (chosen) frequency
B. Unwanted signals that are combined with a 60-Hz hum
C. Unwanted signals caused by sympathetic vibrations from a nearby transmitter
D. Signals which cause skip propagation to occur

N4D08

Why is harmonic radiation from an amateur station not wanted?
A. It may cause interference to other stations and may result in out-of-band signals
B. It uses large amounts of electric power
C. It may cause sympathetic vibrations in nearby transmitters
D. It may cause auroras in the air

N4D08
(A)
Page 10-5

N4D09

What type of interference may come from a multi-band antenna connected to a poorly tuned transmitter?
A. Harmonic radiation
B. Auroral distortion
C. Parasitic excitation
D. Intermodulation

N4D09
(A)
Page 10-6

N4D10

What is the main purpose of shielding in a transmitter?
A. It gives the low-pass filter a solid support
B. It helps the sound quality of transmitters
C. It prevents unwanted RF radiation
D. It helps keep electronic parts warmer and more stable

N4D10
(C)
Page 10-5

N4D11

If you are told that your amateur station is causing television interference, what should you do?
A. First make sure that your station is operating properly, and that it does not cause interference to your own television
B. Immediately turn off your transmitter and contact the nearest FCC office for assistance
C. Connect a high-pass filter to the transmitter output and a low-pass filter to the antenna-input terminals of the television
D. Continue operating normally, because you have no reason to worry about the interference

N4D11
(A)
Page 10-4

SUBELEMENT N5—ELECTRICAL PRINCIPLES
[4 exam questions—4 groups]

Subelement
N5

N5A Metric prefixes, i.e. pico, micro, milli, centi, kilo, mega, giga.

N5A01

If a dial marked in kilohertz shows a reading of 7125 kHz, what would it show if it were marked in megahertz?
A. 0.007125 MHz
B. 7.125 MHz
C. 71.25 MHz
D. 7,125,000 MHz

N5A01
(B)
Page 3-2

N5A02

If a dial marked in megahertz shows a reading of 3.525 MHz, what would it show if it were marked in kilohertz?
A. 0.003525 kHz
B. 35.25 kHz
C. 3525 kHz
D. 3,525,000 kHz

N5A02
(C)
Page 3-2

N5A03
(D)
Page 3-2

N5A03
If a dial marked in kilohertz shows a reading of 3725 kHz, what would it show if it were marked in hertz?
A. 3.725 Hz
B. 37.25 Hz
C. 3,725 Hz
D. 3,725,000 Hz

N5A04
(B)
Page 3-2

N5A04
How long is an antenna that is 400 centimeters long?
A. 0.0004 meters
B. 4 meters
C. 40 meters
D. 40,000 meters

N5A05
(C)
Page 3-2

N5A05
If an ammeter marked in amperes is used to measure a 3000-milliampere current, what reading would it show?
A. 0.003 amperes
B. 0.3 amperes
C. 3 amperes
D. 3,000,000 amperes

N5A06
(B)
Page 3-2

N5A06
If a voltmeter marked in volts is used to measure a 3500-millivolt potential, what reading would it show?
A. 0.35 volts
B. 3.5 volts
C. 35 volts
D. 350 volts

N5A07
(B)
Page 3-2

N5A07
How many farads is 500,000 microfarads?
A. 0.0005 farads
B. 0.5 farads
C. 500 farads
D. 500,000,000 farads

N5A08
(B)
Page 3-2

N5A08
How many microfarads is 1,000,000 picofarads?
A. 0.001 microfarads
B. 1 microfarad
C. 1,000 microfarads
D. 1,000,000,000 microfarads

N5A09
(C)
Page 3-2

N5A09
How many hertz are in a kilohertz?
A. 10
B. 100
C. 1000
D. 1000000

N5A10

How many kilohertz are in a megahertz?

A. 10
B. 100
C. 1000
D. 1000000

N5A10
(C)
Page 3-2

N5A11

If you have a hand-held transceiver which puts out 500 milliwatts, how many watts would this be?

A. 0.02
B. 0.5
C. 5
D. 50

N5A11
(B)
Page 3-2

N5B Concepts of current, voltage, conductor, insulator, resistance, and the measurements thereof.

N5B01

What is the flow of electrons in an electric circuit called?

A. Voltage
B. Resistance
C. Capacitance
D. Current

N5B01
(D)
Page 3-6

N5B02

What is the basic unit of electric current?

A. The volt
B. The watt
C. The ampere
D. The ohm

N5B02
(C)
Page 3-7

N5B03

What is the pressure that forces electrons to flow through a circuit?

A. Magnetomotive force, or inductance
B. Electromotive force, or voltage
C. Farad force, or capacitance
D. Thermal force, or heat

N5B03
(B)
Page 3-6

N5B04

What is the basic unit of voltage?

A. The volt
B. The watt
C. The ampere
D. The ohm

N5B04
(A)
Page 3-6

N5B05

How much voltage does an automobile battery usually supply?

A. About 12 volts
B. About 30 volts
C. About 120 volts
D. About 240 volts

N5B05
(A)
Page 3-6

N5B06
(C)
Page 3-6

N5B06
How much voltage does a wall outlet usually supply (in the US)?
A. About 12 volts
B. About 30 volts
C. About 120 volts
D. About 480 volts

N5B07
(C)
Page 3-7

N5B07
What are three good electrical conductors?
A. Copper, gold, mica
B. Gold, silver, wood
C. Gold, silver, aluminum
D. Copper, aluminum, paper

N5B08
(A)
Page 3-7

N5B08
What are four good electrical insulators?
A. Glass, air, plastic, porcelain
B. Glass, wood, copper, porcelain
C. Paper, glass, air, aluminum
D. Plastic, rubber, wood, carbon

N5B09
(B)
Page 3-7

N5B09
What does an electrical insulator do?
A. It lets electricity flow through it in one direction
B. It does not let electricity flow through it
C. It lets electricity flow through it when light shines on it
D. It lets electricity flow through it

N5B10
(D)
Page 3-9

N5B10
What limits the amount of current that flows through a circuit if the voltage stays the same?
A. Reliance
B. Reactance
C. Saturation
D. Resistance

N5B11
(D)
Page 3-9

N5B11
What is the basic unit of resistance?
A. The volt
B. The watt
C. The ampere
D. The ohm

N5C Ohm's Law {any calculations will be kept to a very low level - no fractions or decimals} and the concepts of energy and power, and open and short circuits.

N5C01
(A)
Page 3-10

N5C01
What formula shows how voltage, current and resistance relate to each other in an electric circuit?
A. Ohm's Law
B. Kirchhoff's Law
C. Ampere's Law
D. Tesla's Law

N5C02

If a current of 2 amperes flows through a 50-ohm resistor, what is the voltage across the resistor?

A. 25 volts
B. 52 volts
C. 100 volts
D. 200 volts

N5C02
(C)
Page 3-10

N5C03

If a 100-ohm resistor is connected to 200 volts, what is the current through the resistor?

A. 1/2 ampere
B. 2 amperes
C. 300 amperes
D. 20000 amperes

N5C03
(B)
Page 3-10

N5C04

If a current of 3 amperes flows through a resistor connected to 90 volts, what is the resistance?

A. 30 ohms
B. 93 ohms
C. 270 ohms
D. 1/30 ohm

N5C04
(A)
Page 3-10

N5C05

What is the word used to describe how fast electrical energy is used?

A. Resistance
B. Current
C. Power
D. Voltage

N5C05
(C)
Page 3-13

N5C06

If you have light bulbs marked 60 watts, 75 watts and 100 watts, which one will use electrical energy the fastest?

A. The 60 watt bulb
B. The 75 watt bulb
C. The 100 watt bulb
D. They will all be the same

N5C06
(C)
Page 3-13

N5C07

What is the basic unit of electrical power?

A. The ohm
B. The watt
C. The volt
D. The ampere

N5C07
(B)
Page 3-13

N5C08

Which electrical circuit can have no current?

A. A closed circuit
B. A short circuit
C. An open circuit
D. A complete circuit

N5C08
(C)
Page 3-12

N5C09
(D)
Page 3-9

N5C09
Which electrical circuit uses too much current?
A. An open circuit
B. A dead circuit
C. A closed circuit
D. A short circuit

N5C10
(B)
Page 3-14

N5C10
What is the name of a current that flows only in one direction?
A. An alternating current
B. A direct current
C. A normal current
D. A smooth current

N5C11
(A)
Page 3-14

N5C11
What is the name of a current that flows back and forth, first in one direction, then in the opposite direction?
A. An alternating current
B. A direct current
C. A rough current
D. A reversing current

N5D Concepts of frequency, including AC vs DC, frequency units, AF vs RF and wavelength.

N5D01
(D)
Page 3-15

N5D01
What term means the number of times per second that an alternating current flows back and forth?
A. Pulse rate
B. Speed
C. Wavelength
D. Frequency

N5D02
(A)
Page 3-15

N5D02
What is the basic unit of frequency?
A. The hertz
B. The watt
C. The ampere
D. The ohm

N5D03
(B)
Page 3-15

N5D03
What frequency can humans hear?
A. 0 - 20 Hz
B. 20 - 20,000 Hz
C. 200 - 200,000 Hz
D. 10,000 - 30,000 Hz

N5D04
(B)
Page 3-15

N5D04
Why do we call signals in the range 20 Hz to 20,000 Hz audio frequencies?
A. Because the human ear cannot sense anything in this range
B. Because the human ear can sense sounds in this range
C. Because this range is too low for radio energy
D. Because the human ear can sense radio waves in this range

N5D05

What is the lowest frequency of electrical energy that is usually known as a radio frequency?

A. 20 Hz
B. 2,000 Hz
C. 20,000 Hz
D. 1,000,000 Hz

N5D05
(C)
Page 3-15

N5D06

Electrical energy at a frequency of 7125 kHz is in what frequency range?

A. Audio
B. Radio
C. Hyper
D. Super-high

N5D06
(B)
Page 3-15

N5D07

If a radio wave makes 3,725,000 cycles in one second, what does this mean?

A. The radio wave's voltage is 3,725 kilovolts
B. The radio wave's wavelength is 3,725 kilometers
C. The radio wave's frequency is 3,725 kilohertz
D. The radio wave's speed is 3,725 kilometers per second

N5D07
(C)
Page 3-15

N5D08

What is the name for the distance an AC signal travels during one complete cycle?

A. Wave speed
B. Waveform
C. Wavelength
D. Wave spread

N5D08
(C)
Page 3-15

N5D09

What happens to a signal's wavelength as its frequency increases?

A. It gets shorter
B. It gets longer
C. It stays the same
D. It disappears

N5D09
(A)
Page 3-16

N5D10

What happens to a signal's frequency as its wavelength gets longer?

A. It goes down
B. It goes up
C. It stays the same
D. It disappears

N5D10
(A)
Page 3-16

N5D11

What does 60 hertz (Hz) mean?

A. 6000 cycles per second
B. 60 cycles per second
C. 6000 meters per second
D. 60 meters per second

N5D11
(B)
Page 3-15

SUBELEMENT N6—CIRCUIT COMPONENTS
[2 exam questions—2 groups]

N6A Electrical function and/or schematic representation of resistor, switch, fuse, or battery.

N6A01
(B)
Page 4-3

N6A01
What can a single-pole, double-throw switch do?
A. It can switch one input to one output
B. It can switch one input to either of two outputs
C. It can switch two inputs at the same time, one input to either of two outputs, and the other input to either of two outputs
D. It can switch two inputs at the same time, one input to one output, and the other input to another output

N6A02
(D)
Page 4-3

N6A02
What can a double-pole, single-throw switch do?
A. It can switch one input to one output
B. It can switch one input to either of two outputs
C. It can switch two inputs at the same time, one input to either of two outputs, and the other input to either of two outputs
D. It can switch two inputs at the same time, one input to one output, and the other input to the other output

N6A03
(A)
Page 4-4

N6A03
Which component has a positive and a negative side?
A. A battery
B. A potentiometer
C. A fuse
D. A resistor

N6A04
(B)
Page 4-1

N6A04
Which component has a value that can be changed?
A. A single-cell battery
B. A potentiometer
C. A fuse
D. A resistor

N6A05
(B)
Page 4-2

N6A05
In Figure N6-1 which symbol represents a variable resistor or potentiometer?
A. Symbol 1
B. Symbol 2
C. Symbol 3
D. Symbol 4

Figure N6–1

N6A06
(C)
Page 4-2

N6A06
In Figure N6-1 which symbol represents a fixed resistor?
A. Symbol 1
B. Symbol 2
C. Symbol 3
D. Symbol 4

N6A07

In Figure N6-1 which symbol represents a fuse?
A. Symbol 1   C. Symbol 3
B. Symbol 2   D. Symbol 4

N6A07
(A)
Page 4-4

N6A08

In Figure N6-1 which symbol represents a single-cell battery?
A. Symbol 1   C. Symbol 3
B. Symbol 2   D. Symbol 4

N6A08
(D)
Page 4-5

N6A09

In Figure N6-2 which symbol represents a single-pole, single-throw switch?
A. Symbol 1   C. Symbol 3
B. Symbol 2   D. Symbol 4

N6A09
(A)
Page 4-2

Figure N6-2

N6A10

In Figure N6-2 which symbol represents a single-pole, double-throw switch?
A. Symbol 1   C. Symbol 3
B. Symbol 2   D. Symbol 4

N6A10
(D)
Page 4-2

N6A11

In Figure N6-2 which symbol represents a double-pole, single-throw switch?
A. Symbol 1   C. Symbol 3
B. Symbol 2   D. Symbol 4

N6A11
(C)
Page 4-2

N6A12

In Figure N6-2 which symbol represents a double-pole, double-throw switch?
A. Symbol 1   C. Symbol 3
B. Symbol 2   D. Symbol 4

N6A12
(B)
Page 4-2

N6B Electrical function and/or schematic representation of a ground, antenna, transistor, or a triode vacuum tube.

N6B01

Which component can amplify a small signal using low voltages?
A. A PNP transistor
B. A variable resistor
C. An electrolytic capacitor
D. A multiple-cell battery

N6B01
(A)
Page 4-7

N6B02

Which component conducts electricity from a negative emitter to a positive collector when its base voltage is made positive?
A. A variable resistor
B. An NPN transistor
C. A triode vacuum tube
D. A multiple-cell battery

N6B02
(B)
Page 4-7

N6B03
(A)
Page 4-5

N6B04
(D)
Page 4-5

N6B05
(A)
Page 4-5

N6B06
(C)
Page 4-5

N6B07
(D)
Page 4-7

N6B08
(A)
Page 4-7

N6B09
(B)
Page 4-8

N6B10
(A)
Page 4-8

N6B11
(C)
Page 4-8

N6B03
Which component is used to radiate radio energy?
A. An antenna   C. A chassis ground
B. An earth ground   D. A potentiometer

N6B04
In Figure N6-3 which symbol represents an earth ground?
A. Symbol 1   C. Symbol 3
B. Symbol 2   D. Symbol 4

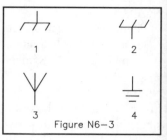

Figure N6-3

N6B05
In Figure N6-3 which symbol represents a chassis ground?
A. Symbol 1   C. Symbol 3
B. Symbol 2   D. Symbol 4

N6B06
In Figure N6-3 which symbol represents an antenna?
A. Symbol 1   C. Symbol 3
B. Symbol 2   D. Symbol 4

N6B07
In Figure N6-4 which symbol represents an NPN transistor?
A. Symbol 1   C. Symbol 3
B. Symbol 2   D. Symbol 4

Figure N6-4

N6B08
In Figure N6-4 which symbol represents a PNP transistor?
A. Symbol 1   C. Symbol 3
B. Symbol 2   D. Symbol 4

N6B09
In Figure N6-4 which symbol represents a triode vacuum tube?
A. Symbol 1   C. Symbol 3
B. Symbol 2   D. Symbol 4

N6B10
What is one reason a triode vacuum tube might be used instead of a transistor in a circuit?
A. It handles higher power   C. It uses less current
B. It uses lower voltages   D. It is much smaller

N6B11
Which component can amplify a small signal but must use high voltages?
A. A transistor   C. A vacuum tube
B. An electrolytic capacitor   D. A multiple-cell battery

SUBELEMENT N7—PRACTICAL CIRCUITS
[2 exam questions—2 groups]

N7A Functional layout of transmitter, transceiver, receiver, power supply, antenna, antenna switch, antenna feed line, impedance-matching device, SWR meter.

N7A01
What would you connect to your transceiver if you wanted to switch it between more than one type of antenna?
A. A terminal-node switch
B. An antenna switch
C. A telegraph key switch
D. A high-pass filter

N7A01
(B)
Page 5-3

N7A02
What device might allow use of an antenna on a band it was not designed for?
A. An SWR meter
B. A low-pass filter
C. An antenna tuner
D. A high-pass filter

N7A02
(C)
Page 5-3

N7A03
What connects your transceiver to your antenna?
A. A dummy load
B. A ground wire
C. The power cord
D. A feed line

N7A03
(D)
Page 5-3

N7A04
What might you connect between your transceiver and an antenna switch connected to several types of antennas?
A. A high-pass filter
B. An SWR meter
C. A key-click filter
D. A mixer

N7A04
(B)
Page 5-3

N7A05
If your SWR meter is connected to an antenna tuner on one side, what would you connect to the other side of it?
A. A power supply
B. An antenna
C. An antenna switch
D. A transceiver

N7A05
(D)
Page 5-3

N7A06
Which of these should never be connected to the output of a transceiver?
A. An antenna switch
B. An SWR meter
C. An antenna
D. A receiver

N7A06
(D)
Page 5-3

N7A07
(A)
Page 5-2

N7A07
If your mobile transceiver works in your car but not in your home, what should you check first?
A. The power supply
B. The speaker
C. The microphone
D. The SWR meter

N7A08
(A)
Page 5-3

N7A08
What does an antenna tuner do?
A. It matches a transceiver to a mismatched antenna system
B. It helps a receiver automatically tune in stations that are far away
C. It switches an antenna system to a transceiver when sending, and to a receiver when listening
D. It switches a transceiver between different kinds of antennas connected to one feed line

N7A09
(B)
Page 5-3

N7A09
In Figure N7-1, if block 1 is a transceiver and block 3 is a dummy antenna what is block 2?
A. A terminal-node switch
B. An antenna switch
C. A telegraph key switch
D. A high-pass filter

Figure N7–1

N7A10
(A)
Page 5-3

N7A10
In Figure N7-2, if block 2 is an SWR meter and block 3 is an antenna switch, what is block 1?
A. A transceiver
B. A high-pass filter
C. An antenna tuner
D. A modem

Figure N7–2

N7A11
(B)
Page 5-3

N7A11
In Figure N7-3, if block 1 is a transceiver and block 2 is an SWR meter, what is block 3?
A. An antenna switch
B. An antenna tuner
C. A key-click filter
D. A terminal-node controller

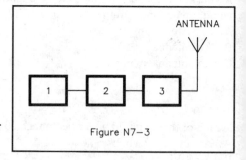

Figure N7–3

N7A12
(C)
Page 5-2

N7A12
What device converts household current to 12 VDC?
A. A catalytic converter
B. A low-pass filter
C. A power supply
D. An RS-232 interface

N7A13
Which of these usually needs a heavy-duty power supply?
A. An SWR meter
B. A receiver
C. A transceiver
D. An antenna switch

N7A13
(C)
Page 5-2

N7B Station layout and accessories for telegraphy, radiotelephone, radioteleprinter or packet

N7B01
What would you connect to a transceiver to send Morse code?
A. A terminal-node controller
B. A telegraph key
C. An SWR meter
D. An antenna switch

N7B01
(B)
Page 5-4

N7B02
Where would you connect a telegraph key to send Morse code?
A. To a power supply
B. To an antenna switch
C. To a transceiver
D. To an antenna

N7B02
(C)
Page 5-4

N7B03
What do many amateurs use to help form good Morse code characters?
A. A key-operated on/off switch
B. An electronic keyer
C. A key-click filter
D. A DTMF keypad

N7B03
(B)
Page 5-4

N7B04
Where would you connect a microphone for voice operation?
A. To a power supply
B. To an antenna switch
C. To a transceiver
D. To an antenna

N7B04
(C)
Page 5-5

N7B05
What would you connect to a transceiver for voice operation?
A. A splatter filter
B. A terminal-voice controller
C. A receiver audio filter
D. A microphone

N7B05
(D)
Page 5-5

N7B06
What would you connect to a transceiver for RTTY operation?
A. A modem and a teleprinter or computer system
B. A computer, a printer and a RTTY refresh unit
C. A terminal voice controller
D. A modem, a monitor and a DTMF keypad

N7B06
(A)
Page 5-5

N7B07
(C)
Page 5-5

N7B07
What would you connect between a transceiver and a computer system or teleprinter for RTTY operation?
A. An RS-232 interface
B. A DTMF keypad
C. A modem
D. A terminal-network controller

N7B08
(A)
Page 5-6

N7B08
What would you connect between a computer system and a transceiver for packet-radio operation?
A. A terminal-node controller
B. A DTMF keypad
C. An SWR bridge
D. An antenna tuner

N7B09
(C)
Page 5-6

N7B09
Where would you connect a terminal-node controller for packet-radio operation?
A. Between your antenna and transceiver
B. Between your computer and monitor
C. Between your computer and transceiver
D. Between your keyboard and computer

N7B10
(D)
Page 5-5

N7B10
In RTTY operation, what equipment connects to a modem?
A. A DTMF keypad, a monitor and a transceiver
B. A DTMF microphone, a monitor and a transceiver
C. A transceiver and a terminal-network controller
D. A transceiver and a teleprinter or computer system

N7B11
(B)
Page 5-6

N7B11
In packet-radio operation, what equipment connects to a terminal-node controller?
A. A transceiver and a modem
B. A transceiver and a terminal or computer system
C. A DTMF keypad, a monitor and a transceiver
D. A DTMF microphone, a monitor and a transceiver

Subelement N8

SUBELEMENT N8—SIGNALS AND EMISSIONS
[2 exam questions—2 groups]

N8A Emission types, key clicks, chirps or superimposed hum.

N8A01
(B)
Page 9-3

N8A01
How is CW usually transmitted?
A. By frequency-shift keying an RF signal
B. By on/off keying an RF signal
C. By audio-frequency-shift keying an oscillator tone
D. By on/off keying an audio-frequency signal

N8A02
(A)
Page 9-16

N8A02
How is RTTY usually transmitted?
A. By frequency-shift keying an RF signal
B. By on/off keying an RF signal
C. By digital pulse-code keying of an unmodulated carrier
D. By on/off keying an audio-frequency signal

N8A03
What is the name for international Morse code emissions?
A. RTTY
B. Data
C. CW
D. Phone

N8A03
(C)
Page 9-3

N8A04
What is the name for narrow-band direct-printing telegraphy emissions?
A. RTTY
B. Data
C. CW
D. Phone

N8A04
(A)
Page 9-17

N8A05
What is the name for packet-radio emissions?
A. RTTY
B. Data
C. CW
D. Phone

N8A05
(B)
Page 9-22

N8A06
What is the name for voice emissions?
A. RTTY
B. Data
C. CW
D. Phone

N8A06
(D)
Page 9-7

N8A07
How can you prevent key clicks?
A. By sending CW more slowly
B. By increasing power
C. By using a better power supply
D. By using a key-click filter

N8A07
(D)
Page 10-7

N8A08
What does chirp mean?
A. An overload in a receiver's audio circuit whenever CW is received
B. A high-pitched tone which is received along with a CW signal
C. A small change in a transmitter's frequency each time it is keyed
D. A slow change in transmitter frequency as the circuit warms up

N8A08
(C)
Page 10-7

N8A09
What can be done to keep a CW transmitter from chirping?
A. Add a low-pass filter
B. Use an RF amplifier
C. Keep the power supply current very steady
D. Keep the power supply voltages very steady

N8A09
(D)
Page 10-7

N8A10
What may cause a buzzing or hum in the signal of an HF transmitter?
A. Using an antenna which is the wrong length
B. Energy from another transmitter
C. Bad design of the transmitter's RF power output circuit
D. A bad filter capacitor in the transmitter's power supply

N8A10
(D)
Page 10-8

N8A11
(A)
Page 9-7

N8A11
Which sideband is commonly used for 10-meter phone operation?
A. Upper-sideband
B. Lower-sideband
C. Amplitude-compandored sideband
D. Double-sideband

N8B Harmonics and unwanted signals, equipment and adjustments to help reduce interference to others.

N8B01
(C)
Page 10-1

N8B01
How does the frequency of a harmonic compare to the desired transmitting frequency?
A. It is slightly more than the desired frequency
B. It is slightly less than the desired frequency
C. It is exactly two, or three, or more times the desired frequency
D. It is much less than the desired frequency

N8B02
(A)
Page 10-1

N8B02
What is the fourth harmonic of a 7160-kHz signal?
A. 28,640 kHz
B. 35,800 kHz
C. 28,160 kHz
D. 1790 kHz

N8B03
(C)
Page 10-1

N8B03
If you are told your station was heard on 21,375 kHz, but at the time you were operating on 7125 kHz, what is one reason this could happen?
A. Your transmitter's power-supply filter capacitor was bad
B. You were sending CW too fast
C. Your transmitter was radiating harmonic signals
D. Your transmitter's power-supply filter choke was bad

N8B04
(D)
Page 10-1

N8B04
If someone tells you that signals from your hand-held transceiver are interfering with other signals on a frequency near yours, what may be the cause?
A. You may need a power amplifier for your hand-held
B. Your hand-held may have chirp from weak batteries
C. You may need to turn the volume up on your hand-held
D. Your hand-held may be transmitting spurious emissions

N8B05
(D)
Page 10-1

N8B05
If your transmitter sends signals outside the band where it is transmitting, what is this called?
A. Off-frequency emissions
B. Transmitter chirping
C. Side tones
D. Spurious emissions

N8B06
(A)
Page 10-5

N8B06
What problem may occur if your transmitter is operated without the cover and other shielding in place?
A. It may transmit spurious emissions
B. It may transmit a chirpy signal
C. It may transmit a weak signal
D. It may interfere with other stations operating near its frequency

N8B07
What may happen if an SSB transmitter is operated with the microphone gain set too high?
A. It may cause digital interference to computer equipment
B. It may cause splatter interference to other stations operating near its frequency
C. It may cause atmospheric interference in the air around the antenna
D. It may cause interference to other stations operating on a higher frequency band

N8B08
What may happen if an SSB transmitter is operated with too much speech processing?
A. It may cause digital interference to computer equipment
B. It may cause splatter interference to other stations operating near its frequency
C. It may cause atmospheric interference in the air around the antenna
D. It may cause interference to other stations operating on a higher frequency band

N8B09
What may happen if an FM transmitter is operated with the microphone gain or deviation control set too high?
A. It may cause digital interference to computer equipment
B. It may cause interference to other stations operating near its frequency
C. It may cause atmospheric interference in the air around the antenna
D. It may cause interference to other stations operating on a higher frequency band

N8B10
What may your FM hand-held or mobile transceiver do if you shout into its microphone?
A. It may cause digital interference to computer equipment
B. It may cause interference to other stations operating near its frequency
C. It may cause atmospheric interference in the air around the antenna
D. It may cause interference to other stations operating on a higher frequency band

N8B11
What can you do if you are told your FM hand-held or mobile transceiver is over deviating?
A. Talk louder into the microphone
B. Let the transceiver cool off
C. Change to a higher power level
D. Talk farther away from the microphone

N8B07
(B)
Page 9-16

N8B08
(B)
Page 9-16

N8B09
(B)
Page 9-16

N8B10
(B)
Page 9-16

N8B11
(D)
Page 9-16

SUBELEMENT N9—ANTENNAS AND FEED LINES
[3 exam questions—3 groups]

N9A Wavelength vs antenna length.

N9A01
(D)
Page 7-9

N9A01
How do you calculate the length (in feet) of a half-wavelength dipole antenna?
A. Divide 150 by the antenna's operating frequency (in MHz) [150/f(in MHz)]
B. Divide 234 by the antenna's operating frequency (in MHz) [234/f (in MHz)]
C. Divide 300 by the antenna's operating frequency (in MHz) [300/f (in MHz)]
D. Divide 468 by the antenna's operating frequency (in MHz) [468/f (in MHz)]

N9A02
(B)
Page 7-17

N9A02
How do you calculate the length (in feet) of a quarter-wavelength vertical antenna?
A. Divide 150 by the antenna's operating frequency (in MHz) [150/f (in MHz)]
B. Divide 234 by the antenna's operating frequency (in MHz) [234/f (in MHz)]
C. Divide 300 by the antenna's operating frequency (in MHz) [300/f (in MHz)]
D. Divide 468 by the antenna's operating frequency (in MHz) [468/f (in MHz)]

N9A03
(A)
Page 7-9

N9A03
If you made a half-wavelength dipole antenna for 3725 kHz, how long would it be (to the nearest foot)?
A. 126 ft
B. 81 ft
C. 63 ft
D. 40 ft

N9A04
(C)
Page 7-9

N9A04
If you made a half-wavelength dipole antenna for 28.150 MHz, how long would it be (to the nearest foot)?
A. 22 ft
B. 11 ft
C. 17 ft
D. 34 ft

N9A05
(D)
Page 7-17

N9A05
If you made a quarter-wavelength vertical antenna for 7125 kHz, how long would it be (to the nearest foot)?
A. 11 ft
B. 16 ft
C. 21 ft
D. 33 ft

N9A06

If you made a quarter-wavelength vertical antenna for 21.125 MHz, how long would it be (to the nearest foot)?

A. 7 ft
B. 11 ft
C. 14 ft
D. 22 ft

N9A06
(B)
Page 7-17

N9A07

If you made a half-wavelength vertical antenna for 223 MHz, how long would it be (to the nearest inch)?

A. 112 inches
B. 50 inches
C. 25 inches
D. 12 inches

N9A07
(C)
Page 7-9

N9A08

If an antenna is made longer, what happens to its resonant frequency?

A. It decreases
B. It increases
C. It stays the same
D. It disappears

N9A08
(A)
Page 7-14

N9A09

If an antenna is made shorter, what happens to its resonant frequency?

A. It decreases
B. It increases
C. It stays the same
D. It disappears

N9A09
(B)
Page 7-14

N9A10

How could you lower the resonant frequency of a dipole antenna?

A. Lengthen the antenna
B. Shorten the antenna
C. Use less feed line
D. Use a smaller size feed line

N9A10
(A)
Page 7-14

N9A11

How could you raise the resonant frequency of a dipole antenna?

A. Lengthen the antenna
B. Shorten the antenna
C. Use more feed line
D. Use a larger size feed line

N9A11
(B)
Page 7-14

N9B Yagi parts, concept of directional antennas, and safety near antennas.

N9B01

In what direction does a Yagi antenna send out radio energy?

A. It goes out equally in all directions
B. Most of it goes in one direction
C. Most of it goes equally in two opposite directions
D. Most of it is aimed high into the air

N9B01
(B)
Page 7-22

N9B02
(C)
Page 7-23

N9B02
About how long is the driven element of a Yagi antenna?
A. 1/4 wavelength
B. 1/3 wavelength
C. 1/2 wavelength
D. 1 wavelength

N9B03
(D)
Page 7-23

N9B03
In Diagram N9-1, what is the name of element 2 of the Yagi antenna?
A. Director
B. Reflector
C. Boom
D. Driven element

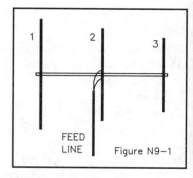

FEED
LINE   Figure N9-1

N9B04
(A)
Page 7-23

N9B04
In Diagram N9-1, what is the name of element 3 of the Yagi antenna?
A. Director
B. Reflector
C. Boom
D. Driven element

N9B05
(B)
Page 7-23

N9B05
In Diagram N9-1, what is the name of element 1 of the Yagi antenna?
A. Director
B. Reflector
C. Boom
D. Driven element

N9B06
(B)
Page 7-23

N9B06
Looking at the Yagi antenna in Diagram N9-1, in which direction on the page would it send most of its radio energy?
A. Left
B. Right
C. Top
D. Bottom

N9B07
(B)
Page 7-18

N9B07
Why is a 5/8-wavelength vertical antenna better than a 1/4-wavelength vertical antenna for VHF or UHF mobile operations?
A. A 5/8-wavelength antenna can handle more power
B. A 5/8-wavelength antenna has more gain
C. A 5/8-wavelength antenna has less corona loss
D. A 5/8-wavelength antenna is easier to install on a car

N9B08
(C)
Page 7-16

N9B08
In what direction does a vertical antenna send out radio energy?
A. Most of it goes in two opposite directions
B. Most of it goes high into the air
C. Most of it goes equally in all horizontal directions
D. Most of it goes in one direction

N9B09

If the ends of a half-wave dipole antenna point east and west, which way would the antenna send out radio energy?

A. Equally in all directions
B. Mostly up and down
C. Mostly north and south
D. Mostly east and west

N9B09
(C)
Page 7-10

N9B10

How should you hold the antenna of a hand-held transceiver while you are transmitting?

A. Away from your head and away from others
B. Pointed towards the station you are contacting
C. Pointed away from the station you are contacting
D. Pointed down to bounce the signal off the ground

N9B10
(A)
Page 7-25

N9B11

Why should your outside antennas be high enough so that no one can touch them while you are transmitting?

A. Touching the antenna might cause television interference
B. Touching the antenna might cause RF burns
C. Touching the antenna might radiate harmonics
D. Touching the antenna might reflect the signal back to the transmitter and cause damage

N9B11
(B)
Page 7-25

N9C Feed lines, baluns and polarization via element orientation.

N9C01

What is a coaxial cable?

A. Two wires side-by-side in a plastic ribbon
B. Two wires side-by-side held apart by insulating rods
C. Two wires twisted around each other in a spiral
D. A center wire inside an insulating material covered by a metal sleeve or shield

N9C01
(D)
Page 7-2

N9C02

Why does coaxial cable make a good antenna feed line?

A. You can make it at home, and its impedance matches most amateur antennas
B. It is weatherproof, and its impedance matches most amateur antennas
C. It is weatherproof, and its impedance is higher than that of most amateur antennas
D. It can be used near metal objects, and its impedance is higher than that of most amateur antennas

N9C02
(B)
Page 7-3

N9C03

Which kind of antenna feed line can carry radio energy very well even if it is buried in the ground?

A. Twin lead
B. Coaxial cable
C. Parallel conductor
D. Twisted pair

N9C03
(B)
Page 7-3

N9C04
(A)
Page 7-3

N9C04
What is the best antenna feed line to use if it must be put near grounded metal objects?
A. Coaxial cable
B. Twin lead
C. Twisted pair
D. Ladder-line

N9C05
(B)
Page 7-3

N9C05
What is parallel-conductor feed line?
A. Two wires twisted around each other in a spiral
B. Two wires side-by-side held apart by insulating rods
C. A center wire inside an insulating material which is covered by a metal sleeve or shield
D. A metal pipe which is as wide or slightly wider than a wavelength of the signal it carries

N9C06
(D)
Page 7-4

N9C06
What are some reasons to use parallel-conductor feed line?
A. It has low impedance, and will operate with a high SWR
B. It will operate with a high SWR, and it works well when tied down to metal objects
C. It has a low impedance, and has less loss than coaxial cable
D. It will operate with a high SWR, and has less loss than coaxial cable

N9C07
(A)
Page 7-4

N9C07
What are some reasons not to use parallel-conductor feed line?
A. It does not work well when tied down to metal objects, and you must use an impedance-matching device with your transceiver
B. It is difficult to make at home, and it does not work very well with a high SWR
C. It does not work well when tied down to metal objects, and it cannot operate under high power
D. You must use an impedance-matching device with your transceiver, and it does not work very well with a high SWR

N9C08
(B)
Page 7-3

N9C08
What kind of antenna feed line is made of two conductors held apart by insulated rods?
A. Coaxial cable
B. Open-conductor ladder line
C. Twin lead in a plastic ribbon
D. Twisted pair

N9C09
(C)
Page 7-2

N9C09
What would you use to connect a coaxial cable of 50-ohms impedance to an antenna of 35-ohms impedance?
A. A terminating resistor
B. An SWR meter
C. An impedance-matching device
D. A low-pass filter

N9C10
What does balun mean?
A. Balanced antenna network
B. Balanced unloader
C. Balanced unmodulator
D. Balanced to unbalanced

N9C11
Where would you install a balun to feed a dipole antenna with 50-ohm coaxial cable?
A. Between the coaxial cable and the antenna
B. Between the transmitter and the coaxial cable
C. Between the antenna and the ground
D. Between the coaxial cable and the ground

N9C10
(D)
Page 7-6

N9C11
(A)
Page 7-6

Element 2 (Novice)

Figure N6-1

Figure N6-2

Figure N6-3

Figure N6-4

ANTENNA

Figure N7-1

DIPOLE

BEAM

DUMMY
LOAD

Figure N7-2

ANTENNA

Figure N7-3

FEED
LINE   Figure N9-1

$$\frac{E}{I \ R} \qquad \frac{984}{F_{MHz}} = Ft \qquad \frac{468}{F_{MHz}} = Ft \qquad \frac{234}{F_{MHz}} = Ft \qquad \frac{300}{F_{MHz}} = M$$

As Prepared by the
Question Pool Committee
for all examinations
administered
July 1, 1993 through June 30, 1997

Chapter 13

Technician (Element 3A) Question Pool—With Answers

DON'T START HERE

This chapter contains the complete question pool for the Element 3A exam. Element 3A is part of the Technician license exam. To earn a Technician license, you must also pass the Element 2 exam. The Element 2 question pool is in Chapter 12. The Technician license does not require a Morse code test.

Before you read the questions and answers printed in this chapter, be sure to read the text in the previous chapters. Use these questions as review exercises, when the text tells you to study them. Don't try to memorize all the questions and answers.

This pool, released by the Volunteer Examiner Coordinators' Question Pool Committee, will be used on exams beginning July 1, 1993. The pool is scheduled to be used until June 30, 1997. Changes to FCC Rules and other factors may result in earlier revisions. Such changes will be announced in *QST* and other Amateur Radio publications.

HOW MANY QUESTIONS?

The FCC specifies that an Element 3A exam must include 25 questions. Nine subelements make up each exam question pool. The FCC also specifies the number of questions from each subelement that must appear on the exam. For example, there must be three questions from the Operat-

ing Procedures subelement and two questions from the Circuit Components subelement. Table 13-1 summarizes the number of questions to be selected from each subelement on the exam. These numbers are also included at the beginning of each subelement in the question pool.

The Volunteer Examiner Coordinators' Question Pool Committee has broken the subelements into smaller groups. Each subelement has the same number of groups as there are questions from that subelement on the exam. This means there are three groups for the Operating Procedures subelement and two groups for the Circuit Components subelement. One question from each group is to be used on the exam. These groups are called out with a list of topics given in bold type in the question pool. They are labeled alphabetically within each subelement. The five groups in the Commission's Rules subelement are labeled T1A through T1E.

The question numbers used in the question pool relate to the syllabus or study guide printed at the end of Chapter 1. The syllabus is an outline of topics covered by the exam. Each question number begins with a T to indicate the question is from the Technician question pool. Next is a number to indicate which of the nine subelements the question is from. A letter following this number indicates which subelement group the question is from. Each question number ends with a two-digit number to specify its position in the set. So question T2A01 is the first question in the A group of the second subelement. Question T9C08 is the eighth question in the C group of the ninth subelement.

This question pool contains more than ten times the number of questions necessary to make up an exam. This ensures that the examiners have sufficient questions to choose from when they make up an exam.

WHO PICKS THE QUESTIONS?

The FCC allows Volunteer-Examiner teams to select the questions that will be used on amateur exams. If your test is coordinated by the ARRL/VEC, however, your test will be prepared by the VEC, or using a computer program supplied by the VEC.

Table 13-1

Technician Exam Content

Subelement	Topic	Number of Questions
T1	Commission's Rules	5
T2	Operating Procedures	3
T3	Radio-Wave Propagation	3
T4	Amateur Radio Practices	4
T5	Electrical Principles	2
T6	Circuit Components	2
T7	Practical Circuits	1
T8	Signals and Emissions	2
T9	Antennas and Feed Lines	3

QUESTION POOL FORMAT

The rest of this chapter contains the entire Element 3A question pool. We have printed the answer key to these questions along the edge of the page. There is a line to indicate where you should fold the page to hide the answer key while you study. After making your best effort to answer the questions, you can look at the answers to check your understanding. We also have included page references along with the answers. These page numbers indicate where you will find the text discussion related to each question. If you have any problems with a question, refer to the page listed for that question. You may have to study beyond the listed page number to review all the related material. We also have included references to sections of Part 97 for the Commission's Rules Subelement, T1. This is to help you identify the particular rule citations for these questions. The complete text of the FCC Rules in Part 97 is included in the ARRL publication, *The FCC Rule Book*.

The Question Pool Committee included all the drawings for the question pool on a single page. We placed a copy of this page at the back of the question pool, for your information. You should expect to receive a page like this as part of your exam papers. Notice there is a representation of the Ohm's Law Circle you learned about in Chapter 3 on the page.

$$\frac{E}{I \ R}$$

This will help you remember the Ohm's Law equations for your exam. You may not need this equation for your exam, but it will be available for you on the Figure sheet. For your convenience, we also placed the individual figures near the questions that refer to them in the question pool.

Good luck with your studies.

Element 3A (Technician Class) Question Pool

ANSWER
KEY
Page numbers tell you where to look in this book for more information.

Subelement T1

Numbers in [square brackets] indicate sections in FCC Part 97, the Amateur Radio rules.

SUBELEMENT T1—COMMISSION'S RULES
[5 exam questions—5 groups]

T1A Station control, frequency privileges authorized to the Technician-class control operator, term of licenses, grace periods and modifications of licenses.

T1A01
What is the control point of an amateur station?
A. The on/off switch of the transmitter
B. The input/output port of a packet controller
C. The variable frequency oscillator of a transmitter
D. The location at which the control operator function is performed

T1A02
What is the term for the location at which the control operator function is performed?
A. The operating desk
B. The control point
C. The station location
D. The manual control location

T1A03
What must you do to renew or change your operator/primary station license?
A. Properly fill out FCC Form 610 and send it to the FCC in Gettysburg, PA
B. Properly fill out FCC Form 610 and send it to the nearest FCC field office
C. Properly fill out FCC Form 610 and send it to the FCC in Washington, DC
D. An amateur license never needs changing or renewing

T1A04
What is the "grace period" during which the FCC will renew an expired 10-year license?
A. 2 years
B. 5 years
C. 10 years
D. There is no grace period

T1A01
(D)
[97.3a12]
Page 2-23

T1A02
(B)
[97.3a12]
Page 2-23

T1A03
(A)
[97.19a/b]
Page 2-24

T1A04
(A)
[97.19c]
Page 2-24

T1A05
(C)
[97.301/
305e]
Page 2-23

T1A05
Which of the following frequencies may a Technician operator who has passed a Morse code test use?
A. 7.1 - 7.2 MHz
B. 14.1 - 14.2 MHz
C. 21.1 - 21.2 MHz
D. 28.1 - 29.2 MHz

T1A06
(C)
[97.301a]
Page 2-23

T1A06
Which operator licenses authorize privileges on 52.525 MHz?
A. Extra, Advanced only
B. Extra, Advanced, General only
C. Extra, Advanced, General, Technician only
D. Extra, Advanced, General, Technician, Novice

T1A07
(B)
[97.301a]
Page 2-23

T1A07
Which operator licenses authorize privileges on 146.52 MHz?
A. Extra, Advanced, General, Technician, Novice
B. Extra, Advanced, General, Technician only
C. Extra, Advanced, General only
D. Extra, Advanced only

T1A08
(A)
[97.301a]
Page 2-23

T1A08
Which operator licenses authorize privileges on 223.50 MHz?
A. Extra, Advanced, General, Technician, Novice
B. Extra, Advanced, General, Technician only
C. Extra, Advanced, General only
D. Extra, Advanced only

T1A09
(B)
[97.301a]
Page 2-23

T1A09
Which operator licenses authorize privileges on 446.0 MHz?
A. Extra, Advanced, General, Technician, Novice
B. Extra, Advanced, General, Technician only
C. Extra, Advanced, General only
D. Extra, Advanced only

T1A10
(D)
[97.301e]
Page 2-23

T1A10
In addition to passing the Technician written examination (Elements 2 and 3A), what must you do before you are allowed to use amateur frequencies below 30 MHz?
A. Nothing special is needed; all Technicians may use the HF bands at any time
B. You must notify the FCC that you intend to operate on the HF bands
C. You must attend a class to learn about HF communications
D. You must pass a Morse code test (either Element 1A, 1B or 1C)

T1A11
(C)
[97.301e]
Page 2-29

T1A11
If you are a Technician licensee, what must you have to prove that you are authorized to use the Novice amateur frequencies below 30 MHz?
A. A certificate from the FCC showing that you have notified them that you will be using the HF bands
B. A certificate from an instructor showing that you have attended a class in HF communications
C. Written proof of having passed a Morse code test
D. No special proof is required before using the HF bands

T1B Emission privileges for Technician-class control operator, frequency selection and sharing, transmitter power.

T1B01

At what point in your station is transceiver power measured?
A. At the power supply terminals inside the transmitter or amplifier
B. At the final amplifier input terminals inside the transmitter or amplifier
C. At the antenna terminals of the transmitter or amplifier
D. On the antenna itself, after the feed line

T1B01
(C)
[97.3b6]
Page 2-24

T1B02

What is the term for the average power supplied to an antenna transmission line during one RF cycle at the crest of the modulation envelope?
A. Peak transmitter power
B. Peak output power
C. Average radio-frequency power
D. Peak envelope power

T1B02
(D)
[97.3b6]
Page 2-24

T1B03

What is the maximum transmitting power permitted an amateur station in beacon operation?
A. 10 watts PEP output
B. 100 watts PEP output
C. 500 watts PEP output
D. 1500 watts PEP output

T1B03
(B)
[97.203c]
Page 2-26

T1B04

If the FCC rules say that the amateur service is a secondary user of a frequency band, and another service is a primary user, what does this mean?
A. Nothing special; all users of a frequency band have equal rights to operate
B. Amateurs are only allowed to use the frequency band during emergencies
C. Amateurs are allowed to use the frequency band only if they do not cause harmful interference to primary users
D. Amateurs must increase transmitter power to overcome any interference caused by primary users

T1B04
(C)
[97.303]
Page 2-24

T1B05

If you are using a frequency within a band assigned to the amateur service on a secondary basis, and a station assigned to the primary service on that band causes interference, what action should you take?
A. Notify the FCC's regional Engineer in Charge of the interference
B. Increase your transmitter's power to overcome the interference
C. Attempt to contact the station and request that it stop the interference
D. Change frequencies; you may be causing harmful interference to the other station, in violation of FCC rules

T1B05
(D)
[97.303]
Page 2-24

T1B06
(C)
[97.303a]
Page 2-24

T1B06
What rule applies if two amateur stations want to use the same frequency?
A. The station operator with a lesser class of license must yield the frequency to a higher-class licensee
B. The station operator with a lower power output must yield the frequency to the station with a higher power output
C. Both station operators have an equal right to operate on the frequency
D. Station operators in ITU Regions 1 and 3 must yield the frequency to stations in ITU Region 2

T1B07
(A)
[97.305a]
Page 2-29

T1B07
What emission type may always be used for station identification, regardless of the transmitting frequency?
A. CW
B. RTTY
C. MCW
D. Phone

T1B08
(B)
[97.305c]
Page 2-23

T1B08
On what frequencies within the 6-meter band may phone emissions be transmitted?
A. 50.0 - 54.0 MHz only
B. 50.1 - 54.0 MHz only
C. 51.0 - 54.0 MHz only
D. 52.0 - 54.0 MHz only

T1B09
(A)
[97.305c]
Page 2-23

T1B09
On what frequencies within the 2-meter band may image emissions be transmitted?
A. 144.1 - 148.0 MHz only
B. 146.0 - 148.0 MHz only
C. 144.0 - 148.0 MHz only
D. 146.0 - 147.0 MHz only

T1B10
(D)
[97.313b]
Page 2-24

T1B10
What is the maximum transmitting power permitted an amateur station on 146.52 MHz?
A. 200 watts PEP output
B. 500 watts ERP
C. 1000 watts DC input
D. 1500 watts PEP output

T1B11
(A)
[97.209b2]
Page 2-25

T1B11
Which band may NOT be used by Earth stations for satellite communications?
A. 6 meters
B. 2 meters
C. 70 centimeters
D. 23 centimeters

T1C Digital communications, station identification, ID with CSCE.

T1C01

If you are a Novice licensee with a Certificate of Successful Completion of Examination (CSCE) for Technician privileges, how do you identify your station when transmitting on 146.34 MHz?

A. You must give your call sign, followed by any suitable word that denotes the slant mark and the identifier "KT"
B. You may not operate on 146.34 until your new license arrives
C. No special form of identification is needed
D. You must give your call sign and the location of the VE examination where you obtained the CSCE

T1C01
(A)
[97.119e1]
Page 2-29

T1C02

What is the maximum frequency shift permitted for RTTY or data transmissions below 50 MHz?

A. 0.1 kHz
B. 0.5 kHz
C. 1 kHz
D. 5 kHz

T1C02
(C)
[97.307f3/4]
Page 2-23

T1C03

What is the maximum frequency shift permitted for RTTY or data transmissions above 50 MHz?

A. 0.1 kHz or the sending speed, in bauds, whichever is greater
B. 0.5 kHz or the sending speed, in bauds, whichever is greater
C. 5 kHz or the sending speed, in bauds, whichever is greater
D. The FCC rules do not specify a maximum frequency shift above 50 MHz

T1C03
(D)
[97.307]
Page 2-23

T1C04

What is the maximum symbol rate permitted for packet transmissions on the 10-meter band?

A. 300 bauds
B. 1200 bauds
C. 19.6 kilobauds
D. 56 kilobauds

T1C04
(B)
[97.307f4]
Page 2-23

T1C05

What is the maximum symbol rate permitted for packet transmissions on the 2-meter band?

A. 300 bauds
B. 1200 bauds
C. 19.6 kilobauds
D. 56 kilobauds

T1C05
(C)
[97.307f5]
Page 2-23

T1C06

What is the maximum symbol rate permitted for RTTY or data transmissions between 28 and 50 MHz?

A. 56 kilobauds
B. 19.6 kilobauds
C. 1200 bauds
D. 300 bauds

T1C06
(C)
[97.307f4]
Page 2-23

T1C07
(B)
[97.307f5]
Page 2-23

T1C07
What is the maximum symbol rate permitted for RTTY or data transmissions between 50 and 222 MHz?
A. 56 kilobauds
B. 19.6 kilobauds
C. 1200 bauds
D. 300 bauds

T1C08
(A)
[97.307f5]
Page 2-23

T1C08
What is the maximum authorized bandwidth of RTTY, data or multiplexed emissions using an unspecified digital code within the frequency range of 50 to 222 MHz?
A. 20 kHz
B. 50 kHz
C. The total bandwidth shall not exceed that of a single-sideband phone emission
D. The total bandwidth shall not exceed 10 times that of a CW emission

T1C09
(D)
[97.307f6]
Page 2-23

T1C09
What is the maximum symbol rate permitted for RTTY or data transmissions above 222 MHz?
A. 300 bauds
B. 1200 bauds
C. 19.6 kilobauds
D. 56 kilobauds

T1C10
(B)
[97.307f6]
Page 2-23

T1C10
What is the maximum authorized bandwidth of RTTY, data or multiplexed emissions using an unspecified digital code within the frequency range of 222 to 450 MHz?
A. 50 kHz
B. 100 kHz
C. 150 kHz
D. 200 kHz

T1C11
(C)
[97.307f6]
Page 2-23

T1C11
What is the maximum authorized bandwidth of RTTY, data or multiplexed emissions using an unspecified digital code within the 70-cm amateur band?
A. 300 kHz
B. 200 kHz
C. 100 kHz
D. 50 kHz

T1D Correct language, Phonetics, Beacons and Radio Control of model craft and vehicles.

T1D01
(A)
[97.3a9]
Page 2-26

T1D01
What is an amateur station called which transmits communications for the purpose of observation of propagation and reception?
A. A beacon
B. A repeater
C. An auxiliary station
D. A radio control station

T1D02
What is the fastest code speed a repeater may use for automatic identification?
A. 13 words per minute
B. 20 words per minute
C. 25 words per minute
D. There is no limitation

T1D03
If you are using a language besides English to make a contact, what language must you use when identifying your station?
A. The language being used for the contact
B. The language being used for the contact, providing the US has a third-party communications agreement with that country
C. English
D. Any language of a country which is a member of the International Telecommunication Union

T1D04
What do the FCC rules suggest you use as an aid for correct station identification when using phone?
A. A speech compressor
B. Q signals
C. A phonetic alphabet
D. Unique words of your choice

T1D05
What minimum class of amateur license must you hold to operate a beacon station?
A. Novice
B. Technician
C. General
D. Amateur Extra

T1D06
If a repeater is causing harmful interference to another repeater and a frequency coordinator has recommended the operation of one station only, who is responsible for resolving the interference?
A. The licensee of the unrecommended repeater
B. Both repeater licensees
C. The licensee of the recommended repeater
D. The frequency coordinator

T1D07
If a repeater is causing harmful interference to another amateur repeater and a frequency coordinator has recommended the operation of both stations, who is responsible for resolving the interference?
A. The licensee of the repeater which has been recommended for the longest period of time
B. The licensee of the repeater which has been recommended the most recently
C. The frequency coordinator
D. Both repeater licensees

T1D02
(B)
[97.119b1]
Page 2-29

T1D03
(C)
[97.119b2]
Page 2-29

T1D04
(C)
[97.119b2]
Page 2-29

T1D05
(B)
[97.203a]
Page 2-26

T1D06
(A)
[97.205c]
Page 2-25

T1D07
(D)
[97.205c]
Page 2-25

T1D08
If a repeater is causing harmful interference to another repeater and a frequency coordinator has NOT recommended either station, who is primarily responsible for resolving the interference?
A. Both repeater licensees
B. The licensee of the repeater which has been in operation for the longest period of time
C. The licensee of the repeater which has been in operation for the shortest period of time
D. The frequency coordinator

T1D09
What minimum information must be on a label affixed to a transmitter used for telecommand (control) of model craft?
A. Station call sign
B. Station call sign and the station licensee's name
C. Station call sign and the station licensee's name and address
D. Station call sign and the station licensee's class of license

T1D10
What are the station identification requirements for an amateur transmitter used for telecommand (control) of model craft?
A. Once every ten minutes
B. Once every ten minutes, and at the beginning and end of each transmission
C. At the beginning and end of each transmission
D. Station identification is not required if the transmitter is labeled with the station licensee's name, address and call sign

T1D11
What is the maximum transmitter power an amateur station is allowed when used for telecommand (control) of model craft?
A. One milliwatt
B. One watt
C. Two watts
D. Three watts

T1E Emergency communications; broadcasting; permissible one-way, satellite and third-party communication; indecent and profane language.

T1E01
What is meant by the term broadcasting?
A. Transmissions intended for reception by the general public, either direct or relayed
B. Retransmission by automatic means of programs or signals from non-amateur stations
C. One-way radio communications, regardless of purpose or content
D. One-way or two-way radio communications between two or more stations

T1E02
Which of the following one-way communications may not be transmitted
in the amateur service?
A. Telecommands to model craft
B. Broadcasts intended for the general public
C. Brief transmissions to make adjustments to the station
D. Morse code practice

T1E02
(B)
[97.3a10]
Page 2-27

T1E03
What kind of payment is allowed for third-party messages sent by an
amateur station?
A. Any amount agreed upon in advance
B. Donation of equipment repairs
C. Donation of amateur equipment
D. No payment of any kind is allowed

T1E03
(D)
[97.113b]
Page 2-27

T1E04
When may you send obscene words from your amateur station?
A. Only when they do not cause interference to other communications
B. Never; obscene words are prohibited in amateur transmissions
C. Only when they are not retransmitted through a repeater
D. Any time, but there is an unwritten rule among amateurs that they
   should not be used on the air

T1E04
(B)
[97.113d]
Page 2-28

T1E05
When may you send indecent words from your amateur station?
A. Only when they do not cause interference to other communications
B. Only when they are not retransmitted through a repeater
C. Any time, but there is an unwritten rule among amateurs that they
   should not be used on the air
D. Never; indecent words are prohibited in amateur transmissions

T1E05
(D)
[97.113d]
Page 2-28

T1E06
When may you send profane words from your amateur station?
A. Only when they do not cause interference to other communications
B. Only when they are not retransmitted through a repeater
C. Never; profane words are prohibited in amateur transmissions
D. Any time, but there is an unwritten rule among amateurs that they
   should not be used on the air

T1E06
(C)
[97.113d]
Page 2-28

T1E07
If you wanted to use your amateur station to retransmit communications
between a space shuttle and its associated Earth stations, what agency
must first give its approval?
A. The FCC in Washington, DC
B. The office of your local FCC Engineer In Charge (EIC)
C. The National Aeronautics and Space Administration
D. The Department of Defense

T1E07
(C)
[97.113e]
Page 2-28

T1E08
When are third-party messages allowed to be sent to a foreign country?
A. When sent by agreement of both control operators
B. When the third party speaks to a relative
C. They are not allowed under any circumstances
D. When the US has a third-party agreement with the foreign country or
   the third party is qualified to be a control operator

T1E08
(D)
[97.115a2]
Page 2-28

T1E09
(A)
[97.115b1]
Page 2-28

T1E09

If you let an unlicensed third party use your amateur station, what must you do at your station's control point?

A. You must continuously monitor and supervise the third-party's participation
B. You must monitor and supervise the communication only if contacts are made in countries which have no third-party communications agreement with the US
C. You must monitor and supervise the communication only if contacts are made on frequencies below 30 MHz
D. You must key the transmitter and make the station identification

T1E10
(A)
[97.401a]
Page 2-25

T1E10

If a disaster disrupts normal communication systems in an area where the amateur service is regulated by the FCC, what kinds of transmissions may stations make?

A. Those which are necessary to meet essential communication needs and facilitate relief actions
B. Those which allow a commercial business to continue to operate in the affected area
C. Those for which material compensation has been paid to the amateur operator for delivery into the affected area
D. Those which are to be used for program production or newsgathering for broadcasting purposes

T1E11
(C)
[97.401c]
Page 2-25

T1E11

What information is included in an FCC declaration of a temporary state of communication emergency?

A. A list of organizations authorized to use radio communications in the affected area
B. A list of amateur frequency bands to be used in the affected area
C. Any special conditions and special rules to be observed during the emergency
D. An operating schedule for authorized amateur emergency stations

**Subelement
T2**

SUBELEMENT T2— OPERATING PROCEDURES
[3 exam questions—3 groups]

T2A Repeater operation, courteous operation.

T2A01
(B)
Page 9-14

T2A01

How do you call another station on a repeater if you know the station's call sign?

A. Say "break, break 79," then say the station's call sign
B. Say the station's call sign, then identify your own station
C. Say "CQ" three times, then say the station's call sign
D. Wait for the station to call "CQ," then answer it

T2A02
(C)
Page 9-14

T2A02

Why should you pause briefly between transmissions when using a repeater?

A. To check the SWR of the repeater
B. To reach for pencil and paper for third-party communications
C. To listen for anyone wanting to break in
D. To dial up the repeater's autopatch

T2A03

Why should you keep transmissions short when using a repeater?

A. A long transmission may prevent someone with an emergency from using the repeater

B. To see if the receiving station operator is still awake

C. To give any listening non-hams a chance to respond

D. To keep long-distance charges down

T2A04

What is the proper way to break into a conversation on a repeater?

A. Wait for the end of a transmission and start calling the desired party

B. Shout, "break, break!" to show that you're eager to join the conversation

C. Turn on an amplifier and override whoever is talking

D. Say your call sign during a break between transmissions

T2A05

What is the purpose of repeater operation?

A. To cut your power bill by using someone else's higher power system

B. To help mobile and low-power stations extend their usable range

C. To transmit signals for observing propagation and reception

D. To make calls to stores more than 50 miles away

T2A06

What causes a repeater to "time out"?

A. The repeater's battery supply runs out

B. Someone's transmission goes on longer than the repeater allows

C. The repeater gets too hot and stops transmitting until its circuitry cools off

D. Something is wrong with the repeater

T2A07

During commuting rush hours, which type of repeater operation should be discouraged?

A. Mobile stations

B. Low-power stations

C. Highway traffic information nets

D. Third-party communications nets

T2A08

What is a courtesy tone (used in repeater operations)?

A. A sound used to identify the repeater

B. A sound used to indicate when a transmission is complete

C. A sound used to indicate that a message is waiting for someone

D. A sound used to activate a receiver in case of severe weather

T2A09

What is the meaning of: "Your signal is full quieting..."?

A. Your signal is strong enough to overcome all receiver noise

B. Your signal has no spurious sounds

C. Your signal is not strong enough to be received

D. Your signal is being received, but no audio is being heard

T2A03
(A)
Page 9-14

T2A04
(D)
Page 9-13

T2A05
(B)
Page 9-12

T2A06
(B)
Page 9-14

T2A07
(D)
Page 9-14

T2A08
(B)
Page 9-14

T2A09
(A)
Page 9-9

T2A10
(C)
Page 9-9

T2A10

How should you give a signal report over a repeater?

A. Say what your receiver's S-meter reads
B. Always say: "Your signal report is five five..."
C. Say the amount of signal quieting into the repeater
D. Try to imitate the sound quality you are receiving

T2A11
(A)
Page 9-13

T2A11

What is a repeater called which is available for anyone to use?

A. An open repeater
B. A closed repeater
C. An autopatch repeater
D. A private repeater

T2A12
(A)
Page 9-12

T2A12

What is the usual input/output frequency separation for repeaters in the 2-meter band?

A. 600 kHz
B. 1.0 MHz
C. 1.6 MHz
D. 5.0 MHz

T2A13
(C)
Page 9-12

T2A13

What is the usual input/output frequency separation for repeaters in the 1.25-meter band?

A. 600 kHz
B. 1.0 MHz
C. 1.6 MHz
D. 5.0 MHz

T2A14

What is the usual input/output frequency separation for repeaters in the 70-centimeter band?

A. 600 kHz
B. 1.0 MHz
C. 1.6 MHz
D. 5.0 MHz

T2A14
(D)
Page 9-12

T2A15
(A)
Page 9-9

T2A15

Why should local amateur communications use VHF and UHF frequencies instead of HF frequencies?

A. To minimize interference on HF bands capable of long-distance communication
B. Because greater output power is permitted on VHF and UHF
C. Because HF transmissions are not propagated locally
D. Because signals are louder on VHF and UHF frequencies

T2A16
(A)
Page 9-13

T2A16

How might you join a closed repeater system?

A. Contact the control operator and ask to join
B. Use the repeater until told not to
C. Use simplex on the repeater input until told not to
D. Write the FCC and report the closed condition

T2A17

How can on-the-air interference be minimized during a lengthy
transmitter testing or loading up procedure?

A. Choose an unoccupied frequency
B. Use a dummy load
C. Use a non-resonant antenna
D. Use a resonant antenna that requires no loading up procedure

T2A18

What is the proper way to ask someone their location when using a
repeater?

A. What is your QTH
B. What is your 20
C. Where are you
D. Locations are not normally told by radio

T2B Simplex operations, Q signals, RST signal reporting, repeater frequency coordination.

T2B01

Why should simplex be used where possible, instead of using a repeater?

A. Signal range will be increased
B. Long distance toll charges will be avoided
C. The repeater will not be tied up unnecessarily
D. Your antenna's effectiveness will be better tested

T2B02

If you are talking to a station using a repeater, how would you find out if
you could communicate using simplex instead?

A. See if you can clearly receive the station on the repeater's input
   frequency
B. See if you can clearly receive the station on a lower frequency band
C. See if you can clearly receive a more distant repeater
D. See if a third station can clearly receive both of you

T2B03

If you are operating simplex on a repeater frequency, why would it be
good amateur practice to change to another frequency?

A. The repeater's output power may ruin your station's receiver
B. There are more repeater operators than simplex operators
C. Changing the repeater's frequency is not practical
D. Changing the repeater's frequency requires the authorization of the
   FCC

T2B04

What is a repeater frequency coordinator?

A. Someone who organizes the assembly of a repeater station
B. Someone who provides advice on what kind of repeater to buy
C. The person whose call sign is used for a repeater's identification
D. A person or group that recommends frequencies for repeater usage

T2A17
(B)
Page 9-6

T2A18
(C)
Page 9-14

T2B01
(C)
Page 9-14

T2B02
(A)
Page 9-14

T2B03
(C)
Page 9-14

T2B04
(D)
Page 9-13

T2B05
(C)
Page 9-3

T2B05
What is the proper Q signal to use to see if a frequency is in use before transmitting on CW?
A. QRV?
B. QRU?
C. QRL?
D. QRZ?

T2B06
(A)
Page 9-3

T2B06
What is one meaning of the Q signal "QSY"?
A. Change frequency
B. Send more slowly
C. Send faster
D. Use more power

T2B07
(B)
Page 9-2

T2B07
What is one meaning of the Q signal "QSO"?
A. A contact is confirmed
B. A conversation is in progress
C. A contact is ending
D. A conversation is desired

T2B08
(B)
Page 9-4

T2B08
What is the proper Q signal to use to ask if someone is calling you on CW?
A. QSL?
B. QRZ?
C. QRL?
D. QRT?

T2B09
(A)
Page 9-5

T2B09
What is the meaning of: "Your signal report is five seven..."?
A. Your signal is perfectly readable and moderately strong
B. Your signal is perfectly readable, but weak
C. Your signal is readable with considerable difficulty
D. Your signal is perfectly readable with near pure tone

T2B10
(C)
Page 9-5

T2B10
What is the meaning of: "Your signal report is three three..."?
A. The contact is serial number thirty-three
B. The station is located at latitude 33 degrees
C. Your signal is readable with considerable difficulty and weak in strength
D. Your signal is unreadable, very weak in strength

T2B11
(D)
Page 9-9

T2B11
What is the meaning of: "Your signal report is five nine plus 20 dB..."?
A. Your signal strength has increased by a factor of 100
B. Repeat your transmission on a frequency 20 kHz higher
C. The bandwidth of your signal is 20 decibels above linearity
D. A relative signal-strength meter reading is 20 decibels greater than strength 9

T2C Distress calling and emergency drills and communications—operations and equipment, Radio Amateur Civil Emergency Service (RACES).

T2C01

What is the proper distress call to use when operating phone?

A. Say "MAYDAY" several times
B. Say "HELP" several times
C. Say "EMERGENCY" several times
D. Say "SOS" several times

T2C02

What is the proper distress call to use when operating CW?

A. MAYDAY
B. QRRR
C. QRZ
D. SOS

T2C03

What is the proper way to interrupt a repeater conversation to signal a distress call?

A. Say "BREAK" twice, then your call sign
B. Say "HELP" as many times as it takes to get someone to answer
C. Say "SOS," then your call sign
D. Say "EMERGENCY" three times

T2C04

With what organization must you register before you can participate in RACES drills?

A. A local Amateur Radio club
B. A local racing organization
C. The responsible civil defense organization
D. The Federal Communications Commission

T2C05

What is the maximum number of hours allowed per week for RACES drills?

A. One
B. Six, but not more than one hour per day
C. Eight
D. As many hours as you want

T2C06

How must you identify messages sent during a RACES drill?

A. As emergency messages
B. As amateur traffic
C. As official government messages
D. As drill or test messages

T2C07

What is one reason for using tactical call signs such as "command post" or "weather center" during an emergency?

A. They keep the general public informed about what is going on
B. They are more efficient and help coordinate public-service communications
C. They are required by the FCC
D. They increase goodwill between amateurs

T2C01
(A)
Page 2-19

T2C02
(D)
Page 2-19

T2C03
(A)
Page 9-14

T2C04
(C)
Page 2-25

T2C05
(A)
Page 2-25

T2C06
(D)
Page 2-25

T2C07
(B)
Page 2-26

T2C08
What type of messages concerning a person's well-being are sent into or out of a disaster area?
A. Routine traffic
B. Tactical traffic
C. Formal message traffic
D. Health and Welfare traffic

T2C09
What are messages called which are sent into or out of a disaster area concerning the immediate safety of human life?
A. Tactical traffic
B. Emergency traffic
C. Formal message traffic
D. Health and Welfare traffic

T2C10
Why is it a good idea to have a way to operate your amateur station without using commercial AC power lines?
A. So you may use your station while mobile
B. So you may provide communications in an emergency
C. So you may operate in contests where AC power is not allowed
D. So you will comply with the FCC rules

T2C11
What is the most important accessory to have for a hand-held radio in an emergency?
A. An extra antenna
B. A portable amplifier
C. Several sets of charged batteries
D. A microphone headset for hands-free operation

T2C12
Which type of antenna would be a good choice as part of a portable HF amateur station that could be set up in case of an emergency?
A. A three-element quad
B. A three-element Yagi
C. A dipole
D. A parabolic dish

SUBELEMENT T3—RADIO-WAVE PROPAGATION
[3 exam questions—3 groups]

T3A Ionosphere, ionospheric regions, solar radiation.

T3A01
What is the ionosphere?
A. An area of the outer atmosphere where enough ions and free electrons exist to propagate radio waves
B. An area between two air masses of different temperature and humidity, along which radio waves can travel
C. An ionized path in the atmosphere where lightning has struck
D. An area of the atmosphere where weather takes place

T3A02
What is the name of the area that makes long-distance radio communications possible by bending radio waves?
A. Troposphere
B. Stratosphere
C. Magnetosphere
D. Ionosphere

T3A03
What causes the ionosphere to form?
A. Solar radiation ionizing the outer atmosphere
B. Temperature changes ionizing the outer atmosphere
C. Lightning ionizing the outer atmosphere
D. Release of fluorocarbons into the atmosphere

T3A04
What type of solar radiation is most responsible for ionization in the outer atmosphere?
A. Thermal
B. Ionized particle
C. Ultraviolet
D. Microwave

T3A05
Which ionospheric region limits daytime radio communications on the 80-meter band to short distances?
A. D region
B. E region
C. F1 region
D. F2 region

T3A06
Which ionospheric region is closest to the earth?
A. The A region
B. The D region
C. The E region
D. The F region

T3A07
Which ionospheric region most affects sky-wave propagation on the 6-meter band?
A. The D region
B. The E region
C. The F1 region
D. The F2 region

T3A08
Which region of the ionosphere is the least useful for long-distance radio-wave propagation?
A. The D region
B. The E region
C. The F1 region
D. The F2 region

T3A02
(D)
Page 9-29

T3A03
(A)
Page 9-29

T3A04
(C)
Page 9-44

T3A05
(A)
Page 9-45

T3A06
(B)
Page 9-44

T3A07
(B)
Page 9-45

T3A08
(A)
Page 9-44

T3A09
(D)
Page 9-45

T3A09
Which region of the ionosphere is mainly responsible for long-distance sky-wave radio communications?
A. D region
B. E region
C. F1 region
D. F2 region

T3A10
(B)
Page 9-45

T3A10
What two sub-regions of ionosphere exist only in the daytime?
A. Troposphere and stratosphere
B. F1 and F2
C. Electrostatic and electromagnetic
D. D and E

T3A11
(C)
Page 9-45

T3A11
Which two daytime ionospheric regions combine into one region at night?
A. E and F1
B. D and E
C. F1 and F2
D. E1 and E2

T3B Ionospheric absorption, causes and variation, maximum usable frequency.

T3B01
(D)
Page 9-44

T3B01
Which region of the ionosphere is mainly responsible for absorbing radio signals during the daytime?
A. The F2 region
B. The F1 region
C. The E region
D. The D region

T3B02
(B)
Page 9-44

T3B02
When does ionospheric absorption of radio signals occur?
A. When tropospheric ducting occurs
B. When long-wavelength signals enter the D region
C. When signals travel to the F region
D. When a temperature inversion occurs

T3B03
(A)
Page 9-44

T3B03
What effect does the D region of the ionosphere have on lower-frequency HF signals in the daytime?
A. It absorbs the signals
B. It bends the radio waves out into space
C. It refracts the radio waves back to earth
D. It has little or no effect on 80-meter radio waves

T3B04
(B)
Page 9-44

T3B04
What causes the ionosphere to absorb radio waves?
A. The weather below the ionosphere
B. The ionization of the D region
C. The presence of ionized clouds in the E region
D. The splitting of the F region

T3B05

What is the condition of the ionosphere just before local sunrise?

A. Atmospheric attenuation is at a maximum
B. The D region is above the E region
C. The E region is above the F region
D. Ionization is at a minimum

T3B05
(D)
Page 9-45

T3B06

When is the ionosphere most ionized?

A. Dusk
B. Midnight
C. Midday
D. Dawn

T3B06
(C)
Page 9-45

T3B07

When is the ionosphere least ionized?

A. Shortly before dawn
B. Just after noon
C. Just after dusk
D. Shortly before midnight

T3B07
(A)
Page 9-45

T3B08

When is the E region most ionized?

A. Dawn
B. Midday
C. Dusk
D. Midnight

T3B08
(B)
Page 9-45

T3B09

What happens to signals higher in frequency than the critical frequency?

A. They pass through the ionosphere
B. They are absorbed by the ionosphere
C. Their frequency is changed by the ionosphere to be below the maximum usable frequency
D. They are reflected back to their source

T3B09
(A)
Page 9-45

T3B10

What causes the maximum usable frequency to vary?

A. The temperature of the ionosphere
B. The speed of the winds in the upper atmosphere
C. The amount of radiation received from the sun, mainly ultraviolet
D. The type of weather just below the ionosphere

T3B10
(C)
Page 9-46

T3B11

What does maximum usable frequency mean?

A. The highest frequency signal that will reach its intended destination
B. The lowest frequency signal that will reach its intended destination
C. The highest frequency signal that is most absorbed by the ionosphere
D. The lowest frequency signal that is most absorbed by the ionosphere

T3B11
(A)
Page 9-46

T3C Propagation, including ionospheric, tropospheric, line-of-sight scatter propagation, and Maximum Usable Frequency.

T3C01
(C)
Page 9-47

T3C01

What kind of propagation would best be used by two stations within each other's skip zone on a certain frequency?

A. Ground-wave
B. Sky-wave
C. Scatter-mode
D. Ducting

T3C02
(C)
Page 9-47

T3C02

If you are receiving a weak and distorted signal from a distant station on a frequency close to the maximum usable frequency, what type of propagation is probably occurring?

A. Ducting
B. Line-of-sight
C. Scatter
D. Ground-wave

T3C03
(B)
Page 9-48

T3C03

How are VHF signals propagated within the range of the visible horizon?

A. By sky wave
B. By direct wave
C. By plane wave
D. By geometric wave

T3C04
(C)
Page 9-49

T3C04

Ducting occurs in which region of the atmosphere?

A. F2
B. Ectosphere
C. Troposphere
D. Stratosphere

T3C05
(A)
Page 9-49

T3C05

What effect does tropospheric bending have on 2-meter radio waves?

A. It lets you contact stations farther away
B. It causes them to travel shorter distances
C. It garbles the signal
D. It reverses the sideband of the signal

T3C06
(D)
Page 9-49

T3C06

What causes tropospheric ducting of radio waves?

A. A very low pressure area
B. An aurora to the north
C. Lightning between the transmitting and receiving stations
D. A temperature inversion

T3C07
(B)
Page 9-49

T3C07

What causes VHF radio waves to be propagated several hundred miles over oceans?

A. A polar air mass
B. A widespread temperature inversion
C. An overcast of cirriform clouds
D. A high-pressure zone

T3C08
In what frequency range does tropospheric ducting most often occur?
A. SW
B. MF
C. HF
D. VHF

T3C08
(D)
Page 9-49

T3C09
In what frequency range does sky-wave propagation least often occur?
A. LF
B. MF
C. HF
D. VHF

T3C09
(D)
Page 9-47

T3C10
What weather condition may cause tropospheric ducting?
A. A stable high-pressure system
B. An unstable low-pressure system
C. A series of low-pressure waves
D. Periods of heavy rainfall

T3C10
(A)
Page 9-49

T3C11
What band conditions might indicate long-range skip on the 6-meter and
2-meter bands?
A. Noise on the 80-meter band
B. The absence of signals on the 10-meter band
C. Very long-range skip on the 10-meter band
D. Strong signals on the 10-meter band from stations about 500-600
   miles away

T3C11
(D)
Page 9-47

SUBELEMENT T4—AMATEUR RADIO PRACTICES
[4 exam questions—4 groups]

Subelement
T4

**T4A Electrical wiring, including switch location, dangerous voltages
and currents.**

T4A01
Where should the green wire in a three-wire AC line cord be connected
in a power supply?
A. To the fuse
B. To the "hot" side of the power switch
C. To the chassis
D. To the white wire

T4A01
(C)
Page 8-11

T4A02
Where should the black (or red) wire in a three-wire AC line cord be
connected in a power supply?
A. To the white wire
B. To the green wire
C. To the chassis
D. To the fuse

T4A02
(D)
Page 8-10

T4A03
(B)
Page 8-10

T4A03
Where should the white wire in a three-wire AC line cord be connected in a power supply?
A. To the side of the power transformer's primary winding that has a fuse
B. To the side of the power transformer's primary winding that does not have a fuse
C. To the chassis
D. To the black wire

T4A04
(D)
Page 8-11

T4A04
What document is used by almost every US city as the basis for electrical safety requirements for power wiring and antennas?
A. The Code of Federal Regulations
B. The Proceedings of the IEEE
C. The ITU Radio Regulations
D. The National Electrical Code

T4A05
(C)
Page 8-11

T4A05
What document would you use to see if you comply with standard electrical safety rules when building an amateur antenna?
A. The Code of Federal Regulations
B. The Proceedings of the IEEE
C. The National Electrical Code
D. The ITU Radio Regulations

T4A06
(D)
Page 8-5

T4A06
Where should fuses be connected on a mobile transceiver's DC power cable?
A. Between the red and black wires
B. In series with just the black wire
C. In series with just the red wire
D. In series with both the red and black wires

T4A07
(B)
Page 8-11

T4A07
Why is the retaining screw in one terminal of a wall outlet made of brass while the other one is silver colored?
A. To prevent corrosion
B. To indicate correct wiring polarity
C. To better conduct current
D. To reduce skin effect

T4A08
(A)
Page 8-12

T4A08
How much electrical current flowing through the human body is usually fatal?
A. As little as 1/10 of an ampere
B. Approximately 10 amperes
C. More than 20 amperes
D. Current flow through the human body is never fatal

T4A09
(A)
Page 8-12

T4A09
Which body organ can be fatally affected by a very small amount of electrical current?
A. The heart
B. The brain
C. The liver
D. The lungs

T4A10

How much electrical current flowing through the human body is usually painful?

A. As little as 1/500 of an ampere
B. Approximately 10 amperes
C. More than 20 amperes
D. Current flow through the human body is never painful

T4A10
(A)
Page 8-12

T4A11

What is the minimum voltage which is usually dangerous to humans?

A. 30 volts
B. 100 volts
C. 1000 volts
D. 2000 volts

T4A11
(A)
Page 8-12

T4A12

Where should the main power switch for a high-voltage power supply be located?

A. Inside the cabinet, to kill the power if the cabinet is opened
B. On the back side of the cabinet, out of sight
C. Anywhere that can be seen and reached easily
D. A high-voltage power supply should not be switch-operated

T4A12
(C)
Page 8-12

T4A13

What precaution should you take when leaning over a power amplifier?

A. Take your shoes off
B. Watch out for loose jewelry contacting high voltage
C. Shield your face from the heat produced by the power supply
D. Watch out for sharp edges which may snag your clothing

T4A13
(B)
Page 8-12

T4A14

What is an important safety rule concerning the main electrical box in your home?

A. Make sure the door cannot be opened easily
B. Make sure something is placed in front of the door so no one will be able to get to it easily
C. Make sure others in your home know where it is and how to shut off the electricity
D. Warn others in your home never to touch the switches, even in an emergency

T4A14
(C)
Page 8-13

T4A15

What should you do if you discover someone who is being burned by high voltage?

A. Run from the area so you won't be burned too
B. Turn off the power, call for emergency help and give CPR if needed
C. Immediately drag the person away from the high voltage
D. Wait for a few minutes to see if the person can get away from the high voltage on their own, then try to help

T4A15
(B)
Page 8-13

T4B Meters, including volt, amp, multi, peak-reading, RF watt and placement, and ratings of fuses and switches.

T4B01
(B)
Page 8-13

T4B01
How is a voltmeter usually connected to a circuit under test?
A. In series with the circuit
B. In parallel with the circuit
C. In quadrature with the circuit
D. In phase with the circuit

T4B02
(C)
Page 8-13

T4B02
How can the range of a voltmeter be increased?
A. By adding resistance in series with the circuit under test
B. By adding resistance in parallel with the circuit under test
C. By adding resistance in series with the meter, between the meter and the circuit under test
D. By adding resistance in parallel with the meter, between the meter and the circuit under test

T4B03
(A)
Page 8-13

T4B03
What happens inside a voltmeter when you switch it from a lower to a higher voltage range?
A. Resistance is added in series with the meter
B. Resistance is added in parallel with the meter
C. Resistance is reduced in series with the meter
D. Resistance is reduced in parallel with the meter

T4B04
(A)
Page 8-13

T4B04
How is an ammeter usually connected to a circuit under test?
A. In series with the circuit
B. In parallel with the circuit
C. In quadrature with the circuit
D. In phase with the circuit

T4B05
(D)
Page 8-13

T4B05
How can the range of an ammeter be increased?
A. By adding resistance in series with the circuit under test
B. By adding resistance in parallel with the circuit under test
C. By adding resistance in series with the meter
D. By adding resistance in parallel with the meter

T4B06
(D)
Page 8-14

T4B06
What does a multimeter measure?
A. SWR and power
B. Resistance, capacitance and inductance
C. Resistance and reactance
D. Voltage, current and resistance

T4B07
(A)
Page 8-14

T4B07
Where should an RF wattmeter be connected for the most accurate readings of transmitter output power?
A. At the transmitter output connector
B. At the antenna feed point
C. One-half wavelength from the transmitter output
D. One-half wavelength from the antenna feed point

T4B08

At what line impedance do most RF wattmeters usually operate?

A. 25 ohms
B. 50 ohms
C. 100 ohms
D. 300 ohms

T4B08
(B)
Page 8-14

T4B09

What does a directional wattmeter measure?

A. Forward and reflected power
B. The directional pattern of an antenna
C. The energy used by a transmitter
D. Thermal heating in a load resistor

T4B09
(A)
Page 8-14

T4B10

If a directional RF wattmeter reads 90 watts forward power and 10 watts reflected power, what is the actual transmitter output power?

A. 10 watts
B. 80 watts
C. 90 watts
D. 100 watts

T4B10
(B)
Page 8-14

T4B11

If a directional RF wattmeter reads 96 watts forward power and 4 watts reflected power, what is the actual transmitter output power?

A. 80 watts
B. 88 watts
C. 92 watts
D. 100 watts

T4B11
(C)
Page 8-14

T4C Marker generator, crystal calibrator, signal generators and impedance-match indicator.

T4C01

What is a marker generator?

A. A high-stability oscillator that generates reference signals at exact frequency intervals
B. A low-stability oscillator that "sweeps" through a range of frequencies
C. A low-stability oscillator used to inject a signal into a circuit under test
D. A high-stability oscillator which can produce a wide range of frequencies and amplitudes

T4C01
(A)
Page 8-15

T4C02

How is a marker generator used?

A. To calibrate the tuning dial on a receiver
B. To calibrate the volume control on a receiver
C. To test the amplitude linearity of a transmitter
D. To test the frequency deviation of a transmitter

T4C02
(A)
Page 8-15

T4C03
(D)
Page 8-15

T4C03
What device is used to inject a frequency calibration signal into a receiver?
A. A calibrated voltmeter
B. A calibrated oscilloscope
C. A calibrated wavemeter
D. A crystal calibrator

T4C04
(B)
Page 8-15

T4C04
What frequency standard may be used to calibrate the tuning dial of a receiver?
A. A calibrated voltmeter
B. Signals from WWV and WWVH
C. A deviation meter
D. A sweep generator

T4C05
(C)
Page 8-15

T4C05
How might you check the accuracy of your receiver's tuning dial?
A. Tune to the frequency of a shortwave broadcasting station
B. Tune to a popular amateur net frequency
C. Tune to one of the frequencies of station WWV or WWVH
D. Tune to another amateur station and ask what frequency the operator is using

T4C06
(C)
Page 8-15

T4C06
What device produces a stable, low-level signal that can be set to a desired frequency?
A. A wavemeter
B. A reflectometer
C. A signal generator
D. An oscilloscope

T4C07
(B)
Page 8-15

T4C07
What is an RF signal generator used for?
A. Measuring RF signal amplitudes
B. Aligning tuned circuits
C. Adjusting transmitter impedance-matching networks
D. Measuring transmission-line impedances

T4C08
(D)
Page 8-16

T4C08
What device can measure an impedance mismatch in your antenna system?
A. A field-strength meter
B. An ammeter
C. A wavemeter
D. A reflectometer

T4C09
(A)
Page 8-16

T4C09
Where should a reflectometer be connected for best accuracy when reading the impedance match between an antenna and its feed line?
A. At the antenna feed point
B. At the transmitter output connector
C. At the midpoint of the feed line
D. Anywhere along the feed line

T4C10

If you use a 3-30 MHz RF power meter for VHF, how accurate will its readings be?

A. They will not be accurate
B. They will be accurate enough to get by
C. If it properly calibrates to full scale in the set position, they may be accurate
D. They will be accurate providing the readings are multiplied by 4.5

T4C11

If you use a 3-30 MHz SWR meter for VHF, how accurate will its readings be?

A. They will not be accurate
B. They will be accurate enough to get by
C. If it properly calibrates to full scale in the set position, they may be accurate
D. They will be accurate providing the readings are multiplied by 4.5

T4D Dummy antennas, S-meter, exposure of the human body to RF.

T4D01

What device should be connected to a transmitter's output when you are making transmitter adjustments?

A. A multimeter
B. A reflectometer
C. A receiver
D. A dummy antenna

T4D02

What is a dummy antenna?

A. An nondirectional transmitting antenna
B. A nonradiating load for a transmitter
C. An antenna used as a reference for gain measurements
D. A flexible antenna usually used on hand-held transceivers

T4D03

What is the main component of a dummy antenna?

A. A wire-wound resistor
B. An iron-core coil
C. A noninductive resistor
D. An air-core coil

T4D04

What device is used in place of an antenna during transmitter tests so that no signal is radiated?

A. An antenna matcher
B. A dummy antenna
C. A low-pass filter
D. A decoupling resistor

T4D05

Why would you use a dummy antenna?

A. For off-the-air transmitter testing
B. To reduce output power
C. To give comparative signal reports
D. To allow antenna tuning without causing interference

T4C10
(A)
Page 8-14

T4C11
(C)
Page 8-16

T4D01
(D)
Page 8-16

T4D02
(B)
Page 8-16

T4D03
(C)
Page 8-17

T4D04
(B)
Page 8-16

T4D05
(A)
Page 8-17

T4D06
(A)
Page 8-17

T4D06
What minimum rating should a dummy antenna have for use with a 100-watt single-sideband phone transmitter?
A. 100 watts continuous
B. 141 watts continuous
C. 175 watts continuous
D. 200 watts continuous

T4D07
(D)
Page 8-17

T4D07
Why might a dummy antenna get warm when in use?
A. Because it stores electric current
B. Because it stores radio waves
C. Because it absorbs static electricity
D. Because it changes RF energy into heat

T4D08
(A)
Page 8-18

T4D08
What is used to measure relative signal strength in a receiver?
A. An S meter
B. An RST meter
C. A signal deviation meter
D. An SSB meter

T4D09
(B)
Page 8-19

T4D09
How can exposure to a large amount of RF energy affect body tissue?
A. It causes radiation poisoning
B. It heats the tissue
C. It paralyzes the tissue
D. It produces genetic changes in the tissue

T4D10
(A)
Page 8-20

T4D10
Which body organ is the most likely to be damaged from the heating effects of RF radiation?
A. Eyes
B. Hands
C. Heart
D. Liver

T4D11
(D)
Page 8-20

T4D11
What organization has published safety guidelines for the maximum limits of RF energy near the human body?
A. The Institute of Electrical and Electronics Engineers (IEEE)
B. The Federal Communications Commission (FCC)
C. The Environmental Protection Agency (EPA)
D. The American National Standards Institute (ANSI)

T4D12
(B)
Page 8-20

T4D12
What is the purpose of the ANSI RF protection guide?
A. It lists all RF frequency allocations for interference protection
B. It gives RF exposure limits for the human body
C. It sets transmitter power limits for interference protection
D. It sets antenna height limits for aircraft protection

T4D13
According to the ANSI RF protection guide, what frequencies cause us the greatest risk from RF energy?
A. 3 to 30 MHz
B. 300 to 3000 MHz
C. Above 1500 MHz
D. 30 to 300 MHz

T4D14
Why is the limit of exposure to RF the lowest in the frequency range of 30 MHz to 300 MHz, according to the ANSI RF protection guide?
A. There are more transmitters operating in this range
B. There are fewer transmitters operating in this range
C. Most transmissions in this range are for a longer time
D. The human body absorbs RF energy the most in this range

T4D15
According to the ANSI RF protection guide, what is the maximum safe power output to the antenna of a hand-held VHF or UHF radio?
A. 125 milliwatts
B. 7 watts
C. 10 watts
D. 25 watts

T4D16
After you have opened a VHF power amplifier to make internal tuning adjustments, what should you do before you turn the amplifier on?
A. Remove all amplifier shielding to ensure maximum cooling
B. Make sure that the power interlock switch is bypassed so you can test the amplifier
C. Be certain all amplifier shielding is fastened in place
D. Be certain no antenna is attached so that you will not cause any interference

SUBELEMENT T5—ELECTRICAL PRINCIPLES
[2 exam questions—2 groups]

T5A Definition of resistance, inductance, and capacitance and unit of measurement, calculation of values in series and parallel.

T5A01
What does resistance do in an electric circuit?
A. It stores energy in a magnetic field
B. It stores energy in an electric field
C. It provides electrons by a chemical reaction
D. It opposes the flow of electrons

T5A02
What is the ability to store energy in a magnetic field called?
A. Admittance
B. Capacitance
C. Resistance
D. Inductance

T4D13
(D)
Page 8-20

T4D14
(D)
Page 8-20

T4D15
(B)
Page 8-20

T4D16
(C)
Page 8-20

Subelement
T5

T5A01
(D)
Page 3-9

T5A02
(D)
Page 3-19

T5A03
(C)
Page 3-20

T5A03
What is the basic unit of inductance?
A. The coulomb
B. The farad
C. The henry
D. The ohm

T5A04
(C)
Page 3-20

T5A04
What is a henry?
A. The basic unit of admittance
B. The basic unit of capacitance
C. The basic unit of inductance
D. The basic unit of resistance

T5A05
(D)
Page 3-20

T5A05
What is the ability to store energy in an electric field called?
A. Inductance
B. Resistance
C. Tolerance
D. Capacitance

T5A06
(A)
Page 3-20

T5A06
What is the basic unit of capacitance?
A. The farad
B. The ohm
C. The volt
D. The henry

T5A07
(B)
Page 3-20

T5A07
What is a farad?
A. The basic unit of resistance
B. The basic unit of capacitance
C. The basic unit of inductance
D. The basic unit of admittance

T5A08
(B)
Page 3-20

T5A08
If two equal-value inductors are connected in series, what is their total inductance?
A. Half the value of one inductor
B. Twice the value of one inductor
C. The same as the value of either inductor
D. The value of one inductor times the value of the other

T5A09
(A)
Page 3-20

T5A09
If two equal-value inductors are connected in parallel, what is their total inductance?
A. Half the value of one inductor
B. Twice the value of one inductor
C. The same as the value of either inductor
D. The value of one inductor times the value of the other

T5A10

If two equal-value capacitors are connected in series, what is their total capacitance?

A. Twice the value of one capacitor
B. The same as the value of either capacitor
C. Half the value of either capacitor
D. The value of one capacitor times the value of the other

T5A10
(C)
Page 3-22

T5A11

If two equal-value capacitors are connected in parallel, what is their total capacitance?

A. Twice the value of one capacitor
B. Half the value of one capacitor
C. The same as the value of either capacitor
D. The value of one capacitor times the value of the other

T5A11
(A)
Page 3-21

T5B Ohm's Law.

T5B01

Ohm's Law describes the mathematical relationship between what three electrical quantities?

A. Resistance, voltage and power
B. Current, resistance and power
C. Current, voltage and power
D. Resistance, current and voltage

T5B01
(D)
Page 3-9

T5B02

How is the current in a DC circuit calculated when the voltage and resistance are known?

A. I = R × E [current equals resistance multiplied by voltage]
B. I = R / E [current equals resistance divided by voltage]
C. I = E / R [current equals voltage divided by resistance]
D. I = P / E [current equals power divided by voltage]

T5B02
(C)
Page 3-10

T5B03

How is the resistance in a DC circuit calculated when the voltage and current are known?

A. R = I / E [resistance equals current divided by voltage]
B. R = E / I [resistance equals voltage divided by current]
C. R = I × E [resistance equals current multiplied by voltage]
D. R = P / E [resistance equals power divided by voltage]

T5B03
(B)
Page 3-10

T5B04

How is the voltage in a DC circuit calculated when the current and resistance are known?

A. E = I / R [voltage equals current divided by resistance]
B. E = R / I [voltage equals resistance divided by current]
C. E = I × R [voltage equals current multiplied by resistance]
D. E = P / I [voltage equals power divided by current]

T5B04
(C)
Page 3-10

T5B05

If a 12-volt battery supplies 0.25 ampere to a circuit, what is the circuit's resistance?

A. 0.25 ohm
B. 3 ohm
C. 12 ohms
D. 48 ohms

T5B05
(D)
Page 3-10

T5B06
(D)
Page 3-10

T5B07
(B)
Page 3-10

T5B08
(D)
Page 3-10

T5B09
(A)
Page 3-10

T5B10
(A)
Page 3-10

T5B11
(A)
Page 3-10

T5B06
If a 12-volt battery supplies 0.15 ampere to a circuit, what is the circuit's resistance?
A. 0.15 ohm
B. 1.8 ohm
C. 12 ohms
D. 80 ohms

$R = 12 \div 0.15$

T5B07
If a 4800-ohm resistor is connected to 120 volts, approximately how much current will flow through it?
A. 4 A
B. 25 mA
C. 25 A
D. 40 MA

$I = 120 \div 4800$

T5B08
If a 48000-ohm resistor is connected to 120 volts, approximately how much current will flow through it?
A. 400 A
B. 40 A
C. 25 mA
D. 2.5 mA

$I = 120 \div 48000$

T5B09
If a 4800-ohm resistor is connected to 12 volts, approximately how much current will flow through it?
A. 2.5 mA
B. 25 mA
C. 40 A
D. 400 A

$I = 12 \div 4800$

T5B10
If a 48000-ohm resistor is connected to 12 volts, approximately how much current will flow through it?
A. 250 µA
B. 250 mA
C. 4000 mA
D. 4000 A

$I = 12 \div 48000$

T5B11
If you know the voltage and current supplied to a circuit, what formula would you use to calculate the circuit's resistance?
A. Ohm's Law
B. Tesla's Law
C. Ampere's Law
D. Kirchhoff's Law

SUBELEMENT T6— CIRCUIT COMPONENTS
[2 exam questions—2 groups]

T6A Resistors, construction types, variable and fixed, color code, power ratings, schematic symbols.

T6A01
Which of the following are common resistor types?
A. Plastic and porcelain
B. Film and wire-wound
C. Electrolytic and metal-film
D. Iron core and brass core

T6A01
(B)
Page 4-10

T6A02
What does a variable resistor or potentiometer do?
A. Its resistance changes when AC is applied to it
B. It transforms a variable voltage into a constant voltage
C. Its resistance changes when its slide or contact is moved
D. Its resistance changes when it is heated

T6A02
(C)
Page 4-12

T6A03
How do you find a resistor's tolerance rating?
A. By using a voltmeter
B. By reading the resistor's color code
C. By using Thevenin's theorem for resistors
D. By reading its Baudot code

T6A03
(B)
Page 4-12

T6A04
What do the first three color bands on a resistor indicate?
A. The value of the resistor in ohms
B. The resistance tolerance in percent
C. The power rating in watts
D. The resistance material

T6A04
(A)
Page 4-12

T6A05
What does the fourth color band on a resistor indicate?
A. The value of the resistor in ohms
B. The resistance tolerance in percent
C. The power rating in watts
D. The resistance material

T6A05
(B)
Page 4-12

T6A06
Why do resistors sometimes get hot when in use?
A. Some electrical energy passing through them is lost as heat
B. Their reactance makes them heat up
C. Hotter circuit components nearby heat them up
D. They absorb magnetic energy which makes them hot

T6A06
(A)
Page 4-10

T6A07
Why would a large size resistor be used instead of a smaller one of the same resistance?
A. For better response time
B. For a higher current gain
C. For greater power dissipation
D. For less impedance in the circuit

T6A07
(C)
Page 4-13

T6A08
(C)
Page 4-12

T6A08
What are the possible values of a 100-ohm resistor with a 10% tolerance?
A. 90 to 100 ohms
B. 10 to 100 ohms
C. 90 to 110 ohms
D. 80 to 120 ohms

T6A09
(B)
Page 4-12

T6A09
How do you find a resistor's value?
A. By using a voltmeter
B. By using the resistor's color code
C. By using Thevenin's theorem for resistors
D. By using the Baudot code

T6A10
(A)
Page 4-13

T6A10
Which tolerance rating would a high-quality resistor have?
A. 0.1%
B. 5%
C. 10%
D. 20%

T6A11
(D)
Page 4-13

T6A11
Which tolerance rating would a low-quality resistor have?
A. 0.1%
B. 5%
C. 10%
D. 20%

T6B Schematic symbols—inductors and capacitors, construction of variable and fixed, factors affecting inductance and capacitance, capacitor construction.

T6B01
(D)
Page 4-13

T6B01
What is an inductor core?
A. The place where a coil is tapped for resonance
B. A tight coil of wire used in a transformer
C. Insulating material placed between the wires of a transformer
D. The place inside an inductor where its magnetic field is concentrated

T6B02
(C)
Page 4-13

T6B02
What does an inductor do?
A. It stores a charge electrostatically and opposes a change in voltage
B. It stores a charge electrochemically and opposes a change in current
C. It stores a charge electromagnetically and opposes a change in current
D. It stores a charge electromechanically and opposes a change in voltage

T6B03
(D)
Page 4-13

T6B03
What determines the inductance of a coil?
A. The core material, the core diameter, the length of the coil and whether the coil is mounted horizontally or vertically
B. The core diameter, the number of turns of wire used to wind the coil and the type of metal used for the wire
C. The core material, the number of turns used to wind the core and the frequency of the current through the coil
D. The core material, the core diameter, the length of the coil and the number of turns of wire used to wind the coil

T6B04

As an iron core is inserted in a coil, what happens to the coil's inductance?

A. It increases
B. It decreases
C. It stays the same
D. It disappears

T6B05

What can happen if you tune a ferrite-core coil with a metal tool?

A. The metal tool can change the coil's inductance and cause you to tune the coil incorrectly
B. The metal tool can become magnetized so much that you might not be able to remove it from the coil
C. The metal tool can pick up enough magnetic energy to become very hot
D. The metal tool can pick up enough magnetic energy to become a shock hazard

T6B06

In Figure T6-1 which symbol represents an adjustable inductor?

A. Symbol 1   C. Symbol 3
B. Symbol 2   D. Symbol 4

Figure T6-1

T6B07

In Figure T6-1 which symbol represents an iron-core inductor?
A. Symbol 1   C. Symbol 3
B. Symbol 2   D. Symbol 4

T6B08

In Figure T6-1 which symbol represents an inductor wound over a toroidal core?
A. Symbol 1   C. Symbol 3
B. Symbol 2   D. Symbol 4

T6B09

In Figure T6-1 which symbol represents an electrolytic capacitor?
A. Symbol 1   C. Symbol 3
B. Symbol 2   D. Symbol 4

T6B10

In Figure T6-1 which symbol represents a variable capacitor?
A. Symbol 1   C. Symbol 3
B. Symbol 2   D. Symbol 4

T6B11

What describes a capacitor?

A. Two or more layers of silicon material with an insulating material between them
B. The wire used in the winding and the core material
C. Two or more conductive plates with an insulating material between them
D. Two or more insulating plates with a conductive material between them

T6B04
(A)
Page 4-13

T6B05
(A)
Page 4-14

T6B06
(B)
Page 4-14

T6B07
(D)
Page 4-14

T6B08
(D)
Page 4-15

T6B09
(A)
Page 4-16

T6B10
(C)
Page 4-17

T6B11
(C)
Page 4-15

T6B12

What does a capacitor do?
A. It stores a charge electrochemically and opposes a change in current
B. It stores a charge electrostatically and opposes a change in voltage
C. It stores a charge electromagnetically and opposes a change in current
D. It stores a charge electromechanically and opposes a change in voltage

T6B13

What determines the capacitance of a capacitor?
A. The material between the plates, the area of one side of one plate, the number of plates and the spacing between the plates
B. The material between the plates, the number of plates and the size of the wires connected to the plates
C. The number of plates, the spacing between the plates and whether the dielectric material is N type or P type
D. The material between the plates, the area of one plate, the number of plates and the material used for the protective coating

T6B14

As the plate area of a capacitor is increased, what happens to its capacitance?
A. It decreases
B. It increases
C. It stays the same
D. It disappears

Subelement T7

SUBELEMENT T7—PRACTICAL CIRCUITS
[1 exam question—1 group]

T7A Practical circuits.

T7A01

Why do modern HF transmitters have a built-in low-pass filter in their RF output circuits?
A. To reduce RF energy below a cutoff point
B. To reduce low-frequency interference to other amateurs
C. To reduce harmonic radiation
D. To reduce fundamental radiation

T7A02

What circuit blocks RF energy above and below a certain limit?
A. A band-pass filter
B. A high-pass filter
C. An input filter
D. A low-pass filter

T7A03

What type of filter is used in the IF section of receivers to block energy outside a certain frequency range?
A. A band-pass filter
B. A high-pass filter
C. An input filter
D. A low-pass filter

T7A04

What circuit is found in all types of receivers?

A. An audio filter
B. A beat-frequency oscillator
C. A detector
D. An RF amplifier

T7A04
(C)
Page 5-13

T7A05

What circuit has a variable-frequency oscillator connected to a driver and a power amplifier?

A. A packet-radio transmitter
B. A crystal-controlled transmitter
C. A single-sideband transmitter
D. A VFO-controlled transmitter

T7A05
(D)
Page 5-11

T7A06

What circuit combines signals from an IF amplifier stage and a beat-frequency oscillator (BFO), to produce an audio signal?

A. An AGC circuit
B. A detector circuit
C. A power supply circuit
D. A VFO circuit

T7A06
(B)
Page 5-14

T7A07

What circuit uses a limiter and a frequency discriminator to produce an audio signal?

A. A double-conversion receiver
B. A variable-frequency oscillator
C. A superheterodyne receiver
D. An FM receiver

T7A07
(D)
Page 5-15

T7A08

What circuit is pictured in Figure T7-1 if block 1 is a variable-frequency oscillator?

A. A packet-radio transmitter
B. A crystal-controlled transmitter
C. A single-sideband transmitter
D. A VFO-controlled transmitter

T7A08
(D)
Page 5-11

Figure T7-1

T7A09

What is the unlabeled block in Figure T7-2?

A. An AGC circuit
B. A detector
C. A power supply
D. A VFO circuit

T7A09
(B)
Page 5-14

Figure T7-2

T7A10
What circuit is pictured in Figure T7-3?
A. A double-conversion receiver
B. A variable-frequency oscillator
C. A superheterodyne receiver
D. An FM receiver

Figure T7–3

T7A11
What is the unlabeled block in Figure T7-4?
A. A band-pass filter
B. A crystal oscillator
C. A reactance modulator
D. A rectifier modulator

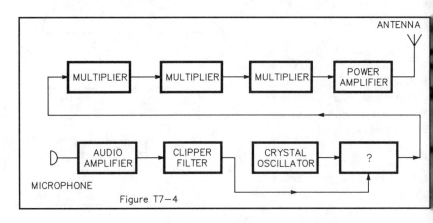

Figure T7–4

**Subelement
T8**

SUBELEMENT T8—SIGNALS AND EMISSIONS
[2 exam questions—2 groups]

T8A Definition of modulation and emission types.

T8A01
What is the name for unmodulated carrier wave emissions?
A. Phone   C. CW
B. Test   D. RTTY

T8A02
What is the name for Morse code emissions produced by switching a transmitter's output on and off?
A. Phone   C. CW
B. Test   D. RTTY

T8A03

What is RTTY?

A. Amplitude-keyed telegraphy
B. Frequency-shift-keyed telegraphy
C. Frequency-modulated telephony
D. Phase-modulated telephony

T8A04

What is the name for packet-radio emissions?

A. CW   C. Phone
B. Data   D. RTTY

T8A05

How is tone-modulated Morse code produced?

A. By feeding a microphone's audio signal into an FM transmitter
B. By feeding an on/off keyed audio tone into a CW transmitter
C. By on/off keying of a carrier
D. By feeding an on/off keyed audio tone into a transmitter

T8A06

What is the name of the voice emission most used on VHF/UHF repeaters?

A. Single-sideband phone
B. Pulse-modulated phone
C. Slow-scan phone
D. Frequency-modulated phone

T8A07

What is the name of the voice emission most used on amateur HF bands?

A. Single-sideband phone
B. Pulse-modulated phone
C. Slow-scan phone
D. Frequency-modulated phone

T8A08

What is meant by the upper sideband (USB)?

A. The part of a single-sideband signal which is above the carrier frequency
B. The part of a single-sideband signal which is below the carrier frequency
C. Any frequency above 10 MHz
D. The carrier frequency of a single-sideband signal

T8A09

What emissions are produced by a transmitter using a reactance modulator?

A. CW
B. Test
C. Single-sideband, suppressed-carrier phone
D. Phase-modulated phone

T8A10

What other emission does phase modulation most resemble?

A. Amplitude modulation
B. Pulse modulation
C. Frequency modulation
D. Single-sideband modulation

T8A03	(B) Page 9-16
T8A04	(B) Page 9-22
T8A05	(D) Page 9-16
T8A06	(D) Page 9-12
T8A07	(A) Page 9-7
T8A08	(A) Page 9-11
T8A09	(D) Page 5-13
T8A10	(C) Page 5-13

T8A11
What is the name for emissions produced by an on/off keyed audio tone?
A. RTTY
B. MCW
C. CW
D. Phone

T8B RF carrier, modulation, bandwidth and deviation.

T8B01
What is another name for a constant-amplitude radio-frequency signal?
A. An RF carrier
B. An AF carrier
C. A sideband carrier
D. A subcarrier

T8B02
What is modulation?
A. Varying a radio wave in some way to send information
B. Receiving audio information from a signal
C. Increasing the power of a transmitter
D. Suppressing the carrier in a single-sideband transmitter

T8B03
What kind of emission would your FM transmitter produce if its microphone failed to work?
A. An unmodulated carrier
B. A phase-modulated carrier
C. An amplitude-modulated carrier
D. A frequency-modulated carrier

T8B04
How would you modulate a 2-meter FM transceiver to produce packet-radio emissions?
A. Connect a terminal-node controller to interrupt the transceiver's carrier wave
B. Connect a terminal-node controller to the transceiver's microphone input
C. Connect a keyboard to the transceiver's microphone input
D. Connect a DTMF key pad to the transceiver's microphone input

T8B05
Why is FM voice best for local VHF/UHF radio communications?
A. The carrier is not detectable
B. It is more resistant to distortion caused by reflected signals
C. It has high-fidelity audio which can be understood even when the signal is somewhat weak
D. Its RF carrier stays on frequency better than the AM modes

T8B06
Why do many radio receivers have several IF filters of different bandwidths that can be selected by the operator?
A. Because some frequency bands are wider than others
B. Because different bandwidths help increase the receiver sensitivity
C. Because different bandwidths improve S-meter readings
D. Because some emission types need a wider bandwidth than others to be received properly

T8B07

Which list of emission types is in order from the narrowest bandwidth to the widest bandwidth?

A. RTTY, CW, SSB voice, FM voice
B. CW, FM voice, RTTY, SSB voice
C. CW, RTTY, SSB voice, FM voice
D. CW, SSB voice, RTTY, FM voice

T8B07
(C)
Page 9-50

T8B08

What is the usual bandwidth of a single-sideband amateur signal?

A. 1 kHz
B. 2 kHz
C. Between 3 and 6 kHz
D. Between 2 and 3 kHz

T8B08
(D)
Page 9-50

T8B09

What is the usual bandwidth of a frequency-modulated amateur signal?

A. Less than 5 kHz
B. Between 5 and 10 kHz
C. Between 10 and 20 kHz
D. Greater than 20 kHz

T8B09
(C)
Page 9-50

T8B10

What is the result of overdeviation in an FM transmitter?

A. Increased transmitter power
B. Out-of-channel emissions
C. Increased transmitter range
D. Poor carrier suppression

T8B10
(B)
Page 9-51

T8B11

What causes splatter interference?

A. Keying a transmitter too fast
B. Signals from a transmitter's output circuit are being sent back to its input circuit
C. Overmodulation of a transmitter
D. The transmitting antenna is the wrong length

T8B11
(C)
Page 9-51

SUBELEMENT T9—ANTENNAS AND FEED LINES
[3 exam questions—3 groups]

Subelement T9

T9A Parasitic beam and non-directional antennas.

T9A01

What is a directional antenna?

A. An antenna which sends and receives radio energy equally well in all directions
B. An antenna that cannot send and receive radio energy by skywave or skip propagation
C. An antenna which sends and receives radio energy mainly in one direction
D. An antenna which sends and receives radio energy equally well in two opposite directions

T9A01
(C)
Page 7-30

Technician (Element 3A) Question Pool—With Answers   13-43

T9A02

How is a Yagi antenna constructed?

A. Two or more straight, parallel elements are fixed in line with each other
B. Two or more square or circular loops are fixed in line with each other
C. Two or more square or circular loops are stacked inside each other
D. A straight element is fixed in the center of three or more elements which angle toward the ground

T9A03

What type of beam antenna uses two or more straight elements arranged in line with each other?

A. A delta loop antenna
B. A quad antenna
C. A Yagi antenna
D. A Zepp antenna

T9A04

How many directly driven elements do most beam antennas have?

A. None   C. Two
B. One   D. Three

T9A05

What is a parasitic beam antenna?

A. An antenna where some elements obtain their radio energy by induction or radiation from a driven element
B. An antenna where wave traps are used to magnetically couple the elements
C. An antenna where all elements are driven by direct connection to the feed line
D. An antenna where the driven element obtains its radio energy by induction or radiation from director elements

T9A06

What are the parasitic elements of a Yagi antenna?

A. The driven element and any reflectors
B. The director and the driven element
C. Only the reflectors (if any)
D. Any directors or any reflectors

T9A07

What is a cubical quad antenna?

A. Four straight, parallel elements in line with each other, each approximately 1/2-electrical wavelength long
B. Two or more parallel four-sided wire loops, each approximately one-electrical wavelength long
C. A vertical conductor 1/4-electrical wavelength high, fed at the bottom
D. A center-fed wire 1/2-electrical wavelength long

T9A08

What is a delta loop antenna?

A. A type of cubical quad antenna, except with triangular elements rather than square
B. A large copper ring or wire loop, used in direction finding
C. An antenna system made of three vertical antennas, arranged in a triangular shape
D. An antenna made from several triangular coils of wire on an insulating form

T9A09

What type of non-directional antenna is easy to make at home and works well outdoors?

A. A Yagi
B. A delta loop
C. A cubical quad
D. A ground plane

T9A10

What type of antenna is made when a magnetic-base whip antenna is placed on the roof of a car?

A. A Yagi
B. A delta loop
C. A cubical quad
D. A ground plane

T9A11

If a magnetic-base whip antenna is placed on the roof of a car, in what direction does it send out radio energy?

A. It goes out equally well in all horizontal directions
B. Most of it goes in one direction
C. Most of it goes equally in two opposite directions
D. Most of it is aimed high into the air

T9B Polarization, impedance matching and SWR, feed lines, balanced vs unbalanced (including baluns).

T9B01

What does horizontal wave polarization mean?

A. The magnetic lines of force of a radio wave are parallel to the earth's surface
B. The electric lines of force of a radio wave are parallel to the earth's surface
C. The electric lines of force of a radio wave are perpendicular to the earth's surface
D. The electric and magnetic lines of force of a radio wave are perpendicular to the earth's surface

T9B02

What does vertical wave polarization mean?

A. The electric lines of force of a radio wave are parallel to the earth's surface
B. The magnetic lines of force of a radio wave are perpendicular to the earth's surface
C. The electric lines of force of a radio wave are perpendicular to the earth's surface
D. The electric and magnetic lines of force of a radio wave are parallel to the earth's surface

T9B03

What electromagnetic-wave polarization does a Yagi antenna have when its elements are parallel to the earth's surface?

A. Circular
B. Helical
C. Horizontal
D. Vertical

T9A09
(D)
Page 7-16

T9A10
(D)
Page 7-18

T9A11
(A)
Page 7-18

T9B01
(B)
Page 7-29

T9B02
(C)
Page 7-29

T9B03
(C)
Page 7-29

T9B04
(D)
Page 7-29

T9B04
What electromagnetic-wave polarization does a half-wavelength antenna have when it is perpendicular to the earth's surface?
A. Circular
B. Horizontal
C. Parabolical
D. Vertical

T9B05
(D)
Page 7-30

T9B05
What electromagnetic-wave polarization does most man-made electrical noise have in the HF and VHF spectrum?
A. Horizontal
B. Left-hand circular
C. Right-hand circular
D. Vertical

T9B06
(D)
Page 7-6

T9B06
What does standing-wave ratio mean?
A. The ratio of maximum to minimum inductances on a feed line
B. The ratio of maximum to minimum resistances on a feed line
C. The ratio of maximum to minimum impedances on a feed line
D. The ratio of maximum to minimum voltages on a feed line

T9B07
(A)
Page 7-6

T9B07
What does forward power mean?
A. The power traveling from the transmitter to the antenna
B. The power radiated from the top of an antenna system
C. The power produced during the positive half of an RF cycle
D. The power used to drive a linear amplifier

T9B08
(B)
Page 7-6

T9B08
What does reflected power mean?
A. The power radiated down to the ground from an antenna
B. The power returned to a transmitter from an antenna
C. The power produced during the negative half of an RF cycle
D. The power returned to an antenna by buildings and trees

T9B09
(C)
Page 7-3

T9B09
What happens to radio energy when it is sent through a poor quality coaxial cable?
A. It causes spurious emissions
B. It is returned to the transmitter's chassis ground
C. It is converted to heat in the cable
D. It causes interference to other stations near the transmitting frequency

T9B10
(C)
Page 7-6

T9B10
What is an unbalanced line?
A. Feed line with neither conductor connected to ground
B. Feed line with both conductors connected to ground
C. Feed line with one conductor connected to ground
D. Feed line with both conductors connected to each other

T9B11
What device can be installed to feed a balanced antenna with an unbalanced feed line?
A. A balun
B. A loading coil
C. A triaxial transformer
D. A wavetrap

T9B11
(A)
Page 7-6

T9C Line losses by line type, length and frequency, RF safety.

T9C01
What common connector usually joins RG-213 coaxial cable to an HF transceiver?
A. An F-type cable connector
B. A PL-259 connector
C. A banana plug connector
D. A binding post connector

T9C01
(B)
Page 7-28

T9C02
What common connector usually joins a hand-held transceiver to its antenna?
A. A BNC connector
B. A PL-259 connector
C. An F-type cable connector
D. A binding post connector

T9C02
(A)
Page 7-28

T9C03
Which of these common connectors has the lowest loss at UHF?
A. An F-type cable connector
B. A type-N connector
C. A BNC connector
D. A PL-259 connector

T9C03
(B)
Page 7-28

T9C04
If you install a 6-meter Yagi antenna on a tower 150 feet from your transmitter, which of the following feed lines is best?
A. RG-213
B. RG-58
C. RG-59
D. RG-174

T9C04
(A)
Page 7-27

T9C05
If you have a transmitter and an antenna which are 50 feet apart, but are connected by 200 feet of RG-58 coaxial cable, what should be done to reduce feed line loss?
A. Cut off the excess cable so the feed line is an even number of wavelengths long
B. Cut off the excess cable so the feed line is an odd number of wavelengths long
C. Cut off the excess cable
D. Roll the excess cable into a coil which is as small as possible

T9C05
(C)
Page 7-27

T9C06
(B)
Page 7-27

T9C07
(B)
Page 7-27

T9C08
(D)
Page 7-27

T9C09
(D)
Page 7-25

T9C10
(C)
Page 7-25

T9C11
(A)
Page 7-28

T9C06
As the length of a feed line is changed, what happens to signal loss?
A. Signal loss is the same for any length of feed line
B. Signal loss increases as length increases
C. Signal loss decreases as length increases
D. Signal loss is the least when the length is the same as the signal's wavelength

T9C07
As the frequency of a signal is changed, what happens to signal loss in a feed line?
A. Signal loss is the same for any frequency
B. Signal loss increases with increasing frequency
C. Signal loss increases with decreasing frequency
D. Signal loss is the least when the signal's wavelength is the same as the feed line's length

T9C08
If your antenna feed line gets hot when you are transmitting, what might this mean?
A. You should transmit using less power
B. The conductors in the feed line are not insulated very well
C. The feed line is too long
D. The SWR may be too high, or the feed line loss may be high

T9C09
Why should you make sure that no one can touch an open-wire feed line while you are transmitting with it?
A. Because contact might cause a short circuit and damage the transmitter
B. Because contact might break the feed line
C. Because contact might cause spurious emissions
D. Because high-voltage radio energy might burn the person

T9C10
For RF safety, what is the best thing to do with your transmitting antennas?
A. Use vertical polarization
B. Use horizontal polarization
C. Mount the antennas where no one can come near them
D. Mount the antenna close to the ground

T9C11
Why should you regularly clean, tighten and re-solder all antenna connectors?
A. To help keep their resistance at a minimum
B. To keep them looking nice
C. To keep them from getting stuck in place
D. To increase their capacitance

Element 3A (Technician)

Figure T6—1

Figure T7—1

Figure T7—2

Figure T7—3

$$\frac{E}{I\ R}$$

Figure T7—4

As Prepared by the
Question Pool Committee
for all examinations
administered
July 1, 1993 through June 30, 1997

Helpful Data Tables

Standard Resistance Values

Numbers in **bold** type are ± 10% values. Others are 5% values.

Ohms

										Megohms				
1.0	3.6	**12**	43	**150**	510	**1800**	6200	**22000**	75000	0.24	0.62	1.6	4.3	11.0
1.1	**3.9**	13	**47**	160	**560**	2000	**6800**	24000	**82000**	**0.27**	**0.68**	**1.8**	**4.7**	**12.0**
1.2	4.3	**15**	51	**180**	620	**2200**	7500	**27000**	91000	0.30	0.75	2.0	5.1	13.0
1.3	**4.7**	16	**56**	200	**680**	2400	**8200**	30000	**100000**	**0.33**	**0.82**	**2.2**	**5.6**	**15.0**
1.5	5.1	**18**	62	**220**	750	**2700**	9100	**33000**	110000	0.36	0.91	2.4	6.2	16.0
1.6	**5.6**	20	**68**	240	**820**	3000	**10000**	36000	**120000**	**0.39**	**1.0**	**2.7**	**6.8**	**18.0**
1.8	6.2	**22**	75	**270**	910	**3300**	11000	**39000**	130000	0.43	1.1	3.0	7.5	20.0
2.0	**6.8**	24	**82**	300	**1000**	3600	**12000**	43000	**150000**	**0.47**	**1.2**	**3.3**	**8.2**	**22.0**
2.2	7.5	**27**	91	**330**	1100	**3900**	13000	**47000**	160000	0.51	1.3	3.6	9.1	
2.4	**8.2**	30	**100**	360	**1200**	4300	**15000**	51000	**180000**	**0.56**	**1.5**	**3.9**	**10.0**	
2.7	9.1	**33**	110	**390**	1300	**4700**	16000	**56000**	200000					
3.0	**10.0**	36	**120**	430	**1500**	5100	**18000**	62000	**220000**					
3.3	11.0	**39**	130	**470**	1600	**5600**	20000	**68000**						

Standard Values for 1000-V Disc-Ceramic Capacitors

pF	pF	pF	pF
3.3	39	250	1000
5	47	270	1200
6	50	300	1500
6.8	51	330	1800
8	56	360	2000
10	68	390	2500
12	75	400	2700
15	82	470	3000
18	100	500	3300
20	120	510	3900
22	130	560	4700
24	150	600	5000
25	180	680	5600
27	200	750	6800
30	220	820	8200
33	240	910	10000

Resistor Color Code

Color	Sig. Figure	Decimal Multiplier	Tolerance (%)	Color	Sig. Figure	Decimal Multiplier	Tolerance (%)
Black	0	1		Violet	7	10,000,000	
Brown	1	10		Gray	8	100,000,000	
Red	2	100		White	9	1,000,000,000	
Orange	3	1,000		Gold	—	0.1	5
Yellow	4	10,000		Silver	—	0.01	10
Green	5	100,000		No color	—		20
Blue	6	1,000,000					

Nomograph of SWR versus forward and reflected power for levels up to 20 watts. Dashed line shows an SWR of 1.5:1 for 10 W forward and 0.4 W reflected.

Nomograph of SWR versus forward and reflected power for levels up to 2000 watts. Dashed line shows an SWR of 2:1 for 90 W forward and 10 W reflected.

Common Values for Small Electrolytic Capacitors

μF	V*	μF	V*
33	6.3	10	35
33	10	22	35
100	10	33	35
220	10	47	35
330	10	100	35
470	10	220	35
10	16	330	35
22	16	470	35
33	16	1000	35
47	16	1	50
100	16	2.2	50
220	16	3.3	50
470	16	4.7	50
1000	16	10	50
2200	16	33	50
4.7	25	47	50
22	25	100	50
33	25	220	50
47	25	330	50
100	25	470	50
220	25	10	63
330	25	22	63
470	25	47	63
1000	25	1	100
2200	25	10	100
4.7	35	33	100

*Working voltage

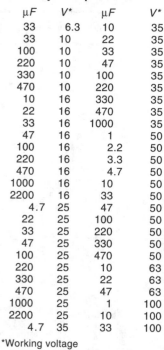

Fractions of an Inch with Metric Equivalents

Fractions Of An Inch		Decimals Of An Inch	Millimeters	Fractions Of An Inch		Decimals Of An Inch	Millimeters
	1/64	0.0156	0.397		33/64	0.5156	13.097
1/32		0.0313	0.794	17/32		0.5313	13.494
	3/64	0.0469	1.191		35/64	0.5469	13.891
1/16		0.0625	1.588	9/16		0.5625	14.288
	5/64	0.0781	1.984		37/64	0.5781	14.684
3/32		0.0938	2.381	19/32		0.5938	15.081
	7/64	0.1094	2.778		39/64	0.6094	15.478
1/8		0.1250	3.175	5/8		0.6250	15.875
	9/64	0.1406	3.572		41/64	0.6406	16.272
5/32		0.1563	3.969	21/32		0.6563	16.669
	11/64	0.1719	4.366		43/64	0.6719	17.066
3/16		0.1875	4.763	11/16		0.6875	17.463
	13/64	0.2031	5.159		45/64	0.7031	17.859
7/32		0.2188	5.556	23/32		0.7188	18.256
	15/64	0.2344	5.953		47/64	0.7344	18.653
1/4		0.2500	6.350	3/4		0.7500	19.050
	17/64	0.2656	6.747		49/64	0.7656	19.447
9/32		0.2813	7.144	25/32		0.7813	19.844
	19/64	0.2969	7.541		51/64	0.7969	20.241
5/16		0.3125	7.938	13/16		0.8125	20.638
	21/64	0.3281	8.334		53/64	0.8281	21.034
11/32		0.3438	8.731	27/32		0.8438	21.431
	23/64	0.3594	9.128		55/64	0.8594	21.828
3/8		0.3750	9.525	7/8		0.8750	22.225
	25/64	0.3906	9.922		57/64	0.8906	22.622
13/32		0.4063	10.319	29/32		0.9063	23.019
	27/64	0.4219	10.716		59/64	0.9219	23.416
7/16		0.4375	11.113	15/16		0.9375	23.813
	29/64	0.4531	11.509		61/64	0.9531	24.209
15/32		0.4688	11.906	31/32		0.9688	24.606
	31/64	0.4844	12.303		63/64	0.9844	25.003
1/2		0.5000	12.700	1		1.0000	25.400

US Customary to Metric Conversions

US Customary—Metric Conversion Factors

International System of Units (SI)—Metric Units

Prefix	Symbol			Multiplication Factor
exa	E	10^{18}	=	1 000 000 000 000 000 000
peta	P	10^{15}	=	1 000 000 000 000 000
tera	T	10^{12}	=	1 000 000 000 000
giga	G	10^{9}	=	1 000 000 000
mega	M	10^{6}	=	1 000 000
kilo	k	10^{3}	=	1 000
hecto	h	10^{2}	=	100
deca	da	10^{1}	=	10
(unit)		10^{0}	=	1
deci	d	10^{-1}	=	0.1
centi	c	10^{-2}	=	0.01
milli	m	10^{-3}	=	0.001
micro	μ	10^{-6}	=	0.000001
nano	n	10^{-9}	=	0.000000001
pico	p	10^{-12}	=	0.000000000001
femto	f	10^{-15}	=	0.000000000000001
atto	a	10^{-18}	=	0.000000000000000001

Linear
1 metre (m) = 100 centimetres (cm) = 1000 millimetres (mm)

Area
$1 \text{ m}^2 = 1 \times 10^4 \text{ cm}^2 = 1 \times 10^6 \text{ mm}^2$

Volume
$1 \text{ m}^3 = 1 \times 10^6 \text{ cm}^3 = 1 \times 10^9 \text{ mm}^3$
1 litre (l) = 1000 cm^3 = 1×10^6 mm^3

Mass
1 kilogram (kg) = 1 000 grams (g)
  (Approximately the mass of 1 litre of water)
1 metric ton (or tonne) = 1 000 kg

US Customary Units

Linear Units
12 inches (in) = 1 foot (ft)
36 inches = 3 feet = 1 yard (yd)
1 rod = 5½ yards = 16½ feet
1 statute mile = 1 760 yards = 5 280 feet
1 nautical mile = 6 076.11549 feet

Area
1 ft^2 = 144 in^2
1 yd^2 = 9 ft^2 = 1 296 in^2
1 rod^2 = 30¼ yd^2
1 acre = 4840 yd^2 = 43 560 ft^2
1 acre = 160 rod^2
1 mile^2 = 640 acres

Volume
1 ft^3 = 1 728 in^3
1 yd^3 = 27 ft^3

Liquid Volume Measure
1 fluid ounce (fl oz) = 8 fluidrams = 1.804 in^3
1 pint (pt) = 16 fl oz
1 quart (qt) = 2 pt = 32 fl oz = 57¾ in^3
1 gallon (gal) = 4 qt = 231 in^3
1 barrel = 31½ gal

Dry Volume Measure
1 quart (qt) = 2 pints (pt) = 67.2 in^3
1 peck = 8 qt
1 bushel = 4 pecks = 2 150.42 in^3

Avoirdupois Weight
1 dram (dr) = 27.343 grains (gr) or (gr a)
1 ounce (oz) = 437.5 gr
1 pound (lb) = 16 oz = 7 000 gr
1 short ton = 2 000 lb, 1 long ton = 2 240 lb

Troy Weight
1 grain troy (gr t) = 1 grain avoirdupois
1 pennyweight (dwt) or (pwt) = 24 gr t
1 ounce troy (oz t) = 480 grains
1 lb t = 12 oz t = 5 760 grains

Apothecaries' Weight
1 grain apothecaries' (gr ap) = 1 gr t = 1 gr a
1 dram ap (dr ap) = 60 gr
1 oz ap = 1 oz t = 8 dr ap = 480 fr
1 lb ap = 1 lb t = 12 oz ap = 5 760 gr

Multiply $\rightarrow$
Metric Unit = Conversion Factor × US Customary Unit

$\leftarrow$ Divide
Metric Unit ÷ Conversion Factor = US Customary Unit

Metric Unit =	Conversion Factor ×	US Unit	Metric Unit =	Conversion Factor ×	US Unit
(Length)			(Volume)		
mm	25.4	inch	mm^3	16387.064	in^3
cm	2.54	inch	cm^3	16.387	in^3
cm	30.48	foot	m^3	0.028316	ft^3
m	0.3048	foot	m^3	0.764555	yd^3
m	0.9144	yard	ml	16.387	in^3
km	1.609	mile	ml	29.57	fl oz
km	1.852	nautical mile	ml	473	pint
			ml	946.333	quart
(Area)			l	28.32	ft^3
mm^2	645.16	$inch^2$	l	0.9463	quart
cm^2	6.4516	in^2	l	3.785	gallon
cm^2	929.03	ft^2	l	1.101	dry quart
m^2	0.0929	ft^2	l	8.809	peck
cm^2	8361.3	yd^2	l	35.238	bushel
m^2	0.83613	yd^2			
m^2	4047	acre	(Mass)	(Troy Weight)	
km^2	2.59	mi^2	g	31.103	oz t
			g	373.248	lb t
(Mass)	(Avoirdupois Weight)				
grams	0.0648	grains	(Mass)	(Apothecaries' Weight)	
g	28.349	oz	g	3.387	dr ap
g	453.59	lb	g	31.103	oz ap
kg	0.45359	lb	g	373.248	lb ap
tonne	0.907	short ton			
tonne	1.016	long ton			

160 METERS

1800 1900 2000 kHz — E,A,G

Amateur stations operating at 1900–2000 kHz must not cause harmful interference to the radiolocation service and are afforded no protection from radiolocation operations.

80 METERS

3675 3725
3525
3850
3775
N,T †
G
A
E
3500 3750 4000 kHz

5167.5 kHz (SSB only): Alaska emergency use only.

40 METERS

7100 7150
7025
7225
N,T †
G *
A *
E *
7000 7150 7300 kHz

* Phone operation is allowed on 7075–7100 kHz in Puerto Rico, US Virgin islands and areas of the Caribbean south of 20 degrees north latitude; and in Hawaii and areas near ITU Region 3, including Alaska.

30 METERS

10,100 10,150 kHz — E,A,G

Maximum power on 30 meters is 200 watts PEP output. Amateurs must avoid interference to the fixed service outside the US.

20 METERS

14,025 14,150 14,225
14,175
G
A
E
14,000 14,150 14,350 kHz

17 METERS

18,068 18,110 18,168 kHz — E,A,G

15 METERS

21,100 21,200
21,025
21,300
21,225
N,T †
G
A
E
21,000 21,200 21,450 kHz

12 METERS

24,890 24,930 24,990 kHz — E,A,G

10 METERS

28,100 28,500
N,T †
E,A,G
28,000 28,300 29,700 kHz

Novices and Technicians are limited to 200 watts PEP output on 10 meters.

6 METERS

50.1
50.0 54.0 MHz — E,A,G,T

2 METERS

144.1
144.0 148.0 MHz — E,A,G,T

1.25 METERS

222.0 225.0 MHz — E,A,G,T,N

Novices are limited to 25 watts PEP output from 222 to 225 MHz.

70 CENTIMETERS **

420.0 450.0 MHz — E,A,G,T

33 CENTIMETERS **

902.0 928.0 MHz — E,A,G,T

23 CENTIMETERS **

1270 1295
N
E,A,G,T
1240 1300 MHz

Novices are limited to 5 watts PEP output from 1270 to 1295 MHz.

US AMATEUR BANDS

February 1, 1994

US AMATEUR POWER LIMITS

At all times, transmitter power should be kept down to that necessary to carry out the desired communications. Power is rated in watts PEP output. Unless otherwise stated, the maximum power output is 1500 W. Power for all license classes is limited to 200 W in the 10,100–10,150 kHz band and in all Novice subbands below 28,100 kHz. Novices and Technicians are restricted to 200 W in the 28,100–28,500 kHz subbands. In addition, Novices are restricted to 25 W in the 222–225 MHz band and 5 W in the 1270–1295 MHz subband.

Operators with Technician class licenses and above may operate on all bands above 50 MHz. For more detailed information see The FCC Rule Book.

KEY

= CW, RTTY and data

= CW, RTTY, data, MCW, test, phone and image

= CW, phone and image

= CW and SSB

= CW, RTTY, data, phone, and image

= CW only

E = EXTRA CLASS
A = ADVANCED
G = GENERAL
T = TECHNICIAN
N = NOVICE

† Only Technician–class licensees who have passed a 5 – WPM code test may use these frequencies.

** Geographical and power restrictions apply to these bands. See The FCC Rule Book for more information about your area.

Above 23 Centimeters:

All licensees except Novices are authorized all modes on the following frequencies:
2300–2310 MHz
2390–2450 MHz
3300–3500 MHz
5650–5925 MHz
10.0–10.5 GHz
24.0–24.25 GHz
47.0–47.2 GHz
75.5–81.0 GHz
119.98–120.02 GHz
142–149 GHz
241–250 GHz
All above 300 GHz

For band plans and sharing arrangements, see The ARRL Operating Manual or The FCC Rule Book.

Schematic Symbols

RESISTORS
FIXED VARIABLE PHOTO
ADJUSTABLE TAPPED THERMISTOR

CAPACITORS
FIXED NON-POLARIZED SPLIT-STATOR
ELECTROLYTIC VARIABLE FEED-THROUGH

INDUCTORS
AIR-CORE ADJUSTABLE PHASING
IRON-CORE OR
FERRITE-BEAD TAPPED

METERS
* = V, mV, A, mA, μA

WIRING
CONDUCTORS NOT JOINED CONDUCTORS JOINED SHIELDED WIRE OR COAXIAL CABLE
TERMINAL ADDRESS OR DATA BUS MULTIPLE CONDUCTOR CABLE

SWITCHES
SPST SPDT NORMALLY OPEN
TOGGLE NORMALLY CLOSED
MULTIPOINT MOMENTARY THERMAL

BATTERIES
SINGLE CELL MULTI CELL

GROUNDS
CHASSIS EARTH A-ANALOG D-DIGITAL

DIODES (D#)
LED (DS#) VOLTAGE VARIABLE CAPACITOR THYRISTOR (SCR) BRIDGE RECTIFIER
DIODE/RECTIFIER
ZENER SCHOTTKY TUNNEL TRIAC (U#)

TRANSFORMERS
AIR CORE ADJUSTABLE INDUCTANCE ADJUSTABLE COUPLING
WITH CORE WITH LINK
3-PIN CERAMIC RESONATOR

MISCELLANEOUS
ANTENNA FUSE
QUARTZ CRYSTAL HAND KEY
MOTOR ASSEMBLY OR MODULE (OTHER THAN IC)

TRANSISTORS
PNP P-CHANNEL P-CHANNEL P-CHANNEL P-CHANNEL P-CHANNEL
NPN N-CHANNEL N-CHANNEL N-CHANNEL N-CHANNEL N-CHANNEL
BIPOLAR UJT JUNCTION FET SINGLE-GATE DUAL-GATE SINGLE-GATE
DEPLETION MODE MOSFET
ENHANCMENT MODE MOSFET

LOGIC (U#)
AND NAND
OR NOR
XOR INVERT
SCHMITT OTHER

RELAYS
SPST SPDT DPDT
THERMAL

INTEGRATED CIRCUITS (U#)
GENERAL AMPLIFIER OTHER
OP AMP

CONNECTORS
COMMON CONNECTIONS LABEL
CONTACTS MALE FEMALE
PHONE JACK PHONE PLUG
COAXIAL CONNECTORS FEMALE MALE
MULTIPLE, MOVABLE
MULTIPLE, FIXED
240 V FEMALE GROUND
FEMALE MALE MALE CHASSIS-MOUNT
HOT 120 V NEUT HOT 120 V HOT 120 V NEUT
GND GND GND

TUBES (V#)
TUBE ELEMENTS
ANODE HEATER OR FILAMENT
GRID GAS FILLED
CATHODE COLD CATHODE
DEFLECTION PLATES
TRIODE PENTODE CRT

LAMPS
INCANDESCENT NEON (AC)

Equations Used in this Book

$$\text{current} = \frac{\text{voltage}}{\text{resistance}} \qquad \text{(Eq 3-1)}$$

$$\text{resistance} = \frac{\text{voltage}}{\text{current}} \qquad \text{(Eq 3-2)}$$

$$\text{voltage} = \text{current} \times \text{resistance} \qquad \text{(Eq 3-3)}$$

$$E = IR \text{ (volts = amperes} \times \text{ohms)} \qquad \text{(Eq 3-4)}$$

$$I = \frac{E}{R} \text{ (amperes = volts divided by ohms)} \qquad \text{(Eq 3-5)}$$

$$R = \frac{E}{I} \text{ (ohms = volts divided by amperes)} \qquad \text{(Eq 3-6)}$$

$$P = IE \text{ (watts = amperes} \times \text{volts)} \qquad \text{(Eq 3-7)}$$

$$I = \frac{P}{E} \text{ (amperes = watts divided by volts)} \qquad \text{(Eq 3-8)}$$

$$E = \frac{P}{I} \text{ (volts = watts divided by amperes)} \qquad \text{(Eq 3-9)}$$

$$c = f\lambda \qquad \text{(Eq 3-10)}$$

$$f = \frac{c}{\lambda} \qquad \text{(Eq 3-11)}$$

$$\lambda = \frac{c}{f} \qquad \text{(Eq 3-12)}$$

$$R_{TOTAL} = R_1 + R_2 + R_3 + \ldots + R_n \qquad \text{(Eq 3-13)}$$

$$R_{TOTAL} = \frac{1}{\dfrac{1}{R_1} + \dfrac{1}{R_2} + \dfrac{1}{R_3} + \ldots + \dfrac{1}{R_n}} \qquad \text{(Eq 3-14)}$$

$$R_{TOTAL} = \frac{R_1 \times R_2}{R_1 + R_2} \qquad \text{(Eq 3-15)}$$

$$L_{TOTAL} = L_1 + L_2 + L_3 + \ldots + L_n \qquad \text{(Eq 3-16)}$$

$$L_{TOTAL} = \frac{1}{\dfrac{1}{L_1} + \dfrac{1}{L_2} + \dfrac{1}{L_3} + \ldots + \dfrac{1}{L_n}} \qquad \text{(Eq 3-17)}$$

$$L_{TOTAL} = \frac{L_1 \times L_2}{L_1 + L_2} \qquad \text{(Eq 3-18)}$$

$$C_{TOTAL} = C_1 + C_2 + C_3 + \ldots + C_n \qquad \text{(Eq 3-19)}$$

$$C_{TOTAL} = \frac{1}{\dfrac{1}{C_1} + \dfrac{1}{C_2} + \dfrac{1}{C_3} + \ldots + \dfrac{1}{C_n}} \qquad \text{(Eq 3-20)}$$

$$C_{TOTAL} = \frac{C_1 \times C_2}{C_1 + C_2} \qquad \text{(Eq 3-21)}$$

$$\lambda \text{ (in feet)} = \frac{984}{f \text{ (in MHz)}} \qquad \text{(Eq 7-1)}$$

$$\tfrac{1}{2}\lambda \text{ (in feet)} = \frac{468}{f \text{ (in MHz)}} \qquad \text{(Eq 7-2)}$$

$$\tfrac{1}{4}\lambda \text{ (in feet)} = \frac{234}{f \text{ (in MHz)}} \qquad \text{(Eq 7-3)}$$

$$B_w = 2 \times (D + M) \qquad \text{(Eq 9-1)}$$

Glossary of Key Words

A1A emission—The FCC emission type designator that describes international Morse code telegraphy (CW) communications without audio modulation of the carrier.

Alternating current (ac)—Electrical current that flows first in one direction in a wire and then in the other. The applied voltage is also changing polarity. This direction reversal continues at a rate that depends on the frequency of the ac.

Amateur operator—A person holding a written authorization to be the control operator of an amateur station.

Amateur service—A radiocommunication service for the purpose of self-training, intercommunication and technical investigations carried out by amateurs, that is, duly authorized persons interested in radio technique solely with a personal aim and without pecuniary interest. (Pecuniary means payment of any type, whether money or other goods.)

Amateur station—A station licensed in the amateur service, including necessary equipment, used for amateur communication.

Ammeter—A test instrument that measures current.

Ampere (A)—The basic unit of electrical current. Current is a measure of the electron flow through a circuit. If we could count electrons, we would find that if there are 6.24×10^{18} electrons moving past a point in one second, we have a current of one ampere.[1] We abbreviate amperes as amps.

Amplifier—A device usually employing electron tubes or transistors to increase the voltage, current or power of a signal. The amplifying device may use a small signal to control voltage and/or current from an external supply. A larger replica of the small input signal appears at the device output.

Amplitude modulation (AM)—A method of combining an information signal and an RF (radio-frequency) carrier. In double-sideband voice AM transmission, we use the voice information to vary (modulate) the amplitude of an RF carrier. Shortwave broadcast stations use this type of AM, as do stations in the Standard Broadcast Band (540-1600 kHz). Few amateurs use double-sideband voice AM, but a variation, known as **single sideband**, is very popular.

ANSI RF protection guide—Standards formulated by the American National Standards Institute that provide RF exposure guidelines.

Antenna—A device that picks up or sends out radio waves.

---

[1]Numbers written as a multiple of some power are expressed in exponential notation. This notation is explained in detail on page 3-2.

Antenna-matching network—A device that matches the antenna system input impedance to the transmitter, receiver or transceiver output impedance. Also called an **antenna tuner** or *Transmatch*.

Antenna switch—A switch used to connect one transmitter, receiver or transceiver to several different antennas.

Antenna tuner—See **antenna-matching network**.

Audio frequency (AF)—The range of frequencies that the human ear can detect. Audio frequencies are usually listed as 20 Hz to 20,000 Hz.

Audio-frequency shift keying (AFSK)—A method of transmitting radioteletype information. Two switched audio tones are fed into the microphone input. AFSK RTTY is most often used on VHF. Also see **frequency-shift keying (FSK)**.

Autopatch—A device that allows repeater users to make telephone calls through a repeater.

Balanced line—Feed line with neither conductor at ground potential, such as open-wire line or ladder line.

Balun—Contraction for balanced to unbalanced. A device to couple a balanced load to an unbalanced source, or vice versa.

Band-pass filter—A circuit that allows signals to go through it only if they are within a certain range of frequencies. It attenuates signals above and below this range.

Bandwidth—A range of associated frequencies (measured in hertz). For example, bandwidth is used to describe the range of frequencies in the radio spectrum that a radio transmission occupies. The range of frequencies that will pass through a given filter. The width of a frequency band outside of which the mean power is attenuated at least 26 dB below the mean power of the total emission, including allowances for transmitter drift or Doppler shift.

Battery—A device that stores electrical energy. It stores the energy in the form of chemical energy, which it converts to electrical energy. It provides excess electrons to produce a **current** and the **voltage** or EMF to push those electrons through a circuit.

Baudot—A code used in radioteletype communications. Each character is represented with a string of five bits of digital information. Each character has a different combination of bits.

Beacon station—An amateur station transmitting communications for the purposes of observation of propagation and reception or other related experimental activities.

Beam antenna—A directional antenna. A beam antenna must be rotated to provide coverage in different directions.

Beat-frequency oscillator (BFO)—A receiver circuit that provides a signal to the product **detector**. The BFO signal mixes with the IF signal to produce an audio tone for **CW** reception. A BFO is needed to copy CW and **SSB** signals.

Block diagram—A drawing using boxes to represent sections of a complicated device or process. The block diagram shows the connections between sections.

Broadcasting—Transmissions intended to be received by the general public, either direct or relayed.

Capacitance—The ability of a capacitor to store energy in an electric field.

Capacitor—An electronic component composed of two or more conductive plates separated by an insulating material. A capacitor stores energy in an electric field.

Centi—The metric prefix for 10^{-2}, or divide by 100.

Characteristic impedance—The opposition to electric current that an antenna feed line presents. Impedance includes factors other than resistance, and applies to alternating currents. Ideally, the characteristic impedance of a feed line is the same as the transmitter output impedance and the antenna input impedance.

Chassis ground—The common connection for all parts of a circuit that connect to the negative side of the power supply.

Chirp—A slight shift in transmitter frequency each time you key the transmitter.

Closed—A repeater that restricts access to those who know a special code. (See **CTCSS**.)

Closed, or complete circuit—An electrical circuit with an uninterrupted path for the current to follow. Turning a switch on, for example, closes or completes the circuit, allowing current to flow.

Coaxial cable—Coax (pronounced kó-aks). A type of feed line with one conductor inside the other.

Code key—A device used as a switch to generate Morse code. Also called a **hand key**, a **straight key** or a **telegraph key**.

Code-practice oscillator—A device that produces an audio tone, used for learning the code.

Color code—A system in which numerical values are assigned to various colors. Colored stripes are painted on the body of **resistors** and sometimes other components to show their value.

Conductor—A material that has a loose grip on its electrons, so an electrical current can pass through it.

Connected—The condition in which two **packet-radio** stations are sending information to each other. Each is acknowledging when the data has been received correctly.

Continuous wave (CW)—Morse code telegraphy.

Control operator—An amateur operator designated by the licensee of a station to be responsible for the transmissions of an amateur station.

Control point—The locations at which the control operator function is performed.

Core—The material used in the center of an **inductor** coil.

CQ—"Calling any station": the general call when requesting a conversation with anyone.

Critical frequency—The highest frequency at which a vertically incident radio wave will return from the **ionosphere**. Above the critical frequency, radio signals pass through the ionosphere instead of returning to Earth.

Crystal calibrator—(See **marker generator**.)

CTCSS—Continuous tone-coded squelch system; a sub-audible tone system used on some repeaters. When added to a carrier, a CTCSS tone allows a receiver to accept a signal. Also called **PL**.

Cubical quad antenna—An antenna built with its elements in the shape of four-sided loops.

Current—A flow of electrons in an electrical circuit.

CW (continuous wave)—The FCC emission type that describes international Morse code telegraphy communication without audio modulation of the carrier. Hams usually produce Morse code signals by interrupting the continuous-wave signal from a transmitter to form the dots and dashes. This is on/off keying of a radio-frequency signal. Another name for international Morse code.

D region—The lowest region of the ionosphere. The D region contributes very little to short-wave radio propagation. It acts mainly to absorb energy from radio waves as they pass through it. This absorption has a significant effect on signals below about 7.5 MHz during daylight.

Dash—The long sound used in Morse code. Pronounce this as "dah" when verbally sounding Morse code characters. Dashes are three times longer than dots.

Data—Computer-based modes, such as RTTY and packet.

DE—The Morse code abbreviation for "from" or "this is."

Deci—The metric prefix for 10^{-1}, or divide by 10.

Delta loop antenna—A variation of the cubical quad with triangular elements.

Detector—The stage in a receiver in which the modulation (voice or other information) is recovered from the RF signal.

Deviation ratio—The ratio between the maximum change in RF-carrier frequency and the highest modulating frequency used in an FM transmitter.

Digipeater—A packet-radio station used to retransmit signals that are specifically addressed to be retransmitted by that station.

Digital communications—Computer-based communications modes. (Also see **Data**.)

Dipole antenna—See **half-wave dipole**. A dipole need not be ½ wavelength long.

Direct current (dc)—Electrical current that flows in one direction only.

Directional wattmeter—(See **wattmeter**.)

Directivity—The ability of an antenna to focus transmitter power into certain directions. Also its ability to enhance received signals from specific directions.

Director—An element in front of the driven element in a Yagi and some other directional antennas.

Direct wave—A (usually) VHF or UHF radio wave that travels in a straight line-of-sight path between the transmitter and receiver.

Dot—The short sound used in Morse code. Pronounce this as "dit" when verbally sounding Morse code characters if the dot comes at the end of the character. If the dot comes at the beginning or in the middle of the character, pronounce it as "di."

Double-pole, double-throw (DPDT) switch—A switch that has six contacts. The DPDT switch has two center contacts. The two center contacts can each be connected to one of two other contacts.

Double-pole, single-throw (DPST) switch—A switch that connects two contacts to another set of contacts. A DPST switch turns two circuits on or off at the same time.

Driven element—The part of an antenna that connects directly to the feed line.

Dummy antenna or **dummy load**—A station accessory that allows you to test or adjust transmitting equipment without sending a signal out over the air.

Duplex operation—Receiving and transmitting on two different frequencies. (Also see **simplex operation**.)

DX—Distance, foreign countries.

E region—The second lowest ionospheric region. Exists only during the day. Under certain conditions, may refract radio waves enough to return them to earth.

Earth ground—A circuit connection to a ground rod driven into the earth or to a cold-water pipe made of copper that goes into the ground.

Earth station—An amateur station located on, or within 50 km of, the Earth's surface intended for communications with space stations or with other Earth stations by means of one or more other objects in space.

Electric field—An invisible force of nature. An electric field exists in a region of space if an electrically charged object placed in the region is subjected to an electrical force.

Electrolytic capacitor—A polarized capacitor that relies on a chemical reaction to form a thin insulating layer between the plates.

Electromotive force (EMF)—The force or pressure that pushes a current through a circuit.

Electron—A tiny, negatively charged particle, normally found in an area surrounding the nucleus of an atom. Moving electrons make up an electrical current.

Electronic keyer—A device that generates Morse code dots and dashes electronically.

Emergency—A situation where there is a danger to lives or property.

Emission privilege—Permission to use a particular emission type (such as Morse code or voice).

Emission types—Term for the different modes authorized for use on the Amateur Radio bands. Examples are CW, phone (such as SSB and FM) and digital (such as RTTY and packet radio).

Emission—RF signals transmitted from a radio station.

Energy—The ability to do work; the ability to exert a force to move some object.

F region—A combination of the two highest ionospheric regions, the F1 and F2 regions. Refracts radio waves and returns them to earth. Its height varies greatly depending on the time of day, season of the year and amount of sunspot activity.

False or deceptive signals—Transmissions that are intended to mislead or confuse those who may receive the transmissions. For example, distress calls transmitted when there is no actual emergency are false or deceptive signals.

Farad—The basic unit of capacitance.

Feed line (feeder)—The wires or cable used to connect a transmitter, receiver or transceiver to an antenna. (Also see **transmission line**.)

Film resistor—A resistor made by depositing a thin layer of resistive material on a ceramic form.

Filter—A circuit that will allow some signals to pass through it but will greatly reduce the strength of others.

Fist—The unique rhythm of an individual amateur's Morse code sending.

Fixed resistor—A resistor whose value cannot be adjusted by the user.

Frequency—The number of complete cycles of an alternating current that occur per second.

Frequency bands—A group of frequencies where amateur communications are authorized.

Frequency coordination—Allocating repeater input and output frequencies to minimize interference between repeaters and to other users of the band.

Frequency coordinators—Individuals or groups that recommend repeater frequencies.

Frequency deviation—The amount the carrier frequency in an FM transmitter changes as it is modulated.

Frequency-modulated (FM) voice—Mode of voice (phone) communications used on repeaters.

Frequency modulation (FM)—A method of combining an information signal with a radio signal. The process of varying the frequency of an RF carrier in response to instantaneous changes in the modulating signal. As you might suspect, we use voice or data to vary the frequency of the transmitted signal. FM broadcast stations and most professional communications (police, fire, taxi) use FM. VHF/UHF FM voice is the most-popular amateur mode. (Also see **amplitude modulation**.)

Frequency privilege—Permission to use a particular group of frequencies.

Frequency-shift keying (FSK)—A method of transmitting radioteletype information by switching an RF carrier between two separate frequencies. FSK RTTY is most often used on HF. (Also see **audio-frequency shift keying**.)

Front-end overload—Interference to a receiver caused by a strong signal that overpowers the receiver RF amplifier or "front end." (Also see **receiver overload**.)

Fuse—A thin metal strip mounted in a holder. When too much current passes through the fuse, the metal strip melts and opens the circuit.

Gain—A measure of the **directivity** of an antenna, as compared with another antenna such as a **dipole**.

General-coverage receiver—A receiver used to listen to a wide range of frequencies. Most general-coverage receivers tune from frequencies below the standard-broadcast band to at least 30 MHz. These frequencies include the shortwave-broadcast bands and the amateur bands from 160 to 10 meters.

Giga—The metric prefix for 10^9, or times 1,000,000,000.

Grace period—The time FCC allows following the expiration of an amateur license to renew that license without having to retake an examination. Those who hold an expired license may not operate an amateur station until the license is reinstated.

Ground connection—A connection made to the earth for electrical safety. This connection can be made inside (to a cold-water pipe) or outside (to a ground rod).

Ground rod—A copper or copper-clad steel rod that is driven into the earth. A heavy copper wire from the ham shack connects all station equipment to the ground rod.

Ground-wave propagation—The method by which radio waves travel along the Earth's surface.

Half-wave dipole—A basic antenna used by radio amateurs. It consists of a length of wire or tubing, opened and fed at the center. The entire antenna is ½ wavelength long at the desired operating frequency.

Ham-bands-only receiver—A receiver designed to cover only the bands used by amateurs. Usually refers to the bands from 80 to 10 meters, sometimes including 160 meters.

Hand key—A simple switch used to send Morse code. Also called a **code key** or a **telegraph key**.

Harmonics—Signals from a transmitter or oscillator occurring on whole-number multiples of the desired operating frequency.

Henry—The basic unit of **inductance**.

Hertz (Hz)—An alternating-current frequency of one cycle per second. The basic unit of **frequency**.

High-pass filter—A filter designed to pass high-frequency signals, while blocking lower-frequency signals. A high-pass filter allows signals above the cutoff frequency to pass through. It attenuates signals below the cutoff frequency.

Horizontally polarized wave—An electromagnetic wave with its electric lines of force parallel to the ground.

Impedance-matching network—A device that matches the impedance of an antenna system to the impedance of a transmitter or receiver. Also called an **antenna-matching network** or *Transmatch*.

Induced EMF—A voltage produced by a change in magnetic lines of force around a conductor. When a magnetic field is formed by current in the conductor, the induced voltage always opposes change in that current.

Inductance—The ability of a coil to store energy in a magnetic field.

Inductor—An electrical component usually composed of a coil of wire wound on a central core. An inductor stores energy in a magnetic field.

Input frequency—A repeater's receiving frequency. To use a repeater, transmit on the input frequency and receive on the output frequency.

Insulator—A material that maintains a tight grip on its electrons, so that an electric current cannot pass through it (within voltage limits).

Intermediate frequency (IF)—The output frequency of a mixing stage in a superheterodyne receiver. The subsequent stages in the receiver are tuned for maximum efficiency at the IF.

Ionosphere—A region of charged (ionized) gases high in the atmosphere. The ionosphere bends radio waves as they travel through it, returning them to Earth. (Also see **sky-wave propagation**.)

K—The Morse code abbreviation for "any station respond."

Key click—A click or thump at the beginning or end of a CW signal.

Key-click filter—A circuit in a transmitter that reduces or eliminates key clicks.

Key-operated on-off switch—A good way to prevent unauthorized persons from using your station is to install a key-operated switch that controls station power.

Kilo—The metric prefix for 10^3, or times 1000.

Lightning protection—There are several ways to help prevent lightning damage to your equipment (and your house), among them unplugging equipment, disconnecting antenna feed lines and using a lightning arrestor.

Line of sight—The term used to describe VHF and UHF propagation in a straight line directly from one station to another.

Low-pass filter—A filter designed to pass low-frequency signals, while blocking higher-frequency signals. A low-pass filter allows signals below the cutoff frequency to pass through and attenuates signals above the cutoff frequency.

Lower sideband (LSB)—The common single-sideband operating mode on the 40, 80 and 160-meter amateur bands.

Major lobe—The shape or pattern of field strength that points in the direction of maximum radiated power from an antenna.

Malicious (harmful) interference—Intentional, deliberate obstruction of radio transmissions.

Marker generator—A high-stability oscillator that produces reference signals at known frequency intervals. It can be used to calibrate receiver and transmitter tuning dials. Also called a **crystal calibrator**.

Matching network—A device that matches one impedance level to another. For example, it may match the impedance of an antenna system to the impedance of a transmitter or receiver. Amateurs also call such devices a Transmatch, impedance-matching network or match box.

Maximum usable frequency (MUF)—The highest-frequency radio signal that will reach a particular destination using E- or F-region skip propagation. The MUF may vary for radio signals sent to different destinations.

MAYDAY—From the French m'aidez (help me), MAYDAY is used when calling for emergency assistance in voice modes.

MCW—The FCC emission type that describes international Morse code telegraphy communication with audio modulation of the carrier.

Mega—The metric prefix for 10^6, or times 1,000,000.

Metric prefixes—A series of terms used in the metric system of measurement. We use metric prefixes to describe a quantity as compared to a basic unit. The metric prefixes indicate multiples of 10.

Metric system—A system of measurement developed by scientists and used in most countries of the world. This system uses a set of prefixes that are multiples of 10 to indicate quantities larger or smaller than the basic unit.

Micro—The metric prefix for 10^{-6}, or divide by 1,000,000.

Microphone—A device that converts sound waves into electrical energy.

Milli—The metric prefix for 10^{-3}, or divide by 1000.

Modem—Short for modulator/demodulator. A modem modulates a radio signal to transmit data and demodulates a received signal to recover transmitted data.

Modulate—To vary the amplitude, frequency, or phase of a radio-frequency signal.

Modulation—The process of varying a radio wave in some way to send information.

Monitor mode—One type of packet radio receiving mode. In monitor mode, everything transmitted on a packet frequency is displayed by the monitoring TNC. This occurs whether or not the transmissions are addressed to the monitoring station.

Morse code—(See **CW**.)

Multimeter—An electronic test instrument used to measure current, voltage and resistance in a circuit. Describes all meters capable of making these measurements, such as the volt-ohm-milliammeter (VOM), vacuum-tube voltmeter (VTVM) and field-effect transistor VOM (FET VOM).

Multimode transceiver—Transceiver capable of SSB, CW and FM operation.

Narrow-band direct-printing telegraphy—The technical term for radioteletype (**RTTY**).

National Electrical Code—A set of guidelines governing electrical safety, including antennas.

Network—A term used to describe several packet stations linked together to transmit data over long distances.

NPN transistor—A transistor that has a layer of P-type semiconductor material sandwiched between layers of N-type semiconductor material.

Offset—The 300- to 1000-Hz difference in CW transmitting and receiving frequencies in a transceiver. For a repeater, offset refers to the difference between its transmitting and receiving frequencies.

Ohm—The basic unit of electrical resistance, used to describe the amount of opposition to current.

Ohm's Law—A basic law of electronics. Ohm's Law gives a relationship between voltage (E), current (I) and resistance (R). The voltage applied to a circuit is equal to the current through the circuit times the resistance of the circuit (E = IR).

Omnidirectional—Antenna characteristic meaning it radiates equal power in all compass directions.

One-way communications—Transmissions that are not intended to be answered. The FCC strictly limits the types of one-way communications allowed on the amateur bands.

Open—A **repeater** that can be used by all hams who have a license that authorizes operation on the repeater frequencies.

Open circuit—An electrical circuit that does not have a complete path, so current can't flow through the circuit.

Open-wire feed line—Parallel-conductor feeder with air as its primary insulation material.

Operator license—The portion of an Amateur Radio license that gives permission to operate an amateur station.

Output frequency—A repeater's transmitting frequency. To use a repeater, transmit on the input frequency and receive on the output frequency.

Overdeviation—Overloading an FM transceiver by speaking too loudly into the microphone or holding it too close to your mouth.

Packet radio—A digital communications system in which information is broken into short bursts. The bursts (packets) also contain addressing and error-detection information.

Parallel circuit—An electrical circuit in which the electrons follow more than one path in going from the negative supply terminal to the positive terminal.

Parallel-conductor feed line—Feed line with two conductors held a constant distance apart.

Parasitic element—Part of a directive antenna that derives energy from mutual coupling with the driven element. Parasitic elements are not connected directly to the feed line.

Peak envelope power (PEP)—The average power of a signal at its largest amplitude peak.

Phase modulation (PM)—Varying the phase of an RF carrier in response to the instantaneous changes in the modulating signal.

Phone—Another name for voice communications.

Phonetic alphabet—Standard words used on voice modes to make it easier to understand letters of the alphabet, such as those in call signs. The call sign KA6LMN stated phonetically is Kilo Alfa Six Lima Mike November.

Pico—The metric prefix for 10^{-12}, or divide by 1,000,000,000,000.

PL—(See **CTCSS**.)

PNP transistor—A transistor that has a layer of N-type semiconductor material sandwiched between layers of P-type semiconductor material.

Polarization—Describes the electrical-field characteristic of a radio wave. An antenna that is parallel to the surface of the earth, such as a dipole, produces horizontally polarized waves. One that is perpendicular to the earth's surface, such as a quarter-wave vertical, produces vertically polarized waves.

Potentiometer—A resistor whose resistance can be varied continuously over a range of values. Another name for a variable resistor. The value of a potentiometer can be changed without removing it from a circuit.

Power—The rate of energy consumption. We calculate power in an electrical circuit by multiplying the voltage applied to the circuit times the current through the circuit (P = IE).

Power supply—An electrical circuit that provides excess electrons to flow into another circuit. The power supply also supplies the voltage or EMF to push the electrons along. Power supplies convert a power source (such as the ac mains) to a form useful for various circuits. A circuit that provides a direct-current output at some desired voltage from an ac input voltage.

Procedural signal (prosign)—One or two letters sent as a single character. Amateurs use prosigns in CW contacts as a short way to indicate the operator's intention. Some examples are K for "Go Ahead," or $\overline{AR}$ for "End of Message." (The bar over the letters indicates that we send the prosign as one character.)

Propagation—The study of how radio waves travel.

Q signals—Three-letter symbols beginning with Q. Used on CW to save time and to improve communication. Some examples are QRS (send slower), QTH (location), QSO (ham conversation) and QSL (acknowledgment of receipt).

QRL?—Ham radio Q signal meaning "Is this frequency in use?"

QSL card—A postcard that serves as a confirmation of communication between two hams.

QSO—A conversation between two radio amateurs.

Quarter-wavelength vertical antenna—An antenna constructed of a quarter-wavelength long radiating element placed perpendicular to the earth.

Radio frequency (RF)—The range of frequencies that can be radiated through space in the form of electro-magnetic radiation. We usually consider RF to be those frequencies higher than the audio frequencies, or above 20 kilohertz.

Radio-frequency (RF) carrier—A constant-amplitude, unmodulated, radio-frequency signal.

Radio-frequency interference (RFI)—Disturbance to electronic equipment caused by radio-frequency signals.

Radio Frequency Protection Guide of the American National Standards Institute (ANSI)—Standards that recommend RF exposure limits for humans.

Radioteletype (RTTY)—Radio signals sent from one teleprinter machine to another machine. Anything that one operator types on his teleprinter will be printed on the other machine. Also known as **narrow-band direct-printing telegraphy.**

Reactance—The property of an inductor or capacitor (measured in ohms) that impedes current in an ac circuit without converting power to heat.

Reactance modulator—A device capable of modulating an ac signal by varying the reactance of a circuit in response to the modulating signal. (The modulating signal may be voice, data, video, or some other kind depending on what type of information is being transmitted.)

Receiver—A device that converts radio waves into signals we can hear or see.

Receiver incremental tuning (RIT)—A transceiver control that allows for a slight change in the receiver frequency without changing the transmitter frequency. Some manufacturers call this a clarifier (CLAR) control.

Receiver overload—Interference to a receiver caused by a strong RF signal that forces its way into the equipment. A signal that overloads the receiver RF amplifier (front end) causes front-end overload. Receiver overload is sometimes called **RF overload.**

Reflection—Signals that travel by **line-of-sight** propagation are reflected by large objects like buildings.

Reflectometer—A test instrument used to indicate standing wave ratio (SWR) by measuring the forward power (power from the transmitter) and reflected power (power returned from the antenna system).

Reflector—An element behind the driven element in a Yagi and some other directional antennas.

Repeater station—An amateur station that automatically retransmits the signals of other stations. Repeaters extend the range of the retransmitted signals.

Resistance—The ability to oppose an electric current.

Resistor—A circuit component that controls current through a circuit. Any material that opposes a current in an electrical circuit. An electronic component especially designed to oppose current.

Resolution—The space between markings on a receiver dial. The greater the frequency resolution, the easier it is to separate signals that are close together.

Resonant frequency—The desired operating frequency of a tuned circuit. In an antenna, the resonant frequency is one where the feed-point impedance contains only resistance.

RF burn—A flesh burn caused by exposure to a strong field of radio-frequency (RF) energy.

RF overload—Another term for receiver overload.

RF safety—Preventing injury or illness to humans from the effects of radio-frequency energy.

Rig—The radio amateur's term for a transmitter, receiver or transceiver.

RST—A system of numbers used for signal reports: R is readability, S is strength and T is tone. (On **single-sideband** phone, only R and S reports are used.)

S meter—A meter in a receiver that shows the relative strength of a received signal.

Safety interlock—A switch that automatically turns off ac power to a piece of equipment when the top cover is removed.

Scatter—A small amount of signal that is reflected from the Earth's surface after traveling through the ionoshpere. The reflected signals may go back into the ionosphere along several paths and be refracted to Earth again. Scatter can help provide communications into a station's **skip zone.**

Schematic symbol—A drawing used to represent a circuit component on a wiring diagram.

Selectivity—The ability of a receiver to separate two closely spaced signals.

Sensitivity—The ability of a receiver to detect weak signals.

Series circuit—An electrical circuit in which all the electrons must flow through every part of the circuit. There is only one path for the electrons to follow.

Shack—The room where an Amateur Radio operator keeps his or her station equipment.

Short circuit—An electrical circuit in which the current does not take the desired path, but finds a shortcut instead. Often the current goes directly from the negative power-supply terminal to the positive one, bypassing the rest of the circuit.

Sidebands—The sum or difference frequencies generated when an RF carrier is mixed with an audio signal. **Single-sideband** phone **(SSB)** signals have an **upper sideband** (USB—that part of the signal above the carrier) and a **lower sideband** (LSB—the part of the signal below the carrier). SSB transceivers allow operation on either USB or LSB.

Signal generator—A test instrument that produces a stable low-level radio-frequency signal. The signal can be set to a specific frequency and used to troubleshoot RF equipment.

Simplex operation—Receiving and transmitting on the same frequency. (Also see **duplex operation**.)

Single sideband (SSB)—A common mode of voice operation on the amateur bands. SSB is a form of amplitude modulation.

Single-pole, double-throw (SPDT) switch—A switch that connects one center contact to one of two other contacts.

Single-pole, single-throw (SPST) switch—A switch that only connects one center contact to another contact.

Skip zone—A "dead" area too distant for ground waves and too close for sky waves.

Sky-wave propagation—The method by which radio waves travel through the ionosphere and back to Earth. Sometimes called skip, sky-wave propagation has a far greater range than line-of-sight and ground-wave propagation.

SOS—A Morse code call for emergency assistance.

Space station—An amateur station located more than 50 km above the Earth's surface.

Splatter—A type of interference to nearby stations. Occurs when a transmitter is overmodulated.

Spurious emissions—Signals from a transmitter on frequencies other than the operating frequency.

SSB—Emission mode that describes the type of voice emission used on the HF bands. Abbreviation for **single sideband**.

Standing-wave ratio (SWR)—Sometimes called VSWR. A measure of the impedance match between the feed line and the antenna. Also, with a Transmatch in use, a measure of the match between the feed line from the transmitter and the antenna system. The system includes the Transmatch and the line to the antenna. VSWR is the ratio of maximum voltage to minimum voltage along the feed line. Also the ratio of antenna impedance to feed-line impedance when the antenna is a purely resistive load.

Station grounding—Connecting all station equipment to a good earth ground improves both safety and station performance.

Station license—The portion of an Amateur Radio license that authorizes an amateur station at a specific location. The station license also lists the call sign of that station.

Straight key—Another name for a **code key**.

Sunspot cycle—The number of sunspots increases and decreases in a predictable cycle that lasts about 11 years.

Sunspots—Dark spots on the surface of the sun. When there are few sunspots, long-distance radio propagation is poor on the higher-frequency bands. When there are many sunspots, long-distance HF propagation improves.

Superimposed hum—A low-pitched buzz or hum on a radio signal.

Switch—A device used to connect or disconnect electrical contacts.

SWR meter—A device used to measure SWR. A measuring instrument that can indicate when an antenna system is working well. (Also see **reflectometer**.)

Telegraph key—Another name for a **code key**.

Teleprinter—A machine that can convert keystrokes (typing) into electrical impulses. The teleprinter can also convert the proper electrical impulses back into text. Computers have largely replaced teleprinters for amateur radioteletype work.

Television interference (TVI)—Interruption of television reception caused by another signal.

Temperature inversion—A condition in the atmosphere in which a region of cool air is trapped beneath warmer air.

Temporary state of communication emergency—When a disaster disrupts normal communications in a particular area, the FCC can declare this type of emergency. Certain rules may apply for the duration of the emergency.

Terminal node controller (TNC)—A TNC accepts information from a computer and converts the information into packets. The TNC also receives packets and extracts information to be displayed by a computer.

Third-party communications—Messages passed from one amateur to another on behalf of a third person.

Third-party communications agreement—An official understanding between the United States and another country that allows amateurs in both countries to participate in third-party communications.

Third-party participation—The way an unlicensed person can participate in amateur communications. A control operator must ensure compliance with FCC rules.

Ticket—A commonly used name for an Amateur Radio license.

Time-out timer—A device that limits the amount of time any one person can talk through a repeater.

Toroidal core—A donut-shaped form used to hold the turns of an inductor. Toroidal cores are usually made from ferrite or powdered-iron materials.

Traffic net—An on-the-air meeting of amateurs, for the purpose of relaying messages.

Transceiver—A radio transmitter and receiver combined in one unit.

Transistor—A solid-state device comprising three layers of semiconductor material. An NPN bipolar transistor has a layer of P-type semiconductor material sandwiched between two layers of N-type material. A PNP bipolar transistor has a layer of N-type semiconductor material sandwiched between two layers of P-type material.

Transmission line—The wires or cable used to connect a transmitter or receiver to an antenna.

Transmit-receive (TR) switch—A device used for switching between transmit and receive operation. The TR switch allows you to share one antenna between a receiver and a transmitter.

Transmitter—A device that produces radio-frequency signals.

Triode—A vacuum tube with three active elements: cathode, plate and control grid.

Troposphere—The region in Earth's atmosphere just above the Earth's surface and below the ionosphere.

Tropospheric bending—When radio waves are bent in the troposphere, they return to earth farther away than the visible horizon.

Tropospheric ducting—A type of VHF **propagation** that can occur when warm air overruns cold air (a temperature inversion).

Unbalanced line—Feed line with one conductor at ground potential, such as coaxial cable.

Unidentified communications or signals—Signals or radio communications in which the transmitting station's call sign is not transmitted.

Upper sideband (USB)—The common single-sideband operating mode on the 20, 17, 15, 12 and 10-meter HF amateur bands, and all the VHF and UHF bands.

Variable capacitor—A capacitor that can have its value changed within a certain range.

Variable-frequency oscillator (VFO)—An oscillator used to control the frequency in receivers and transmitters. The frequency is set by a tuned circuit using capacitors and inductors. The frequency can be changed by adjusting the components in the tuned circuit.

Variable resistor—A resistor whose value can be adjusted over a certain range. You can change the value without removing it from a circuit.

Vertical antenna—A common amateur antenna, usually made of metal tubing. The radiating element is vertical. There are usually four or more radial elements parallel to or on the ground.

Vertically polarized wave—A radio wave that has its electric lines of force perpendicular to the surface of the earth.

Visible horizon—The most distant point one can see by line of sight.

Voice communications—Hams can use several voice modes, including FM and SSB.

Volt (V)—The basic unit of electrical pressure or EMF.

Voltage—The EMF or pressure that causes electrons to move through an electrical circuit.

Voltage source—Any source of excess electrons. A voltage source produces a current and the force to push the electrons through an electrical circuit.

Voltmeter—A test instrument used to measure voltage.

W1AW—The headquarters station of the American Radio Relay League. This station is a memorial to the League's cofounder, Hiram Percy Maxim. The station provides daily on-the-air code practice and bulletins of interest to hams.

Watt (W)—The unit of power in the metric system. The watt describes how fast a circuit uses electrical energy.

Wattmeter—Also called a power meter, a test instrument used to measure the power output (in watts) of a transmitter. A directional wattmeter measures both forward and reflected power.

Wavelength—The distance an ac signal will travel during the time it takes the signal to go through one complete cycle. Often abbreviated λ. The distance a radio wave travels in one RF cycle. The wavelength relates to frequency. Higher frequencies have shorter wavelengths.

Wire-wound resistor—A resistor made by winding a length of wire on an insulating form.

WWV and WWVH—Radio stations that broadcast the precise time on precise frequencies. Hams tune in to these stations to calibrate transceiver dials (and to set their clocks).

Yagi antenna—The most popular type of amateur directional (beam) antenna. It has one driven element and one or more additional elements.

73—Ham lingo for "best regards." Used on both phone and CW toward the end of a contact.

WORLD TIME FINDER

DATE LINE

UTC

Index

About the American Radio Relay League

The seed for Amateur Radio was planted in the 1890s, when Guglielmo Marconi began his experiments in wireless telegraphy. Soon he was joined by dozens, then hundreds, of others who were enthusiastic about sending and receiving messages through the air—some with a commercial interest, but others solely out of a love for this new communications medium. The United States government began licensing Amateur Radio operators in 1912.

By 1914, there were thousands of Amateur Radio operators—hams—in the United States. Hiram Percy Maxim, a leading Hartford, Connecticut, inventor and industrialist saw the need for an organization to band together this fledgling group of radio experimenters. In May 1914 he founded the American Radio Relay League (ARRL) to meet that need.

Today ARRL, with more than 170,000 members, is the largest organization of radio amateurs in the United States. The League is a not-for-profit organization that:

- promotes interest in Amateur Radio communications and experimentation
- represents US radio amateurs in legislative matters, and
- maintains fraternalism and a high standard of conduct among Amateur Radio operators.

At League headquarters in the Hartford suburb of Newington, the staff helps serve the needs of members. ARRL is also International Secretariat for the International Amateur Radio Union, which is made up of similar societies in more than 100 countries around the world.

ARRL publishes the monthly journal *QST*, as well as newsletters and many publications covering all aspects of Amateur Radio. Its headquarters station, W1AW, transmits bulletins of interest to radio amateurs and Morse code practice sessions. The League also coordinates an extensive field organization, which includes volunteers who provide technical information for radio amateurs and public-service activities. ARRL also represents US amateurs with the Federal Communications Commission and other government agencies in the US and abroad.

Membership in ARRL means much more than receiving *QST* each month. In addition to the services already described, ARRL offers membership services on a personal level, such as the ARRL Volunteer Examiner Coordinator Program and a QSL bureau.

Full ARRL membership (available only to licensed radio amateurs) gives you a voice in how the affairs of the organization are governed. League policy is set by a Board of Directors (one from each of 15 Divisions). Each year, half of the ARRL Board of Directors stands for election by the full members they represent. The day-to-day operation of ARRL HQ is managed by an Executive Vice President and a Chief Financial Officer.

No matter what aspect of Amateur Radio attracts you, ARRL membership is relevant and important. There would be no Amateur Radio as we know it today were it not for the ARRL. We would be happy to welcome you as a member! (An Amateur Radio license is not required for Associate Membership.) For more information about ARRL and answers to any questions you may have about Amateur Radio, write or call:

ARRL Educational Activities Dept
225 Main Street
Newington CT 06111-1494
(203) 666-1541
Prospective new amateurs call:
800-32-NEW HAM (800-326-3942)

BEST OF MFJ

The Virginia Beach HamFest and Computer Fair™

TM TRC, Inc

For your first real HamFest and Convention, plan a long, Fall weekend in Virginia Beach, VA, and attend the Virginia Beach HamFest.

Relax on the beach, or perhaps try some surf or deep sea fishing, or even visiting Colonial WIlliamsburg, Yorktown or Jamestown; or perhaps play in Busch Gardens or visit the world's largest Naval Base... all within an hour's drive.

Take your first exam or upgrade your license at our show. Most major manufacturers attend the show, as well as many major dealers of radio and computer equipment. Get your best price on ham and computer gear right here! So come to the largest and only real HamFest and Computer Show in the Mid-Atlantic Region! Write today to be placed on the mailing list.

Tidewater Radio Conventions	Tidewater Radio Conventions
Lewis Steingold W4BLO - Marketing Director	Lynn Lilla, WA9DJQ - President
1008 Crabbers Cove Lane - Va. Beach, VA 23452	848 Stacey Place - Va. Beach, VA 23464

Just when you thought you had everything ever published on antennas...

The ARRL presents the **Antenna Compendium Series**—an antenna-lover's feast of never-before-published articles on antennas and transmission lines you can build! Whether you have a casual or a serious interest in antenna design or construction, you'll find something to spark your curiosity in a **Compendium.**

Here's some of what you'll find:

Volume 3 is packed with a mix of practical construction information, ideas for development, and theoretical and tutorial discussions.

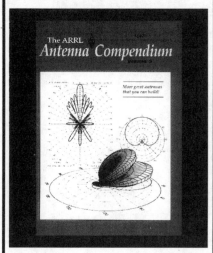

Volume 2 presents articles on selecting antennas, computer modeling, baluns and matching networks and many types of popular and esoteric antennas.

Volume 1 is packed with articles on verticals, quads, loops, Yagis, reduced-size antennas, baluns, Smith Charts, antenna polarization and other interesting subjects.

So warm up the soldering iron, stretch that antenna wire and get set for lots of antenna construction and experimentation because the antenna you've always wanted is in the **ARRL Antenna Compendium series**.

Companion software is available for Volumes 2 and 3. See ordering information in the full-color ARRL Publications Catalog elsewhere in this issue.

The DXCC Companion: *How to Work Your First Hundred Countries*, by Jim Kearman, KR1S, is filled with practical, easy to understand information for the beginning DXer. Follow the advice given by KR1S and you'll have your first hundred countries confirmed in no time. You'll learn about equipment, antennas, propagation and the rights and wrongs of QSLing.

See Ordering information in the ARRL Publications Catalog elsewhere in this issue.

Help for Beginners

Now You're Talking!: All You Need to Get Your First Ham License covers everything you need to know to earn your first Amateur Radio license. More than a study guide, it will help you select equipment for your ham radio station and even explain how to set it up!

Whether you're interested in the new codeless Technician license or the traditional Novice ticket, with *Now You're Talking!*, it's easier than ever to become a ham radio operator! If you're starting with the Novice license you should also purchase audio cassettes or computer software to learn Morse code (described below). 2nd ed, ©1993 ....#4173 **$19**

W1FB's Help for New Hams by Doug DeMaw offers sound advice on getting started in Amateur Radio after you get your license. This book acts as your personal Elmer, covering such topics as selecting equipment, station layout and accessories, constructing and using antennas, and operating. 2nd ed, ©1993 ....#4432 **$10**

Operating an Amateur Radio Station. This booklet answers the basic Amateur Radio questions often posed by newcomers: How do I decide what equipment to buy? What kind of antenna do I need? What procedures do I use, and what do I say when I make my first contact? How do I QSL? and many others. 65th ed, ©1991 ....#226X **$2**

Novice Notes: The Book is a selection of articles for the beginner from the popular *QST* series. It's filled with useful information: What you should do before your license arrives; how to buy used gear; all about antenna tuners and antennas, logging and QSLing; awards chasing; tips on phone, Morse code and digital operating and much more. 1st ed, ©1989 ....#2561 **$6**

Written in an easy-to-understand style for electronics beginners, *Understanding Basic Electronics* is also for those who want to brush up on electronics principles. Starting with math skills, the book progresses to dc and ac electronics principles. It concludes with clear, simple explanations of how components like diodes, transistors and integrated circuits work. Perfect for students, hams and everyone with an interest in the electronics side of ham radio. 1st ed, ©1992 ....#3983 **$17**

Upgrade Your License the Easy Way!

After you receive your license and get on the air, you'll probably want to explore additional operating privileges. The *ARRL License Manual Series* represents the best study material for the Technician, General, Advanced and Extra Class Amateur Radio exams. Each book is carefully revised and updated as new exam questions are released by the VEC Question Pool Committee. The appropriate examination question pool, complete with an answer key, is included for easy reference. The answer key contains page references so you can locate appropriate text explanations as you review the questions before your exam. Our *FCC Rule Book* should be used along with each publication in the series.

ARRL License Manual Series

Technician Class for Novice Class Licensees, 2nd ed, ©1993 ....#4181 **$6**
*General Class, 1st ed, ©1989 ....#2383 **$6**
*Advanced Class, 3rd ed, ©1990 ....#3274 **$8**
*Extra Class, 5th ed, ©1990 ....#3282 **$8**
*(Current editions are good at least until Spring 1994)

FCC Rule Book, 9th ed, ©1993 ....#4270 **$9**

Code Proficiency

When it comes to the code, whether you're just starting out, or you're working on your Extra Class upgrade, practice makes perfect!

The ARRL produces five sets of Morse code tapes to get you from 0 to 22 words per minute. Each set includes two C-90 cassettes. *GGTE Morse Tutor* software for IBM PCs and compatibles teaches you the code, provides plenty of practice for exams and helps keep your code speeds sharp in easy, self-paced lessons. Features include code speeds from 1 to more than 100 words per minute, standard or Farnsworth modes and random QSOs. The *Advanced Morse Tutor* has even more features. You can send code from an ASCII text file that you create on the computer, save to disk the random QSOs created by the program for later replay, and even create your own practice text with special emphasis on problem characters, and more.

Your *Introduction to Morse Code*, our cassette program for beginners, makes learning the code fun. It teaches you all the characters and provides plenty of practice ....#3487 **$10**

ARRL Code Practice Cassettes. Each set of two C-90 tapes provides three hours of practice.
Set 1: 5 to 10 WPM ....#2227 **$10**
Set 2: 10 to 15 WPM ....#2235 **$10**
Set 3: 15 to 22 WPM ....#2243 **$10**
Set 4: 13 to 14 WPM ....#2251 **$10**

GGTE Morse Tutor (5.25-inch) ....#2081 **$20**
GGTE Morse Tutor (3.5-inch)..#2936 **$22**
Advanced Morse Tutor (5.25-inch)....................................#3231 **$30**
Advanced Morse Tutor (3.5-inch) ....#3258 **$30**

Handy References

Every chapter of the 4th edition of *The ARRL Operating Manual* has been updated to include the latest information about every aspect of our dynamic hobby. It's simply the best book available covering on-the-air amateur operating practices. How do I operate on a repeater or on PacketCluster? How can I

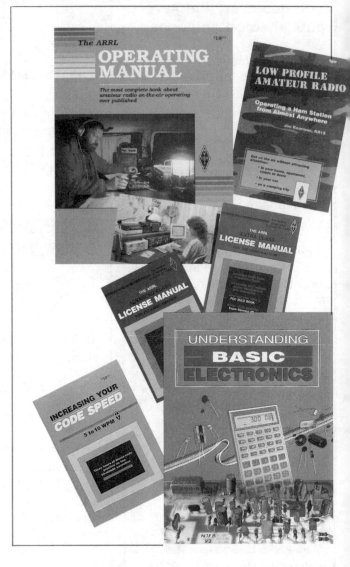

snare a contact through a DXpedition pileup? What satellites are available and how can I use them? You'll find the answers to all of these questions and many more in *The ARRL Operating Manual*!

One impressive and colorful section features dozens of US and overseas operating awards, and a handy reference section includes an ARRL DXCC Countries List, beam-heading information, a series of maps, US counties, sunrise/sunset tables, and much, much more. No shack is complete without this valuable reference. ©1991 ....#1086 **$18**

The 1993-1994 ARRL Repeater Directory includes more than 19,000 listings for voice and digital repeaters and propagation beacons located in North, Central and South America, and Europe. This edition also lists more than 500 beacons from 14 MHz to 24 GHz. You'll also find band plans, a CTCSS (PL™) tone chart, a list of frequency coordinators and a new user-friendly list of ARRL Special Service Clubs. The *Repeater Directory* comes in a handy pocket size for your operating convenience. 22nd ed, ©1993 ....#4246 **$6**

The ARRL DXCC Countries List is the ideal way to record the DXCC countries you've worked and QSLed. The latest printing includes DXCC Advisory Committee members, an expanded cross reference for prefixes and exotic countries and more. (Free shipping). June 1993 ed, ©1993 ....#0291 **$2**

The ARRL Net Directory lists hundreds of Amateur Radio nets of interest to North American hams—DX, ragchew, special-interest, fun and public service nets—they're all here. Updated annually. (Free shipping). 1993-1994 ed, ©1993 ....#4262 **$2**

Low-Profile Amateur Radio is for the ham who lives where antennas are frowned upon. You'll see that you don't need a house with acreage to enjoy your favorite hobby. One practical solution: hide your antennas. Another: operate with low power. This book tells you how to get on the air using these techniques—and others—without calling attention to yourself. 1st ed, ©1993 ....#4114 **$8**

Welcome to Amateur Radio!

Congratulations! You've taken your first step into a hobby—and a service—that knows no limits. Amateur Radio is a worldwide network of people from various cultures, united by a common love of wireless communication. Amateur Radio is as old as radio itself, and its future is no less fantastic than its past.

For most people, Amateur Radio is a lifelong pursuit. We want to make sure you get a good start, which is why we've published this book. But first, who are "we" ?

The American Radio Relay League: What's in it for You?

The seed for Amateur Radio was planted in the 1890s, when radio's pioneers began experimenting with wireless telegraphy. Soon, there were dozens, then hundreds, of experimenters who were enthusiastic about sending and receiving messages through the air—some with a commercial interest, but others solely out of a love for this new and mysterious communications medium. To coordinate all of this activity, the United States government began licensing Amateur Radio operators in 1912.

By 1914, there were thousands of Amateur Radio operators—hams, as they were called—in the United States. Hiram Percy Maxim, a leading Hartford, Connecticut, inventor and industrialist, saw the need for an organization to band together this fledgling group of radio experimenters. In May of 1914 he founded the American Radio Relay League (ARRL) to meet that need.

Today, the League has more than 170,000 members, and is the national organization of radio amateurs and a leading ham radio organization worldwide. By joining the ARRL, you'll receive many special benefits. Here are just a few examples:

◆ **Help for New Hams**: Are you a beginning ham looking for help in getting started in your new hobby? The Educational Activities Department at ARRL HQ in Newington, Connecticut, will be glad to assist you. Call EAD at 800-32-NEWHAM. EAD maintains a computer data base of ham clubs and ham radio "helpers" from across the country who've told us they're interested in helping beginning hams. There are probably several clubs in your area! Contact EAD for more information.

◆ **Licensing Classes**: If you're going to become a ham, you'll need to find a local license exam opportunity sooner or later. ARRL Registered Instructors teach licensing classes all around the country, and ARRL-sponsored Volunteer Examiners are right there to administer your exams. To find the locations and dates of Amateur Radio Licensing classes and test sessions in your area, call the Educational Activities Department at 800-32-NEWHAM.

◆ **Clubs**: As a beginning ham, one of the best moves you can make is to join a local ham club. Whether you join an all-around group or a special-interest club (repeaters, DXing, and so on), you'll make new friends, have a lot of fun, and you can tap into a ready reserve of ham radio knowledge and experience. To find the ham clubs in your area, call HQ's Educational Activities Department at 800-32-NEWHAM.

◆ **Technical Information Service**: Do you have a question of a technical nature? (What new ham doesn't?) Contact the Technical Information Service (TIS) at HQ. Our resident technical experts will help you over the phone, send you specific information on your question (antennas, interference and so on) or refer you to your local ARRL Technical Coordinator or Technical Specialist. It's expert information—and it doesn't cost Members an extra cent!

◆ **Regulatory Information**: Need help with a thorny antenna zoning problem? Having trouble understanding an FCC regulation? Vacationing in a faraway place and want to know how to get permission to operate your ham radio there? HQ's Regulatory Information Branch has the answers you need!

◆ **Operating Awards**: Like to collect "wallpaper"? The ARRL sponsors a wide variety of certificates and Amateur Radio achievement awards. For information on awards you can qualify for, contact the Membership Services Department at HQ.

◆ **Equipment Insurance**: When it comes to protecting their Amateur Radio equipment investments, ARRL Members travel First Class. A. H. Wohlers Company provides the League's "all-risk" equipment insurance plan. (It can protect your ham radio computer, too.) It's comprehensive and cost effective, and it's available only to ARRL Members. Why worry about losing your valuable radio equipment when you can protect it for only a few dollars a year?

◆ **Amateur Radio Emergency Service**: If you're interested in providing public service and emergency communications for your community, you can join more than 25,000 other hams who have registered their communications capabilities with local Emergency Coordinators. Your EC will call on you and other ARES members for vital assistance if disaster should strike your community. Contact the Field Services Department at HQ for information.

◆ **Audio-Visual Library**: Need a program for your next ham club meeting or informal get-together? ARRL affiliated clubs can borrow a tape from the League's video library at HQ! There are many to choose from, including popular titles on Amateur Radio's role in Operation Desert Storm and space shuttle activities. Contact the Educational Activities Department at HQ to check out your tape.

◆ **Blind, Disabled Ham Help**: For a list of available resources and information on the Courage HANDI-HAM System, contact the ARRL Program for the Disabled at HQ.

With your membership you also receive the monthly journal *QST*. Each 200-page issue is packed with valuable information you can use—including a special *New Ham Companion* section! And *QST* Product Reviews are the most respected source of information to help you get the most for your Amateur Radio equipment dollar. (For many hams, *QST* alone is worth far more than the cost of ARRL membership.)

The ARRL also publishes newsletters and dozens of books covering all aspects of Amateur Radio. Our Headquarters station, W1AW, transmits bulletins of interest to radio amateurs and Morse Code practice sessions.

When it comes to representing Amateur Radio's best interests in our nation's capital, the League's team in Washington, DC, is constantly working with the FCC, Congress and industry to protect and foster your privileges as a ham operator.

Regardless of your Amateur Radio interests, ARRL Membership is relevant and important. We will be happy to welcome you as a Member. Use the coupon on the next page to **join today**. And don't hesitate to contact us if you have any questions!

Invitation to Membership

Class of License   Call Sign   Date of Birth

Name

A member of the immediate family of a League member, living at the same address, may become a League member without *QST* at the special rate of $4 per year. Family membership must run concurrently with that of the member receiving *QST*. Blind amateurs may join without *QST* for $4 per year.

Address

City, State, ZIP

If you are 21 or younger, a special rate may apply. Write HQ for details.

Membership Class	☐ Regular		☐ Family ☐ Blind	☐ 65 or older with Proof of Age	
	US	ELSEWHERE		US	ELSEWHERE
☐ 1 year	$30	$42	$4	$24	$36
☐ 2 years	57	81	8	45	69
☐ 3 years	80	116	12	65	101

If you do not wish your name and address made available for non-ARRL related mailings please check this box. ☐

NYT94

THE AMERICAN RADIO RELAY LEAGUE, INC

225 MAIN STREET   NEWINGTON, CONNECTICUT 06111 USA   (203) 666-1541

YOU'VE GOT QUESTIONS? WE'VE GOT ANSWERS!

Now you've got what you need to go after your own Amateur Radio license and call sign! But you've probably still got a question or two. Does anyone in my area teach classes? Where can I find a person who can give me my exam? Where in my area can I buy equipment? Which is better from my location, a dipole antenna or a vertical antenna? If I decide to learn Morse code, where can I find someone to help me practice? Can someone check my station to see if I've set up everything correctly? Who do I turn to for answers? The American Radio Relay League's Educational Activities Department can send you a list of Amateur Radio clubs, instructors, examiners and Elmers who live in your area and enjoy helping newcomers. Here's your first question and answer: What's an Elmer? An Elmer is a person who helps you with whatever you need; an Elmer is another Amateur Radio tradition of hams helping others. For your list, send a postcard to ARRL EAD, 225 Main St, Newington, CT 06111, or call us at 1-800-32NEWHAM.

Approved by OMB
3060-0003
Expires 08/31/96
See instructions for
information regarding
public burden estimate.

APPLICATION FORM 610 FOR
AMATEUR OPERATOR/PRIMARY STATION LICENSE

SECTION 1 - TO BE COMPLETED BY APPLICANT (See instructions)

1. Print or type last name	Suffix	First name	Middle initial	2. Date of birth
				— — month day year

3. Mailing address (Number and street)	City	State code	ZIP code

4.   I.HEREBY APPLY FOR (make an X in the appropriate box(es)):

4A. ☐ **EXAMINATION** for a new license

4B. ☐ **EXAMINATION** for upgrade of my operator license class

4C. ☐ **CHANGE** my name on my license to my new name in Item 1. My former name was:

_____
(Last name)   (Suffix)   (First name)   (MI)

4D. ☐ **CHANGE** my mailing address on my license to my new address in Item 3

4E. ☐ **CHANGE** my station call sign systematically (See instructions) Applicant's Initials _____

4F. ☐ **RENEWAL** of my license

5. Unless you are requesting a new license, attach the original or a photocopy of your license to the back of this Form 610 and complete Items 5A and 5B.	5A. Call sign shown on license	5B. Operator class shown on license

6. Would an FCC grant of your request be an action that may have a significant environmental effect?   ☐ NO   ☐ YES (Attach required statement)

7. If you have filed another Form 610 that we have not acted upon, complete Items 7A and 7B.	7A. Purpose of other form	7B. Date filed — — month day year

WILLFUL FALSE STATEMENTS MADE ON THIS FORM ARE PUNISHABLE BY FINE AND/OR IMPRISONMENT, (U.S. CODE, TITLE 18, SECTION 1001), AND/OR REVOCATION OF ANY STATION LICENSE OR CONSTRUCTION PERMIT (U.S. CODE, TITLE 47, SECTION 312(A)(1)) AND/OR FORFEITURE (U.S. CODE, TITLE 47, SECTION 503).

I CERTIFY THAT ALL STATEMENTS AND ATTACHMENTS ARE TRUE, COMPLETE, AND CORRECT TO THE BEST OF MY KNOWLEDGE AND BELIEF AND ARE MADE IN GOOD FAITH; THAT I AM NOT A REPRESENTATIVE OF A FOREIGN GOVERNMENT; THAT I WAIVE ANY CLAIM TO THE USE OF ANY PARTICULAR FREQUENCY REGARDLESS OF PRIOR USE BY LICENSE OR OTHERWISE; AND THAT THE STATION TO BE LICENSED WILL BE INACCESSIBLE TO UNAUTHORIZED PERSONS.

8. Signature of applicant (Do not print, type, or stamp.) (Must match name in Item 1.)	9. Date signed
✘_____	— — month day year

SECTION 2 - TO BE COMPLETED BY ALL ADMINISTERING VE's

A. Applicant is qualified for operator license class:

☐ NOVICE	(Elements 1(A), 1(B), or 1(C) and 2)
☐ TECHNICIAN	(Elements 2 and 3(A))
☐ TECHNICIAN PLUS	(Elements 1(A), 1(B), or 1(C), 2 and 3(A))
☐ GENERAL	(Elements 1(B) or 1(C), 2, 3(A) and 3(B))
☐ ADVANCED	(Elements 1(B) or 1(C), 2, 3(A), 3(B) and 4(A))
☐ AMATEUR EXTRA	(Elements 1(C), 2, 3(A), 3(B), 4(A) and 4(B))

B. VEC receipt date:

C. Name of Volunteer-Examiner Coordinator (VEC):

D. Date of VEC coordinated examination session:	E. Examination session location:

I CERTIFY THAT I HAVE COMPLIED WITH THE ADMINISTERING VE REQUIREMENTS IN PART 97 OF THE COMMISSION'S RULES AND WITH THE INSTRUCTIONS PROVIDED BY THE COORDINATING VEC AND THE FCC

1st VE's name (Print First, MI, Last, Suffix)	VE's station call sign	VE's signature (must match name)	Date signed
2nd VE's name (Print First, MI, Last, Suffix)	VE's station call sign	VE's signature (must match name)	Date signed
3rd VE's name (Print First, MI, Last, Suffix)	VE's station call sign	VE's signature (must match name)	Date signed

ATTACH ORIGINAL OR A PHOTOCOPY OF YOUR LICENSE HERE:

SECTION 3 - TO BE COMPLETED BY PHYSICIAN

**PHYSICIAN'S CERTIFICATION
OF DISABILITY**

Please see notice below

Print, type, or stamp physician's name: _____

Street address: _____

City, State, ZIP code: _____

Office telephone number: (___) _____

I CERTIFY THAT I have read the Notice to Physician Certifying to a Disability, and that the person named in Item 1 on the reverse is severely handicapped, the duration of which will extend for more than 365 days beyond this date. Because of this severe handicap, this person is unable to pass a 13 or 20 words per minute telegraphy examination. I am licensed to practice in the United States or its Territories as a doctor of medicine (M.D.) or doctor of osteopathy (D.O.). I have considered the accommodations that could be made for this person's disability and have determined that, even with accommodations, this person would be unable to pass a 13 or 20 words per minute telegraphy examination.

WILLFUL FALSE STATEMENT IS PUNISHABLE BY FINE AND IMPRISONMENT (U.S. CODE TITLE 18, SECTION 1001)

➡ _____   _____   _____
PHYSICIAN'S SIGNATURE (DO NOT PRINT, TYPE, OR STAMP)   M.D. or D.O.   DATE SIGNED

PATIENT'S RELEASE

Authorization is hereby given to the physician named above, who participated in my care, to release to the Federal Communications Commission any medical information deemed necessary to process my application for an amateur operator/primary station license.

➡ _____   _____
APPLICANT'S SIGNATURE (DO NOT PRINT, TYPE, OR STAMP)   DATE SIGNED

NOTICE TO PHYSICIAN CERTIFYING TO A DISABILITY

You are being asked by a person who has already passed a 5 words per minute telegraphy examination to certify that, because of a severe handicap, he/she is unable to pass a 13 or 20 words per minute telegraphy examination. If you sign the certification, the person will be exempt from the examination. Before you sign the certification, please consider the following:

THE REASON FOR THE EXAMINATION - Telegraphy is a method of electrical communication that the Amateur Radio Service community strongly desires to preserve. We support their objective by authorizing additional operating privileges to amateur operators who increase their skill to 13 and 20 words per minute. Normally, to attain these levels of skill, intense practice is required. Annually, thousands of amateur operators prove by passing examinations that they have acquired the skill. These examinations are prepared and administered by amateur operators in the local community who volunteer their time and effort.

THE EXAMINATION PROCEDURE - The volunteer examiners (VEs) send a short message in the Morse code. The examinee must decipher a series of audible dots and dashes into 43 different alphabetic, numeric and punctuation characters used in the message. To pass, the examinee must correctly answer questions about the content of the message. Usually, a fill-in-the-blanks format is used. With your certification, they will give the person credit for passing the examination, even though they do not administer it.

MUST A PERSON WITH A HANDICAP SEEK EXEMPTION?

No handicapped person is required to request exemption from the higher speed telegraphy examinations, nor is anyone denied the opportunity to take the examinations because of a handicap. There is available to all otherwise qualified persons, handicapped or not, the Technician Class operator license that does not require passing a telegraphy examination. Because of international regulations, however, any handicapped applicant requesting exemption from the 13 or 20 words per minute examination must have passed the 5 words per minute examination.

ACCOMMODATING A HANDICAPPED PERSON - Many handicapped persons accept and benefit from the personal challenge of passing the examination in spite of their hardships. For handicapped persons without an exemption who have difficulty in proving that they can decipher messages sent in the Morse code, the VEs make exceptionally accommodative arrangements. They will adjust the tone in frequency and volume to suit the examinee. They will administer the examination at a place convenient and comfortable to the examinee, even at bedside. For a deaf person, they will send the dots and dashes to a vibrating surface or flashing light. They will write the examinee's dictation. Where warranted, they will pause in sending the message after each sentence, each phrase, each word, or each character to allow the examinee additional time to absorb and interpret what was sent. They will even allow the examinee to send the message, rather than receive it.

YOUR DECISION - The VEs rely upon you to make the necessary medical determination for them using your professional judgement. You are being asked to decide if the person's handicap is so severe that he/she cannot pass the examination even when the VEs employ their accommodative procedures. The impairment, moreover, will last more than one year. This procedure is not intended to exempt a person who simply wants to avoid expending the effort necessary to acquire greater skill in telegraphy. The person requesting that you sign the certification will give you names and addresses of VEs and other amateur operators in your community who can provide you with more information on this matter.

DETAILED INSTRUCTIONS - If you decide to execute the certification, you should complete and sign the Physician's Certification of Disability on the person's FCC Form 610. You must be an M.D. or D.O. licensed to practice in the United States or its Territories. The person must sign a release permitting disclosure to the FCC of the medical information pertaining to the disability.

FCC Form 610
November 1993

FEEDBACK

Please use this form to give us your comments on this book and what you'd like to see in future editions.

Please check the box that best answers these questions:

How well did this book prepare you for your exam? □ Very well □ Fairly well □ Not very well

How well did this book teach you about ham radio? □ Very well □ Fairly well □ Not very well

Which exam did you take (or will you be taking)? □ Novice □ Technician Did you pass? □ Yes □ No

If you checked Technician: Do you expect to learn Morse code at some point? □ Yes □ No

Which operating modes do you plan to use first?

□ SSB □ FM □ Packet □ RTTY □ Image □ Morse code □ Other_____

Where did you purchase this book? □ From ARRL directly □ From an ARRL dealer

Is there a dealer who carries ARRL publications within: □ 5 miles □ 15 miles □ 30 miles of your location? □ Not sure.

Name

_____ Call sign_____

Address _____

City, State/Province, ZIP/Postal Code _____ . _____

Daytime Phone (   ) _____ Age _____

If licensed, how long? _____ ARRL member? □ Yes □ No

Other hobbies _____

_____

Occupation _____

From _____

_____

_____

EDITOR, NOW YOU'RE TALKING!
AMERICAN RADIO RELAY LEAGUE
225 MAIN ST
NEWINGTON CT 06111-1494

············· please fold and tape ·············